Cioffari's Experiments in College Physics

SEVENTH EDITION

Dean S. Edmonds, Jr.
Boston University

D. C. HEATH AND COMPANY   LEXINGTON, MASSACHUSETTS / TORONTO

Cover Photograph: Stuart L. Craig, Jr./Bruce Coleman, Inc.

International Standard Book Number: 0-669-04492-X

Library of Congress Catalog Card Number: 82-81624

Preface to the Seventh Edition

The seventh edition of Cioffari's manual features a rewriting of several experiments and the addition of extra questions in many of the Questions sections. For example, the former experiment on ray tracing has been rewritten as *Geometrical Optics*, with a small laser that produces a narrow beam, or ray, and with mirrors and lenses from a blackboard optics kit that show how rays are bent by reflection and refraction. A laser has also been introduced in the diffraction grating experiment, which has been redone to show more directly the way a grating works. The actual use of a grating to measure wavelengths in a spectrum has been left to the study of the hydrogen spectrum in Experiment 49.

Two new experiments have been added, namely, Experiment 22, *The Ratio of the Specific Heats of Gases*, and Experiment 38, *Operational Amplifiers*. The former presents a student laboratory version of the well-known method of Rüchhardt, and the latter extends the series of experiments in electronics to a study of some applications of a common type of operational amplifier.

Finally, a new feature of the present edition is the addition of a section entitled *Apparatus Notes* after the Appendix Tables. These notes include specifications of suggested apparatus for the various experiments, details on items that have to be prepared in the laboratory, and comments on any features requiring special explanation. It is hoped that instructors will find this material useful whether they are responsible for the purchase of new equipment or not, and students with a penchant for laboratory work are also encouraged to read it. The actual main text of the manual, however, is now not cluttered with these details and contains only the theory and instructions necessary for the performance of experiments for which the preparation of equipment has already been carried out. It is hoped that such an arrangement will be helpful to instructor and student alike.

Acknowledgments

The author wishes to acknowledge the following users of previous editions of *Experiments in College Physics* whose reviews and evaluations were helpful in improving this seventh edition.

Robert P. James	Grossmont College
Richard W. Mitchell	University of South Florida
Thomas P. Bernat	Louisiana State University and Agricultural and Mechanical College
R. G. Goodrich	Louisiana State University and Agricultural and Mechanical College
Jack B. Greene	Marquette University
Frank G. Karioris	Marquette University
Charles A. Heller	Blinn College
John T. Ritter	Richland College
Edgar J. Rook	West Valley College
Audrey Balentine	Eastern Oklahoma State College
Charles Gosselin	Penn Valley Community College
Lt. Col. R. A. MacArthur	New Mexico Military Institute
David Pierce	El Camino College
David G. Willey	University of Pittsburgh at Johnstown
Philip H. Beatty	Le Tourneau College
Frank J. Falsetto	Gonzaga University
W. C. Wyatt	Middle Georgia College
Physics Faculty	Indiana University—Purdue University at Fort Wayne
Kwok-Leung Li	Bergen Community College

In addition, deserving of special mention are the contributions of Professors J. Gordon Stipe, Bernard Chasan, and George Zimmerman of the Boston University Physics Department, who were always ready with helpful discussions and suggestions and who in fact originated some of the experimental procedures presented herein. Special mention should also be made of Tachisto, Inc. of Needham, Massachusetts, which supplied the operational amplifiers used in the new Experiment 38.

Contents

Introduction

The purpose of the physics laboratory is to supply the practical knowledge necessary for a well-rounded understanding of physics and the physicist's way of looking at the universe. A further aim is to develop familiarity with the experimental method of scientific investigation and to give the student experience in the actual handling of laboratory apparatus. It is one thing to study a certain model of some physical phenomenon and deduce that certain results should be observed. It is quite another to set up an experiment in which these observations can be made and thus produce data on the basis of which the model's validity may be tested. In particular, the obtaining of experimental results depends on the student's ability to make accurate measurements of physical quantities in the real world. A major purpose of every experiment in this book is to provide practice in doing so.

INSTRUCTIONS

The instructions for each experiment include some basic theory on the phenomenon to be investigated and a description of the procedure to be used. These should be studied carefully before the student comes to the laboratory to avoid waste of valuable laboratory time figuring out what should be done. The student will be told well in advance which experiments are to be performed and the date for which each is scheduled so that there will be time for proper preparation. The necessary equipment will be laid out at each assigned place in the laboratory. Missing or defective apparatus should be reported to the laboratory instructor immediately. The instructor should also be consulted if the student has any questions about the experiment.

All observations and data should be recorded in the blank tables provided for this purpose in each experiment. Columns in these tables are already suitably labeled, but the student must be careful to note the units in which each of the observed quantities is measured. Instruments should be read to the limit of their possibilities by estimating the last figure of the reading, that is, the fraction of the smallest scale division. Each measurement should be recorded directly on the data sheet exactly in the form in which it is made without any mental calculation. Do not use "scratch" data sheets from which data are to be transcribed onto the blank ones provided in this book. Very neat data sheets can be made out this way, but mistakes can also creep in. The instructor is interested in an *original* data sheet and is willing to put up with a certain amount of sloppy penmanship in order to see the direct recording of the actual data taken in the laboratory.

Calculations are to be made in the space provided in the manual whenever this is feasible. When the required computations are too long they should be completed on a separate sheet of paper, which must be included in the report. Each set of calculations should be headed by the pertinent equation so that anyone reading the report can see what mathematical operations are being performed and why. A table of logarithms is provided in the appendix, but students are urged to use electronic calculators or slide rules for all algebraic manipulations. Practice in doing arithmetic by hand is *not* one of the aims of the physics laboratory.

The questions at the end of each experiment are to be answered in the spaces provided for this purpose. Proper English should be used so that communication between the student and a reader will not be impeded. The student is urged to complete his or her calculations and as much of the remainder of the report as possible during the laboratory period. If this is not possible, reports may be completed outside the laboratory and should be handed in at the time and place specified by the instructor.

CARE OF APPARATUS

The apparatus provided with each experiment has been set up to work properly in the arrangement described for that experiment and is in some cases very delicate. Extreme care should be used in handling it. The instructions for each experiment include a list of the required equipment, and the student should check this list against the items on the work bench to make sure everything he or she needs is there and in good condition. Anything missing or broken should be reported to the instructor. At the end of the period the student should again check the apparatus and leave it neatly arranged.

Whenever an experimental setup has been assembled, it should be checked before being placed in operation so that any mistakes that might keep it from working properly or that might cause actual damage can be found and corrected. In particular, electrical circuits should be examined carefully for proper wiring. Application of power to a circuit containing wiring errors can cause serious damage. The source of power (battery or power supply) should always be connected last, and the circuit should be checked and approved by the instructor before this final connection is made. Special care should be exercised in setting meters to the proper range, as these items are expensive and easily destroyed if excessive current is allowed to pass through the movement. Whenever the range of a meter or any wiring in a circuit is to be changed, the source of power should always be disconnected first. The possibility of electrical shock or damage due to a temporary wrong connection will in this way be eliminated.

THE REPORT

A report of the work done in each experiment must be prepared by the student and handed in at the beginning of the next laboratory period or at some other time designated by the instructor. The report will be graded and returned as soon as possible, after which it may be kept in a folder or binder for future reference. The report should include:

1. A title page. This should carry the student's name, the date, and the name and number of the experiment.
2. The instruction sheets. These are the pages describing the object of the experiment, the theory, the apparatus, and the procedure. Perforations allow these pages to be easily torn out of this book for inclusion in the report.
3. All original data and observations. As already noted, these are entered in the blank data tables provided in each experiment. The data table sheets are also perforated so that they can be easily removed from this book.
4. All the required calculations. These are to be made in the space provided. The calculation sheets are then detached along their perforations for inclusion in the report. If extra calculation pages are used, they must be included in the report in the proper order.
5. Graphs and diagrams, whenever these are required. Graph paper pages are provided as needed in this manual and are also perforated so that they can be easily detached and inserted in the report.
6. A summary and discussion of the results. The summary is included in tabular form under the data. It usually involves a comparison of the computed results with the accepted values together with the percentage errors involved. The student is encouraged to add a brief discussion of the sources of these errors and any comments he or she would like to make about the working of the experiment.
7. Answers to the questions at the end of the experiment. The answers are written in the space provided after each question. The question sheets are then torn out along the perforations and added to the report. Care should be taken to use complete sentences and in general to make the answers as clear and readable as possible. A separate sheet may be used for this purpose.

PLOTTING OF CURVES

Graphs are of particular importance in physics because they display in a readily visualized form the relationship between pairs of interdependent quantities. Thus if two quantities x and y have the linear relation

$$y = ax + b \qquad (1)$$

a graph of y against x will be a straight line whose slope is a and whose y intercept (the value of y at which the line crosses the y axis, that is, at $x = 0$) is b. Conversely, if a given theory predicts that a certain physical quantity y depends linearly on another physical quantity x, this conclusion can be tested experimentally by measuring corresponding values of x and y and plotting these results. The plotted points will readily show whether a straight line can be drawn through them, even if various errors cause them to have a "scatter" instead of all lying right on a line. If there is a scatter, the straight line that represents the best average should be drawn as shown in Fig. 1. There are numerous rules for obtaining the line that is a true "best fit" to a given set of experimental points, but a simple determination by visual inspection using a transparent (plastic) straight edge is usually good enough and is all that will be required in these experiments. If the

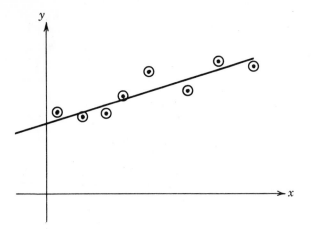

Figure 1   A "Best Fit" Straight Line Drawn through a Set of Experimental Points

scatter of points is so large that a good decision as to where to draw a straight line through them cannot be made, the conclusion that y depends linearly on x should be seriously questioned. If, on the other hand, the points readily define a straight line, not only does the resulting graph supply evidence that x and y are indeed linearly related but the value of a in Equation 1 may be obtained by finding the line's slope. Note that this is a convenient method of getting the average value of a. Moreover, a value for b representing an average result of all the plotted data can be read directly off the graph. This simple procedure is equivalent to the much more tedious one of determining the best values of a and b by a "best fit" calculation.

A linear relation between x and y is easily recognized when the points are plotted, but other relationships are not so obvious. Thus, suppose the relationship to be investigated were

$$y = ax^2 + b \qquad (2)$$

Plotting y against x would give a curve, but it would be very difficult to distinguish this curve from the curve resulting from, say, $y = ax^3 + b$. In fact, the straight line is the only graph that is really obvious. However, if the validity of Equation 2 is to be tested graphically, a new variable $u \equiv x^2$ can be introduced so that Equation 2 becomes

$$y = au + b \qquad (3)$$

Then, if Equation 2 is valid, a plot of y against u (that is, against x^2) will yield a readily recognizable straight line. This procedure may be used in many cases where a new variable u may be substituted for a function of x to produce a linear relation such as Equation 3.

A special case of this procedure arises when y depends exponentially on x, so that

$$y = Ae^{ax} \qquad (4)$$

where e is the base of natural logarithms and is approximately equal to 2.718. This situation occurs often enough to merit special treatment. The first step is to take the natural logarithm of both sides of Equation 4:

$$\ln y = \ln Ae^{ax} = \ln A + \ln e^{ax} = \ln A + ax \qquad (5)$$

Equation 5 is just like Equation 1 except that $\ln y$ rather than y is to be plotted against x. This plot will thus be linear if Equation 4 is valid, and the y intercept (b in Equation 1) will be $\ln A$. Hence A can be found by taking the antilog of the intercept.

Because this situation arises often, special graph paper is printed on which the graduations along the ordinate (y axis) are logarithmically rather than linearly spaced. This means that if a value of y is plotted on the given ordinate scale, the actual position of the point along the y axis will be proportional to the logarithm of y. In other words, plotting a value of y on this special graph paper automatically takes the logarithm, making a separate calculation of $\ln y$ for each value of y unnecessary. However, it should be noted that commercial logarithmic graph paper is set up for common (base 10) logarithms rather than natural (base e) logarithms. Taking the common log of both sides of Equation 4 yields

$$\log y = \log Ae^{ax} = \log A + \log e^{ax}$$
$$= \log A + ax \log e$$
$$= \log A + a(0.4343)x \qquad (6)$$

The slope of the resulting straight line is now $0.4343a$ rather than just a. Notice also that because the y axis is graduated logarithmically, the value of A may be read off from the intercept directly, the paper having also automatically done the job of the taking the antilog.

Another popular special case arises when a functional relation of the form

$$y = ax^n \qquad (7)$$

is to be investigated. Although this can be handled by introducing $u \equiv x^n$ as already discussed, such a procedure requires calculating x^n for each value of x. Since n may be any number, positive or negative, such calculations can get tedious unless a reasonably sophisticated calculator is available. Another very convenient method is again to take the logarithm of both sides of Equation (7). Then

$$\log y = \log a + n \log x \qquad (8)$$

and a plot of $\log y$ against $\log x$ will produce a straight line with slope n^*. The student may object that since he must look up the logarithms of all his values of both x and y, things haven't been simplified much for him, but

---

*Notice that in this case there is no intercept in the usual sense. $\log y = \log a$ and $y = a$ when $x = 1$.

again special graph paper is available that makes this calculation unnecessary. Since log y is now to be plotted against log x rather than x, this graph paper has both the ordinate and the abscissa graduated logarithmically. It is therefore called "full log" or "log-log" paper, whereas paper with the ordinate graduated logarithmically and the abscissa linearly is called "semilog." The logarithmic scales are called "1-cycle," "2-cycle," etc., depending on the number of powers of ten covered on the axis in question. Thus an axis graduated logarithmically from 1 to 10 is called "1-cycle"; from 1 to 100 "2-cycle"; etc. Scales of up to 5 cycles are available commercially, and in the case of full log paper there are various standard combinations of numbers of cycles along the ordinate and the abscissa. Appropriate graph paper pages are included in this book as needed; the student may want to refer to Experiment 32 for an example of semilog paper and to Experiment 35 for the full log type.

In drawing graphs, the student should choose scales for the coordinate axes so that the curve extends over most of the graph sheet and so that decimal parts of units are easily determined. This can be done if each small division is made equal to one, two, five, or ten units. The same scale need not be used for both axes. The independent variable should be plotted along the x axis and the dependent variable along the y axis. Each axis should be labeled with the name of the quantity being plotted and the scale divisions used. The numbers should increase from left to right and from bottom to top. Each graph should have a title indicating what the curve is intended to show.

Each point should be plotted as a dot surrounded by a small circle, which shows where the point is located even if the dot is obscured by the curve drawn through it. A straight line (or smooth curve if a straight-line plot is not being sought) should then be drawn through the dots. The curve need not pass through all the dots but should be drawn so as to fit them as closely as possible, as already mentioned. In general, as many points will lie on one side of the curve as on the other. The extent to which the plotted points coincide with the curve is a measure of the accuracy of the results.

SIGNIFICANT FIGURES

The numbers dealt with in mathematics are exact numbers. That is, when a mathematician writes 2 he means 2.00000 . . . , and all his subsequent calculations assume that his 2 means exactly two, not the tiniest fraction more or less. In physics the situation is very different. Many of the numbers dealt with come from measurements of physical quantities, and these can never be exact. For example, suppose that a distance is measured with an ordinary centimeter rule and found to be 5.23 cm. In this measurement the 3 is an estimate, for the smallest divisions on a centimeter rule are millimeters (tenths of a centimeter). The 3 represents a guess as to where between the 5.2 and 5.3 cm divisions the end of the measured distance lies. The statement that the distance was found to be 5.23 cm does not mean that it is exactly 5.23 cm but merely that it is probably not less than 5.22 cm nor more than 5.24 cm. If a high-quality micrometer had been used, the distance might have been found to be 5.2347 cm, where the 7 represents a guess as to where the micrometer's index line fell between the .234 and the .235 divisions. Thus the micrometer yields a much more precise value of the measured length than does the centimeter rule, but it too is not exact. More precise measurement methods might give further decimal places which cannot be determined with the micrometer any more than the 4 and the 7 could be found with the centimeter rule. Thus, when the result of the centimeter rule measurement has been written as 5.23 cm, it doesn't mean that the distance is exactly 5.23 cm nor can zeros be written after the 3. Nothing can be written after the 3 because the instrument being used gives no information as to what to write there.

The 5, the 2, and the 3 in the centimeter rule measurement are called *significant figures* because they each give trustworthy information about the size of the physical quantity being measured. The centimeter rule is quite good enough as a length-measuring instrument to determine that the length in question lies between 5.2 and 5.3 cm, and the 3 in the next place represents a significant guess as to where between 5.2 and 5.3 cm the actual length lies. The centimeter rule measurement is thus *good to three significant figures,* whereas the micrometer measurement gave five significant figures, the micrometer being a much more precise length-measuring instrument than the centimeter rule. The 3 in the centimeter rule measurement and the 7 in the micrometer measurement are less significant than the other figures but are still considered significant because they give some real information about the desired length even though there is some doubt about their actual values. Clearly, however, if there is some doubt about them, any figures that might get written to the right of them in the respective cases would be meaningless. In particular, the student should be careful not to write zeros there. If the length measurement made with the centimeter rule were recorded as 5.230 cm, the zero would be a significant figure and would mean that somehow the student was able to interpolate between the 5.2 and 5.3 divisions to 1/100 rather than just 1/10 of the space between them. Indeed, the micrometer measurement shows that the figure to the right of the 3 should be 4, so that putting a zero there says something that isn't true. Care should always be taken to distinguish between zeros that are significant and those that are not. In general, zeros that merely serve to place the decimal point are not significant. Thus, if the length measurement were to be stated in meters, the two zeros in 0.0523 meters would not be significant. They merely place the decimal point appropriately in the three-significant-figure measurement. However, if in measuring the distance with the centimeter rule the end of this distance appeared to fall right opposite the .2 cm division following the 5 cm mark, it would be recorded as 5.20 cm and the zero would be significant. In general, zeros appearing to the right of figures that are already to the right of the decimal point must be regarded as significant, for if they weren't they wouldn't be there. Zeros between other figures and the decimal point should usually be regarded as serving only to place the decimal point. The example of the length of 0.0523 meters is typical, there being no doubt that the zeros are not significant. There are some ambiguous cases, however. Suppose that a certain race course is found to be 1.2 km long. As written, this is a two-significant-figure measurement. The same result may be given as 1200 meters. Here again the zeros are not significant but must be present in order to properly located the decimal point. Without the knowledge that the original measurement of 1.2 km contained only two significant figures, however, there is no way to tell whether these zeros are significant or not. In such cases the experimenter must rely on his knowledge of the measuring instrument to tell him how many significant figures are justified.

There is usually no problem in deciding how many significant figures a given measurement should contain, but difficulties arise when these numbers are used in calculations. This is because mathematics assumes that all numbers are exact and thus automatically fills all places to the right of the last significant figure with zeros even though this is physically wrong. The calculations then often produce a great many figures that look as if they were significant but really are not, for clearly no mathematical manipulation can give a result whose precision is greater than that of the quantities put into it. Some examples may serve to show how this problem should be handled.

1. Addition and subtraction: When carrying out addition or subtraction by hand, do not carry the result beyond the first column that contains a doubtful figure. This means that all figures lying to the right of the last column

in which all figures are significant should be dropped. Thus in obtaining the sum of these numbers

806.5		806.5
32.03	they should be written as	32.0
0.0652		0.1
125.0		125.0
		963.6

Note that, in dropping nonsignificant figures, the last figure retained should be unchanged if the first figure dropped is less than 5 and should be increased by 1 if the first figure dropped is 5 or greater. This is a normal convention to which this book will adhere.

If an electronic calculator is used, the numbers to be added may be entered without the bother of determining which figures to drop, in which case all figures will appear in the sum. This is like adding the numbers as given on the left in the example above. The result will be 963.5952. The student must then look at his data and observe that in two of the numbers he is adding he has no knowledge of what the figure in the second decimal place should be. He must therefore round his result off to one decimal place by dropping the 952. Since the 9 is equal to or greater than 5, he raises the figure in the first decimal place by one and gets 963.6 as before.

2. Multiplication and division: The operations of multiplication and division usually produce many more figures than can be justified as significant, so that results must be properly rounded off. The rule is to retain in the result only as many figures as the number of significant figures in the least precise quantity in the data. Suppose the area of a plate is to be measured. A centimeter rule is used to find that the plate has a length of 7.62 cm and a width of 3.81 cm. As in the earlier example with the centimeter rule, these measurements each contain three significant figures, of which the third is doubtful. If the area is now found by multiplying 3.81×7.62 either by hand or with a calculator, 29.0322 cm² will be obtained. This number appears to have six significant figures, but the two original quantities have only three each. Therefore only three significant figures should be retained in the result, which should be written as 29.0 cm². Note that multiplication with a slide rule automatically produces this result. Slide rules, like centimeter rules, can only be read to a certain precision no matter what the numbers involved are; therefore they round off the result automatically. With some exceptions, the measurements to be made in the experimental work covered by this book will contain two or three significant figures, so that the standard foot-long slide rule, which is good to three significant figures, is recommended for all computations involving multiplication and division. Electronic calculators, which allow all calculations (including, in some

cases, trigonometric and logarithmic ones) to be made with great convenience, are also recommended with the proviso that the student must remember to round off his final result to the number of figures that can be justified by the data as being significant.

THEORY OF ERRORS

All measurements are affected by errors; this means that measurements are always subject to some uncertainty. There are many different types of errors, such as personal, systematic, instrumental, and accidental errors. Personal errors include blunders, such as mistakes in arithmetic, in recording an observation, or in reading scale divisions. Another important kind of personal error is known as personal bias, such as trying to fit the measurements to some preconceived idea, or being prejudiced in favor of the first observation. Systematic errors are characterized by their tendency to be in one direction only, either positive or negative. For example, if a meter stick is slightly worn at one end, and measurements are taken from this end, then a constant error will occur in all these measurements. Instrumental errors are those introduced by slight imperfections in the manufacture or calibration of the instrument. The worn meter stick just mentioned or an electrical meter that has not been properly set to zero with no input are examples of instrumental errors. Note that such errors are usually systematic. Accidental errors are deviations beyond the control of the observer. These errors are due to jarring, noise, fluctuations in temperature, variations in atmospheric pressure, and the like. Included in this category are variations in observed data due to inherently random processes such as the intersurface actions that produce the force we call friction and the radioactive decay of atomic nuclei. Since the causes just listed for accidental errors are essentially random in nature, all these causes of data variation are subject to treatment by statistical methods as will be discussed below.

It will be assumed in these experiments that instrumental errors due to improper calibration, zeroing, etc., have been prevented by proper inspection and adjustment of the equipment and that care has been taken to eliminate systematic errors, personal errors, and personal bias. There remain accidental errors, which make themselves known by causing a spread in the values obtained when a given measurement is repeated several times. Two examples may serve to illustrate how this comes about.

Consider first the distance measurement with the centimeter rule discussed in connection with significant figures. In the measurement of 5.23 cm, the 3 was doubtful, being an interpolation between the 5.2 and 5.3 cm divisions, which are the smallest divisions on the centimeter rule. If a two-significant-figure result were adequate, the distance could have been quoted as 5.2 cm.

In this case, if the measurement were repeated many times, even by different experimenters, the likelihood is that 5.2 cm would be obtained each time. No accidental error is revealed because the measuring instrument is not being pushed to the limit of its precision and random processes in the experiment (such as small variations in the length of the rule and/or the distance being measured due to temperature fluctuations) are negligible compared to the smallest scale unit in the measuring instrument (the millimeter divisions on the centimeter rule in this case). The precision here is said to be limited by the scale of the instrument.

However, if the distance measurement is repeated with an estimated interpolation made each time between the 5.2 and 5.3 cm divisions, the same estimate may not always be made. This would be especially true if each measurement were made by a different experimenter who had no knowledge of his or her colleagues' results. Thus one might guess 5.22 cm, another 5.21, another 5.24, etc. To handle this situation, a mean or average of the various measured values is calculated. As will be discussed below, this average is more accurate than any one of the measurements alone and can in fact be shown to improve in accuracy as the square root of the number of individual measurements made. Clearly a way of improving the accuracy of experimental data is to measure each quantity many times, and an important matter of judgment in experimental work is to decide on how many times a given quantity is to be measured. In this regard it must be remembered that to measure something N times takes N times as long as measuring it once, but the accuracy obtained by doing so is only $\sqrt{N}$ times as great. Thus, if a certain measurement takes one minute, making it ten times will take only ten minutes but will yield over three times the accuracy. However, making it a hundred times will take an hour and forty minutes, but this investment in time will only yield another threefold increase in accuracy. Clearly a compromise based on the accuracy required, the time needed for a particular measurement, and the time available must be reached in each case.

A second example, this being one dominated by random processes inherent in the experiment itself, is that of the range of a spring-operated gun. The experimental setup is shown in Fig. 2. The spring gun consists of a tube containing a spring-loaded plunger. A small steel ball is placed in the tube against the plunger. The plunger

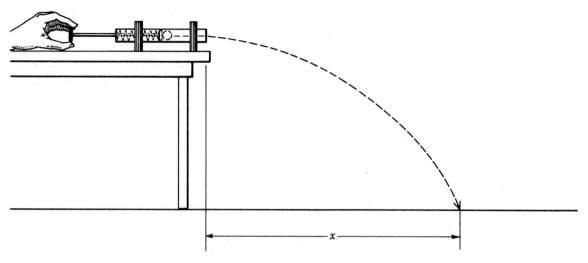

Figure 2   The Range of a Spring Gun

is pulled back a given distance, compressing the spring by a known amount, and is then released sharply, propelling the ball out of the tube. The ball strikes the floor at a horizontal distance x from the end of the tube (the gun's muzzle), this distance being the range in question.

If this experiment is repeated under conditions made as identical as possible to those in effect on the first try (the ball is carefully put back in the tube, the plunger is pulled back by a distance made as closely equal as possible to that used the first time, and care is taken to release the plunger in the same way), will the ball strike the floor at exactly the same point? Simple theory predicts that it will, but small variations in the distance the plunger was pulled back, in the state of the spring, and in the condition of the surfaces of the ball and the inside of the tube — all random, uncontrollable effects — will cause the measured range to vary somewhat on subsequent shots. Indeed, no one would really expect successive shots from a gun to all land in precisely the same spot even though the gun were clamped in a fixed position and given the same charge each time. Instead, a spread of impact points would be expected, as shown in the plan view of the spring-gun experiment in Fig. 3. The extent of the spread may be reduced by using great care in the experimental technique (wiping off the ball after each shot, handling it with plastic gloves to prevent getting fingerprints on it, taking care in the measurement of how far the plunger is retracted, and releasing the plunger smartly each time), but the spread can never be reduced to zero. The size of the spread is a measure of the precision of the experiment. An estimate of this precision is very desirable in all experimental work, and the following discussion will show how the extent of the spread can be used to express such an estimate quantitatively.

According to statistical theory, the arithmetic mean or

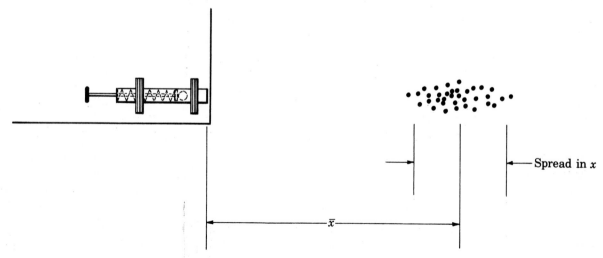

Figure 3   The Range of a Spring Gun. Plan View

average of a number of observations will give the most probable result. This is clear from the results of the range experiment illustrated in Fig. 3. If a single number is to be quoted as the range of the spring gun, it should be the distance from the gun muzzle to the center of the distribution of impact points. In the absence of any peculiar experimental effects, we expect the distribution of points to be densest near the center, to thin out as we go away from the center, and to be symmetrical (to show as many impact points beyond the center as short of it). Hence in this normal case the *average* range $\bar{x}$ is also the *median* (the midpoint of the distribution with as many points with bigger x as with smaller x) and the *most probable* value (the point near which there is the greatest density of points). Thus the first step in data analysis is to find the average of the distances from the gun muzzle to all the individual impact points. This is shown on the left in Table 1. Note that these measurements are made with a meter stick and could therefore be given to one more significant figure by interpolation between the millimeter divisions on the stick. If this order of precision were wanted, each value of x would be measured several times by different investigators, each of whom would make his own interpolation, and an average value obtained for each x. Then every entry in the left-hand column of Table 1 could be quoted to two decimal places (five significant figures). However, this would take a great deal of time and effort all of which would be wasted because the spread in the data is several centimeters, making the fifth significant figure in each measurement nonsignificant in the final result. In other words, the random effects in the experiment dominate the picture and limit the useable precision of the measuring instrument.

One obvious way of expressing the extent of the spread in a set of experimental data is to note the deviation of each measurement from the average or arithmetic mean just found. In the example of the spring-gun range experiment, these deviations (differences between each measurement and the average) are tabulated in the middle column of Table 1 and their average is then computed. Note that, in computing this average, no account is taken of the algebraic signs of the deviations. A deviation represents an error — a difference between a particular measurement and the average of all the measurements, this average being the closest available approximation to the true value of the quantity being measured. Which way the deviation lies makes no difference; it is still an error. The average error is a measure of the scatter of the observed values about their average. The average deviation thus found is therefore often called the average error, and for the purposes of the elementary laboratory, it may be taken as the possible error in the mean value. Con-

sequently the result of the range measurement should be written as 133.9 ± 3.1 cm to show that the true value of the range has a high probability of lying between $133.9 - 3.1 = 130.8$ cm and $133.9 + 3.1 = 137.0$ cm. Actually, a statistical analysis shows that if a very large number of range measurements were made, 57.5% of them would lie inside this interval. That is, 57.5% of the impact points would be between 130.8 and 137.0 cm from the gun muzzle.

TABLE I   The Range of a Spring Gun

Range, cm	Deviations, cm	Deviations Squared, cm²
134.2	+0.3	0.09
139.5	+5.6	31.36
133.0	−0.9	0.81
136.6	+2.7	7.29
129.4	−4.5	20.25
127.8	−6.1	37.21
130.6	−3.3	10.89
136.5	+2.6	6.76
135.3	+1.4	1.96
131.9	−2.0	4.00
138.1	+4.2	17.64
11 ⟌ 1472.9	11 ⟌ 33.6	11 ⟌ 138.26
$\bar{x} = 133.9$ cm	a.d. = 3.1 cm	12.57

$$\sigma = \sqrt{12.57} = 3.5 \text{ cm}$$

Statistical theory also presents some other useful ways of stating the accuracy of an experimental result. For example, the fact that the average of a set of measurements gets more and more accurate in proportion to the square root of the number of measurements made can be reflected in the stated error by dividing the average error by $\sqrt{N}$, where N is the number of measurements. The result is called the *average deviation of the mean* (A.D.). Thus

$$\text{A.D.} = \frac{\text{a.d.}}{\sqrt{N}} \tag{9}$$

where a.d. stands for the average deviation *from the* mean, that is, the average error already discussed. The A.D. is a measure of the deviation of the arithmetical mean from the true value and is in this context generally known as the probable error. The significance of the A.D., from probability theory, is that the chances are 50% that the true value of the quantity being observed will lie within ±A.D. of the mean. Thus, in the example of the spring gun the mean of the measured ranges is

133.9 cm and the average deviation from the mean (the a.d.) is 3.1 cm, which says that on the average the readings differ from the mean (133.9 cm) by ±3.1 cm. The average deviation *of* the mean (the A.D.) is $3.1/\sqrt{11} = 0.9$ cm, which says that the chances that the true value of the range will lie in the interval 133.9 ± 0.9 cm are 50%, while the chances that it will lie outside this interval are also 50%.

Another (and, from the standpoint of statistical theory, most important) measure of the so-called *dispersion* (scatter of experimental points) is the *standard deviation*. This is defined as the square root of the average of the squares of the individual deviations, or, mathematically, by

$$\sigma = \sqrt{\frac{(x_1 - \bar{x})^2 + (x_2 - \bar{x})^2 + \cdots + (x_N - \bar{x})^2}{N}} \quad (10)$$

where σ is the standard deviation, $x_1, x_2, \ldots, x_N$ are the N individual measurements, and $\bar{x}$ is their average. Note that the signs of the various deviations make no difference in calculating σ since each is squared. An example of a standard deviation is given in the right-hand column in Table 1. Like other measures of dispersion, the standard deviation gives information about how closely the distribution is grouped about the mean. Statistical analysis shows that for a large number of normally distributed measurements, 68.3% of them will fall within the interval $x \pm \sigma$. In the results of our hypothetical range experiment, this is 133.9 ± 3.5 cm, and after a large number of firings we would expect to find that about 68% of the impact points lay between 130.4 and 137.4 cm from the gun muzzle.

When a very large number of measurements of a given quantity are made and variations between the different values obtained are due to truly random effects, a *normal distribution* of these values will be found. The word "distribution" as used here means an expression of the relative frequency with which the different observed values occur. Such an expression often takes the form of a graph in which the number of observed values in a small interval centered on a particular value of x is plotted against x. Thus, suppose in the experiment with the spring gun a very large number of range observations were made. We could, for example, count the number of such observations falling in the interval 127.0 ± 0.5 cm and plot this number as the ordinate of a point whose abscissa was 127 cm. Another point would be the number of observations falling between 127.5 and 128.5 cm plotted with an abscissa of 128 cm, and this process could be continued until we found somewhere beyond 140 cm that there were no more observed points to plot. A smooth curve is then drawn through the plotted points.

According to our earlier discussion, this curve should show a maximum at the mean value $\bar{x} = 133.9$ cm and should fall off symmetrically on either side. Such a curve, called the *normal curve*, was first discovered by a famous French mathematician, De Moivre, while working on certain problems in games of chance. It was also derived independently by Laplace and Gauss, who made statistical use of it and found that it accurately represents the errors of observation in scientific measurements. The curve is also known as the *normal probability* curve because of its use in the theory of probability, as the *normal curve of error,* and as the *Gaussian curve.* Here error is used to mean a deviation from the true value. Whenever any measurements are made in which there are random fluctuations, the results predicted by the normal curve are found to be valid. Thus it has been found that this curve describes very well many distributions that arise in the fields of the physical sciences, biology, education, and the social sciences.

The mathematical representation of the normal curve is given by the equation

$$y = \frac{N}{\sigma\sqrt{2\pi}} e^{-(x - \bar{x})^2/2\sigma^2} \quad (11)$$

where y represents the distribution function. In accordance with the foregoing discussion of how the normal curve is obtained in a physical case, the number $y\Delta x$ is the number of measurements of x (out of a very large total number N) that fall within the very small interval Δx centered on the value of x for which y was computed. Thus suppose we want to know how many measurements of x will fall in the vicinity of a certain value x_0. Putting x_0 into Equation 11 for x, we calculate the corresponding value y_0 for y. Then $y_0\Delta x$ is the number of measurements of x expected to be found in the interval $x_0 - \frac{1}{2}\Delta x$ to $x_0 + \frac{1}{2}\Delta x$. Thus y_0 is to be interpreted as the number of measurements *per unit interval in x* falling in the neighborhood of x_0. Note that $y_0\Delta x$ is the area of a tall, thin rectangle Δx wide located at x_0 on the x axis of the coordinate system in which the normal curve is plotted and extending up to the curve (a height y_0). Because this area represents the number of measurements of x falling within Δx, the area under the complete curve should be equal to N, the total number of measurements made. The factor $1/\sigma\sqrt{2\pi}$ in Equation 11, called a normalizing factor, is chosen to bring this about. It follows that the area under the curve between ordinates erected at some pair of values $x = a$ and $x = b$ is the number of measurements falling between a and b.

The curve is bell-shaped and symmetrical about the line $x = \bar{x}$. It has a maximum for this value of x (bearing out the idea that the mean value of x is also the most probable value) and falls quite rapidly toward the x axis on both sides. For different normal distributions, the curve has the same general shape, but its steepness, height, and location along the x axis will depend on the values of N, $\bar{x}$, and σ. The characteristic properties of the normal curve can be studied very readily by representing it in a new set of variables by means of a mathematical transformation. The first step is to divide Equation 11 through by N. The quantity $(y/N)\Delta x$ is then the fraction of the total number of measurements or the probability of obtaining a measurement in the narrow interval Δx centered on the value of x for which y is computed. Put another way, y/N is the probability *per unit interval in x* of getting a value of x lying in that interval whenever a measurement of x is made. Clearly, the area under the curve of y/N is unity, corresponding to the fact that the probability of getting *some* value of x is 1 or 100%. Next, we take the origin of our transformed coordinate system to be at the arithmetic mean and use the standard deviation as the unit of measurement along our new horizontal axis. This is done by choosing a new variable t related to x by the equation $t = (x - \bar{x})/\sigma$. When this is substituted into Equation 11 with N divided out and the normalizing factor appropriately modified, we obtain in place of $y(x)$ a new function $\phi(t)$ given by

$$\phi(t) = \frac{1}{\sqrt{2\pi}}\, e^{-t^2/2} \qquad (12)$$

This is the *standard form* of the normal curve, where $\phi(t)$ is the distribution and t is the variable. For this distribution, the average value of t is zero and the standard deviation is equal to one. The advantage of representing the normal curve in the standard form is that the area under the curve between any two values of t may be calculated once and for all. These values are tabulated and can be obtained from tables of probability integrals. Any normal distribution may then be expressed in the standard form and the required calculations of the characteristics of the distribution can easily be made.

It is important to know the *properties of the normal curve* to be able to use it in practical applications. The curve is bell-shaped and symmetrical about the ordinate at $t = 0$. The ordinates for negative values of t are the same as for the corresponding positive values of t. The curve has a maximum at $t = 0$, falls quite rapidly toward the axis on both sides, and approaches very close to the horizontal axis at both extremities. Theoretically, it extends from $-\infty$ to $+\infty$, but practically it is so close to the axis beyond $t = \pm 3$ that the area under the curve beyond these points is almost negligible.

The mean, the median, and the most probable value all coincide at the origin, where $t = 0$. The distances along the horizontal axis are measured in units of σ, the stan-

dard deviation. The total area under the curve is equal to one. Hence the area under any portion of the curve represents the relative frequency (expressed as a fraction of unity or as a percent) with which measurements in that interval occur. Numerical values of such areas may be obtained from tables and can be changed into the actual frequencies of occurrence by multiplying by N.

The *percentage distribution* of area under the normal curve is given in Fig. 4, where σ is the unit of measurement. The significance of the values given in the figure is that if the values of x are normally distributed, the probability that a value chosen at random will fall within the range $\bar{x} \pm \sigma$ is 0.683. The probability that it will lie within the range $\bar{x} \pm 2\sigma$ is 0.9545. The probability that it will lie within the range $x \pm 3\sigma$ is 0.9973. Thus the probability that it will lie outside of this range is only 0.0027 or 0.27%. Hence a deviation of 3σ on both sides of the arithmetic mean includes practically the whole of a normal distribution. The probability that any particular value of x will lie in the range $\bar{x} \pm$ a.d. is also easily found from Equation 12 and Fig. 4. The average deviation from the mean (a.d.) is simply the average $\bar{t}$ of t expressed in units of σ and can be calculated from Equa-

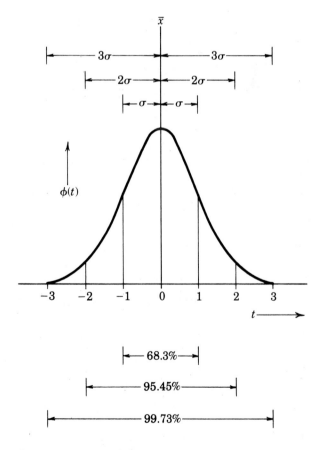

Figure 4   The Percentage Distribution of Area Under the Normal Curve

tion 12. The result is

$$\bar{t} = \sqrt{\frac{2}{\pi}} = 0.798$$

$$\text{a.d.} = \sigma\bar{t} = \sigma\sqrt{\frac{2}{\pi}} = 0.798\sigma \qquad (13)$$

The area under the curve of Fig. 4 between -0.798 and $+0.798$ is then found to be 0.575 or 57.5% of the total area. This means that, as noted earlier, 57.5% of a large number of measurements will fall within the interval $\bar{x} \pm \text{a.d.}$ Similarly the area under the curve between -1 and $+1$ is found to be 0.683 of the total area so that, also as noted earlier, 68.3% of a large number of measurements will fall within the interval $\bar{x} \pm \sigma$.

A somewhat different concept is involved in the so-called *errors of measurement,* by which we mean deviations of our result (the average of our observed values) from the true value of the quantity being measured. Since the true value is not known, the errors must be expressed in terms of the probability that the true value lies within a certain range of the experimental result. Different types of errors can be defined. For example, we have already

defined the probable error as a deviation such that the probability is 50% that the true value lies within a distance of the probable error from the arithmetic mean of the observations. Since the curve of Fig. 4 may be considered as the error curve for a normal distribution, we may find the value of t for which 50% of the area under the curve lies between $\pm t$ and define the probable error as σ times this value. Reference to a probability integral table shows that this value of t is 0.6745, hence the probable error is 0.6745σ. Another type of error is the standard error, which is equal to σ itself, the standard deviation. Again treating Fig. 4 as an error curve, we see that the probability that the true value lies within a distance σ of the arithmetic mean, that is, within the interval $\bar{x} \pm \sigma$, is 68.3%. The *reliable error* is equal to 1.6949σ and carries a 90% probability that the true value lies within this distance of the mean. Of course, the larger the value of the acceptable error, the higher will be the probability that the true value lies within that range of measurement. Conversely, if the measurements show a small spread, the numerical value of the standard deviation will be small and hence the errors will be small, even with a high degree of probability.

PERCENT ERROR

The error in a measured quantity is often conveniently expressed as a percent of the quantity itself. Since the true value is usually not known, one of the errors discussed above must be used in calculating this percent. For this purpose the probable error is usually chosen because it is the same as the A.D. and therefore the easiest to calculate from the tabulated data. This percent error, also called the percent deviation of the mean, or the percent A.D., is equal to the A.D. divided by the arithmetic mean of the measured values and multiplied by 100 to give the result in terms of a percent as desired. That is

$$\% \text{ A.D.} = \frac{\text{A.D.}}{M} \times 100\%$$

where M is the arithmetic mean. This is the quantity usually considered in judging the accuracy of a series of measurements. In the example of the range of the spring gun, the percent A.D. is $\dfrac{0.9}{133.9} \times 100\% = 0.67\%$.

If the true or accepted value of a quantity is known, the actual error can be calculated as the difference between

the result obtained from the experiment (the mean value M of the measurements) and the true value M_t. The relative error is then the ratio of the error to the true value, and the percent error is this ratio times 100%. Thus

$$\% \text{ error} = \frac{M - M_t}{M_t} \times 100\%$$

For example, suppose that a student measures the velocity of sound in dry air at 0°C and finds it to be 333.1 meters per second, while the accepted value is 331.4 meters per second. The error is 1.7 meters per second. The relative error is is $\dfrac{1.7}{331.4}$ or 0.005. The percent error is $\dfrac{1.7}{331.4} \times 100\% = 0.5\%$

There is no definite value for the allowable percent error to be expected in the following experiments. In many cases it is reasonable to expect results within 1%, while in some cases the error may be 5% or more, depending on the apparatus used. However, all measurements should be made with the greatest care, so as to reduce the error as much as possible.

CALCULATING WITH ERRORS

Whenever an experimental result is used in a calculation, account must be taken of the fact that an error is associated with it. Suppose we have two results x_1 and x_2 with respective standard deviations σ_1 and σ_2. We quote these results as $x_1 \pm \sigma_1$ and $x_2 \pm \sigma_2$ on the basis that the 68% chance that the true values lie within these ranges is good enough.* If the theory of the experiment requires these two quantities to be added, the sum is $x_1 + x_2 \pm \sigma_1 \pm \sigma_2$. It would, of course, be very nice if the error in one measurement was in the opposite direction from and therefore cancelled the error in the other, but this happy event can hardly be counted on. It is much safer to assume the worst, that is, that the errors are in the same direction so that the sum becomes $x_1 + x_2 \pm (\sigma_1 + \sigma_2)$. We conclude that to be on the safe side we should add the errors in the individual quantities to obtain the error in the sum. Statistics show, however, that this approach is unduly pessimistic and that in fact, when standard deviations are being used, the standard deviation σ_s in the sum is the square root of the sum of the squares of the individual standard deviations. Thus

$$\sigma_s = \sqrt{\sigma_1{}^2 + \sigma_2{}^2}$$

The procedure is identical in the case of a subtraction, but care should be taken to note that the percent error can increase tremendously when two quantities of about the same value are to be subtracted one from the other. In such a case we get $x_1 - x_2 \pm \sqrt{\sigma_1{}^2 + \sigma_2{}^2}$, and if x_1 and x_2 are nearly equal their difference may be smaller than the error. This means that the errors associated with x_1 and x_2 are large enough so that we cannot tell whether the quantities are equal or slightly unequal and hence whether or not a difference actually exists. Consider, for example, two automobiles driving down the highway with one slowly passing the other. The problem is to measure the passing speed, that is, the difference between the speeds of the individual automobiles. The best way to do this (since we are not interested in the individual speeds) is to measure the relative speed directly, but suppose experimental difficulties made this impossible so that the only data obtainable was the speedometer reading in each car. We find that one reads 61 miles per hour and the other 62 miles per hour. Can we conclude that one car is passing the other at the rate of 1 mph? Not really, for automobile speedometers are good to only two significant figures with the last one in doubt. There is

therefore an error of about 1 mph in each speedometer, so that their readings should be reported as 61 ± 1 and 62 ± 1 mph respectively. This being the case, the difference must be given as $1 \pm \sqrt{1^2 + 1^2} = 1 \pm 1.4$ mph. From this result alone we cannot tell whether the cars are proceeding side by side or even, if they are not, which is passing the other. Certainly there seems to be a bias in favor of one car's doing the passing, but on the basis of the available data a reasonably certain conclusion simply cannot be drawn. Thus, if we really want to know which car is passing the other and how fast, a method must be developed for measuring their relative velocity directly.* Although this does not seem to be a difficult task in our example of the cars, the progress of research in physics has often been marked by breakthroughs due to someone's inventing a method for direct measurement of a quantity that had been formerly obtainable only as the difference between two other quantities.

If the two quantities $x_1 \pm \sigma_1$ and $x_2 \pm \sigma_2$ are to be multiplied, we have

$$(x_1 \pm \sigma_1)(x_2 \pm \sigma_2) = x_1 x_2 \pm \sigma_1 x_2 \pm \sigma_2 x_1 + \sigma_1 \sigma_2$$

On the assumption that the errors are reasonably small, $\sigma_1 \sigma_2$ can be dropped on the basis that the product of two small quantities is negligibly small. Assuming the worst case of errors that are in the same direction and therefore add, we get

$$(x_1 \pm \sigma_1)(x_2 \pm \sigma_2) \approx x_1 x_2 \pm (\sigma_1 x_2 + \sigma_2 x_1)$$

This result is inconvenient to handle in computation, but note that the percent error (the percent standard error in this case) is

$$\% \text{ error} = \frac{\sigma_1 x_2 + \sigma_2 x_1}{x_1 x_2} \times 100\%$$
$$= \left(\frac{\sigma_1}{x_1} + \frac{\sigma_2}{x_2} \right) \times 100\% = \frac{\sigma_1}{x_1} \times 100\% + \frac{\sigma_2}{x_2} \times 100\%$$

In other words, the percent error in the product is the sum of the percent errors in each factor. It is easy to show that the same holds true for division.

Raising an experimental quantity to an integer power is treated just like multiplication, for the integer power merely tells how many times the quantity is to be multi-

---

*For the experiments covered in this manual the a.d., which is easier to find than σ, will usually be good enough.

*We could also substitute speedometers of much higher precision, but this is an inelegant approach, as most of the precision will be wasted in getting the desired result.

plied by itself. Thus in calculating the area x^2 of a square of measured side x, the exponent 2 is simply an instruction to multiply x by x and has no error associated with it. If the error in x is σ, we have

$$(x \pm \sigma)^2 = x^2 \pm 2\sigma x + \sigma^2 \approx x^2 \pm 2\sigma x$$

and the percent error is $\pm \dfrac{2\sigma x}{x^2} \times 100\% = 2\dfrac{\sigma}{x} \times 100\%$ or twice the percent error in the measurement of x. The situation may also be regarded as the product of $x \pm \sigma$ with itself, with error $\pm \sigma x \pm \sigma x$, but note that in this case the two errors σx *must* be considered to add since they are not separate but identical and hence necessarily in the same direction. In general it can be shown that if a measured quantity is raised to any power n, integer or fractional, the percent error in the result will be n times the percent error in the quantity. Notice that this is not n times the error but n times the *percent* error. Consequently, great care must be taken in the measurement of a quantity that is to be raised to a power greater than one, for the percent error in the result will be greater than that in the measurement by a factor of that power. A quantity raised to a high power may be thought of as "dominating" the result in that the error in this quantity will be much more important than the error in other quantities that may be multiplied or added in without being raised to such a power.

Conversely, the percent error in quantities raised to a power less than one is less important because only the fraction of that error represented by the less-than-unity value of n appears in the result. For example, in the experiment on the simple pendulum, in which the theory says that the period is proportional to the square root of the pendulum's length, only half the percent error in the length measurement will show up in the calculated period. But in an investigation of the Stefan-Boltzmann law, which states that the power radiated by a hot surface is proportional to the fourth power of the surface temperature, the temperature measurements will be extremely critical.

Finally, the case of the exponential (a mathematical number such as the base e of natural logarithms, with which no error is associated, being raised to a power given by a measured quantity) must be considered. Suppose that the theory under investigation calls for e to be raised to the power x where x is a measurement containing standard error σ. We have

$$e^{x \pm \sigma} = e^x e^{\pm \sigma} = e^x \left(1 \pm \sigma + \frac{\sigma^2}{2} \pm \cdots \right)$$

where we have used the property $e^{a+b} = e^a e^b$ of exponentials and have expanded $e^{\pm \sigma}$ in its power series. Since σ is a reasonably small quantity, terms in σ^2, σ^3, etc. can be neglected to give

$$e^{x \pm \sigma} \approx e^x (1 \pm \sigma) = e^x \pm \sigma e^x$$

The percent error is then $\pm \sigma \times 100\%$, which means that here *the error in the measured quantity is numerically equal to the relative error in the result and hence to the percent error after multiplication by 100%.* Thus, in any experiment in which e is to be raised to a measured quantity, particular care must be exercised in making the measurement in order to keep the error as small as possible. There is, in fact, a famous case (that of the thermionic emission of electrons from a heated metal surface) in which the theory, which predicts that the current will be proportional to some power of the temperature multiplied by an exponential having the temperature in the exponent, has never been really satisfactorily checked. There is no question about the exponential dependence, but it so dominates the result that even very careful experimental data has not been able to verify what, if any, power of the temperature the exponential should be multiplied by. Such are the challenges, the frustrations, and the fascination of experimental physics. It is hoped that the student will in some measure taste all three as he performs the work described in this manual.

Measurement of Length, Mass, Volume, and Density 1

Physics is a quantitative experimental science, and as such it is largely a science of measurement. Many measuring instruments of great accuracy have been developed to meet the requirements of the physics laboratory. The measurement of length is of fundamental importance in scientific work, hence it is fitting to begin experimental work with this type of measurement.

Mass is another fundamental quantity which must often be measured, and mass determinations will also be done in this experiment. Finally, if the dimensions of a regular geometrical object have been measured, its volume may be calculated; and if its mass is also known, its density may be found. Density is one of the important characteristics of materials. The engineer uses tables of densities in calculating the weights of bridges and other structures that cannot be weighed. From the dimensions the volume can be computed and then from tables of densities the total weight can be calculated.

In this experiment the dimensions of various objects will be measured by means of a meter ruler, vernier caliper, and micrometer caliper, and the probable error in these measurements will be determined. There will also be an opportunity to compare the English and metric units of length. The mass of each object will be found with either an equal-arm balance or a triple-beam balance, and the density of the material of which each object is made will be obtained.

THEORY

In measuring various physical quantities it is important to be able to use the proper instruments. The simplest way to measure length is to use an ordinary meter scale or an English ruler. For ordinary measurements, a length is determined by a rough comparison with the scale — the result is given at a glance and in round numbers. For a more precise determination, the scale must be very accurately made and must be read to a fraction of its smallest scale division.

As discussed in the Introduction, when an instrument is used to the limit of its precision, certain errors occur which cannot be eliminated; these are called accidental errors. Thus, a series of measurements of a physical quantity usually differ among themselves, because of the accidental errors involved. The best value of the quantity measured, that is the most probable value, is the arithmetic mean (the average of the values obtained).

The precision of measurement can usually be increased by using more complex and more accurate equipment and by taking precautions to eliminate errors as much as possible. However, the student should always try to make his measurements with the greatest accuracy attainable with the given apparatus.

● **The Vernier Caliper** When a measurement is made with an ordinary meter stick, it is necessary to estimate the tenths of a millimeter, the smallest division on the scale. A vernier helps in reading accurately the fractional part of a scale division. It is a small auxiliary scale that slides along the main scale. The graduations of the vernier are different than those of the main scale, but they are simply related.

The vernier caliper, which is illustrated in Fig. 5, consists of a fixed part and a movable jaw. The fixed part

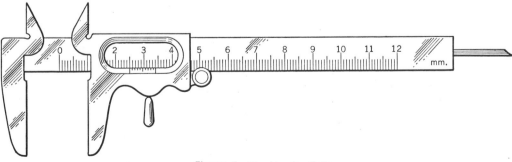

Figure 5 The Vernier Caliper

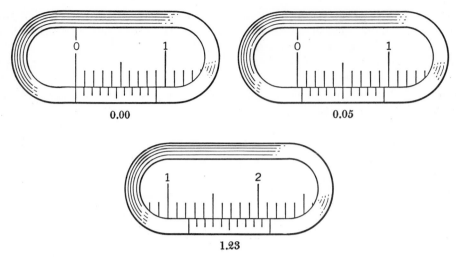

Figure 6   Vernier Readings

includes a stem, on which is engraved the main scale, and a fixed jaw attached to the stem. The movable jaw is free to slide on the fixed stem and has a vernier scale engraved on it. The fixed scale is divided into centimeters and millimeters. The vernier is divided so that ten divisions on it cover the same interval as nine divisions on the main scale. Hence the length of each vernier division is $\frac{9}{10}$ the length of a main-scale division. When the jaws are closed, the first line at the left end of the vernier, called the zero line or the index, coincides with the zero line on the main scale. However, the first vernier division is 0.1 mm away from the first main-scale division, the second vernier division is 0.2 mm away from the second main-scale division, and so on. If the jaws are slightly opened, it is easy to tell what fraction of a main-scale division the vernier index has moved by noting which vernier division coincides with a main-scale division.

A measurement is made with the vernier caliper by closing the jaws on the object to be measured and then reading the position where the zero line of the vernier falls on the main scale. This setting of the zero line of the vernier on the main scale is an incomplete measurement however, because no attempt was made to estimate to a fraction of a main-scale division. The fractional part of a main-scale division is obtained by noting which line on the vernier coincides with a line on the main scale. Several possible vernier readings are illustrated in Fig. 6.

Frequently a vernier does not read zero when the jaws are completely closed. In such cases a zero correction must be applied to every reading; this correction may be either positive or negative. If the errors in the direction of increasing main-scale readings are called positive, then the zero correction is always made by subtracting the zero reading from the final reading.

Vernier scales are attached to many different kinds of instruments, and not all of them divide a main-scale division into ten equal parts. In any given case, the number of vernier divisions equals the number of equal parts into which a main-scale division is to be divided. The vernier works because the total number of vernier divisions spans a lesser number (usually one less) of main-scale divisions.

● **The Micrometer Caliper**   The micrometer caliper (see Fig. 7) is an instrument used for the accurate measurement of short lengths. Essentially, it consists of a carefully machined screw mounted in a strong frame. The object to be measured is placed between the end of the screw and the projecting end of the frame, called the

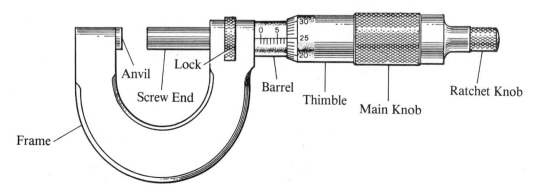

Figure 7   The Micrometer Caliper

anvil. The screw is then advanced until the object is gripped gently between the two jaws of the instrument. Most micrometers are provided with a ratchet arranged to slip on the screw as soon as a light and constant force is exerted on the object. By using the ratchet it is possible to tighten up the screw by the same amount each time and thus avoid using too great a force. If this arrangement is absent, great care should be taken not to force the screw, for the instrument may be easily damaged.

The micrometer caliper used in this experiment consists of a screw with a pitch of 0.5 mm, a longitudinal scale engraved along a barrel containing the screw, and a circular scale engraved around a thimble which rotates with the screw and moves along the scale on the barrel. The longitudinal scale is divided into millimeters. The circular scale has 50 divisions. Since the pitch of the screw is 0.5 mm, which is the distance advanced by the screw on turning through one revolution, it is clear that rotating the thimble through one scale division will cause the screw to move a distance of $\frac{1}{50}$ of 0.5 mm, or 0.01 mm. Hence readings may be taken directly to one hundredth of a millimeter, and by estimating tenths of a thimble-scale division, they may be taken to one thousandth of a millimeter.

The micrometer is read by noting the position of the edge of the thimble on the longitudinal scale and the position of the axial line of the barrel on the circular scale, and adding the two readings. The reading of the main scale gives the measurement to the nearest whole main-scale division; the fractional part of a main-scale division is read on the circular scale. Since it requires two revolutions of the screw to make it advance a distance of one millimeter, it is necessary always to be careful to note whether the reading on the circular scale refers to the first half or the second half of a millimeter. (In Fig. 7 the reading is 7.75 mm.)

The micrometer should be checked for a zero error, for

it may not read zero when the jaws are completely closed. In such cases a zero correction has to be applied to every reading; this correction may be either positive or negative. The value of the zero reading is obtained by rotating the screw until it comes in contact with the anvil and then noting the reading on the circular scale.

The masses of the various objects are measured with a balance. Two kinds of balances are used in this experiment. The first, the equal-arm balance, is provided with a set of weights that can be placed in any combination on one of the two trays. The object to be weighed is placed on the other tray and its mass determined by noting the mass of the weights needed to balance it. Most equal-arm balances have a slider that can be moved on a scale placed along the balance arms to interpolate between the masses of the smallest available weights. A typical scale of this sort may have divisions corresponding to a .5-gram change in mass, which represents about the limit to which the standard equal-arm balance can be read.

The other type of balance used in this experiment is called a triple-beam balance because it has three beams, each provided with a sliding weight. These weights have different sizes, the largest sliding along a scale with a notch every 100 grams, the next on a scale with a notch every 10 grams, and the smallest on a scale with 1-gram main divisions and .1-gram subdivisions. The unknown mass is placed on the tray, the 100-gram and 10-gram weights set in the appropriate notches (determined by trial), and the smallest slider adjusted for balance. The triple-beam balance can thus be read to 0.1 gram, or, if one is willing to interpolate between the subdivisions, to about .05 gram. However, great care is required to make the second decimal place really significant. Figure 8 shows a typical triple-beam balance set to 371.4 grams.

In this and all further work in physics, the student should be very careful to distinguish between mass and weight, which are not the same despite the very common

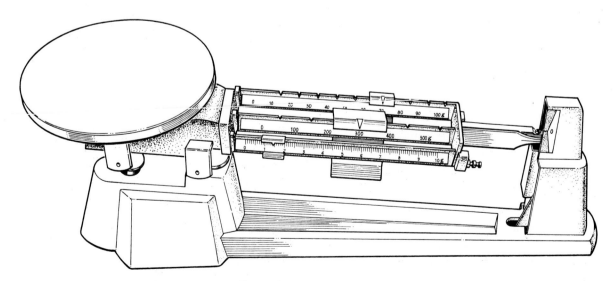

Figure 8   The Triple-Beam Balance

confusion between the two. Mass is the property of material bodies that makes them hard to accelerate. Even if a body is taken out into deep space far from the earth, the sun, and the other planets, it will not become massless — it will still have inertia; that is, a force will still be required to change its velocity. Weight, on the other hand, is simply the force of gravity with which the earth attracts a body. Since this force depends on the distance between the earth and the body, the body will indeed become weightless if taken far away (meaning many thousands of miles). The confusion arises because apparently the property of material bodies called mass is responsible for their mutual gravitational attraction, the two being in fact proportional. Thus the weight of a body (that is, the force with which it is attracted to the earth) is proportional to its mass; if its weight is measured, its mass is also known. Scales and balances measure weight. The balances in this experiment do so by comparing the weight of the unknown object with the weight of known masses. At balance, these weights are equal and we conclude from the proportionality of weight and mass that the masses must be equal also. The scales on the balance beams can therefore be calibrated in grams, which is a unit of mass, rather than in units of force such as pounds or newtons or dynes, even though the weight (the force of gravity) is actually being measured. There is nothing wrong with this — a measuring instrument's scale can always be graduated in units proportional to those it is actually measuring — but a physicist is nevertheless careful not to forget the distinction between mass and weight.

In this experiment the mass of several objects will be determined and their volumes calculated from the measurements of their dimensions. The density of the material of which each is made, defined as the material's mass per unit volume, is then calculated from

$$D = \frac{M}{V}$$

where M is the mass in grams, V is the volume in cubic centimeters, and D is the density, whose units are thus seen to be grams per cubic centimeter.

For the metal cylinders and the wire (which may be considered a long, thin cylinder) the volume is given by

$$V = \pi r^2 L = \frac{\pi d^2 L}{4}$$

where r is the cylinder's radius, d is the diameter ($2r$), and L is the cylinder's length. For the irregular body the volume will be determined by measuring the volume of liquid that it displaces.

APPARATUS

1. Equal-arm balance
2. Set of weights
3. Triple-beam balance
4. English ruler
5. Metric ruler
6. Vernier caliper
7. Micrometer caliper
8. Three cylinders of different metals
   a. aluminum
   b. brass
   c. iron
9. Length of copper wire
10. An irregular solid
11. A 250-ml graduated cylinder

PROCEDURE

1. Measure the length of the piece of copper wire with the metric ruler. Read the position of both ends of the wire, estimating to the nearest tenth of a millimeter. Record both readings in centimeters and read to 0.01 cm. Make four independent measurements, that is, use a different part of the ruler for each measurement, and record all the readings.

2. Measure the length of the wire with the English ruler. Read the position of both ends of the wire, estimating to the nearest tenth of the smallest scale division. Record both readings, expressed in inches, to the nearest 0.01 inch. Make four independent measurements, that is, use a different part of the ruler for each measurement, and record all the readings.

3. Determine the zero reading of the vernier caliper. This is the reading when the jaws of the instrument are in contact with each other. Be sure that the jaws are clean so that no grit or other foreign matter gets between them and prevents true contact. Record the zero reading to 0.01 cm. Make four independent determinations of this reading, that is, open and close the jaws before each setting, and record your results.

4. Measure the length and diameter of each cylinder with the vernier caliper. These measurements are made by closing the jaws of the caliper on the length or diameter of the cylinder being measured and reading the position where the zero line of the vernier falls on the main scale. The fractional part of a main-scale division is obtained by noting which line on the vernier coincides with a line of the main scale. Record the reading in centimeters and read to 0.01 cm. Make four independent measurements, that is, open and close the jaws before each setting. This is most conveniently done by measuring the diameter of each cylinder in turn once, then measuring the length of each in turn once, and repeating the complete sequence four times. Record all readings.

5. Determine the zero reading of the micrometer caliper, *i.e.*, the reading when the surfaces of the anvil and the screw end are in contact. Be sure these surfaces are very clean, since even a small speck of dust can give a false reading. The screw end may be brought almost into contact with the anvil by turning the thimble directly, but actual contact must always be made by turning the ratchet slowly until it clicks several times. If the ratchet is absent, great care must be taken not to force the screw; merely let the screw end approach the anvil very slowly and stop turning the screw as soon as the two surfaces touch. Record the value of this reading in centimeters and read to 0.0001 cm, estimating to one tenth of the smallest scale division. Make four independent determinations of the zero reading, that is, open and close the instrument before each setting, and record the readings.

6. Measure the diameter of the copper wire with the micrometer caliper. The measurement is made by placing the wire between the screw end and the anvil and advancing the screw until the wire is gripped between the anvil and screw-end surfaces. Again the ratchet should be used. If the instrument has no ratchet, care should be taken not to force the screw. Record the reading in centimeters and read to 0.0001 cm, estimating to one tenth of the smallest scale division. Make six independent measurements of the diameter of the wire, that is, open and close the caliper before each setting, and make the measurements at different points along the wire's length. Also, spin the wire between measurements so that different diameters are measured each time. Record the six readings so obtained.

7. Determine the mass of each cylinder and the irregular solid, using the equal-arm balance.

8. Determine the mass of the copper wire, using the triple-beam balance.

9. Determine the volume of the irregular solid by the displacement method. Partly fill the graduated cylinder with water and read the water level. Filling the cylinder to an exact graduation division will prove convenient in the subsequent calculations. Record the water level. Suspend the solid by means of a thread and lower it into the cylinder until it is completely submerged. Read and record the water level now observed. When the first reading is subtracted from the second, the volume of the irregular solid is obtained in cubic centimeters.

DATA

● **Measurement of Wire Length Using a Metric Ruler**

Ruler Readings		Lengths	Deviations
Left End	**Right End**		
Average value of the wire length			

● Measurement of Wire Length Using an English Ruler

Ruler Readings		Lengths	Deviations
Left End	**Right End**		
Average Value of the Wire Length			

Number of centimeters in one inch, calculated value _____
                                  accepted value _____

Percent error _____

● Length and Diameter of Metal Cylinders Using a Vernier Caliper

	Vernier Caliper Readings				Average
	1	**2**	**3**	**4**	
Zero Reading					
Length, Aluminum Cylinder					
Length, Brass Cylinder					
Length, Iron Cylinder					
Diameter, Aluminum Cylinder					
Diameter, Brass Cylinder					
Diameter, Iron Cylinder					

● Diameter of Copper Wire Using a Micrometer Caliper

	Micrometer Readings						Average
	1	**2**	**3**	**4**	**5**	**6**	
Zero Reading							
Reading with Wire							
Diameter of Wire							

● Determination of Density

Object Used	Mass in gm	Length in cm	Diameter in cm	Volume in cc (cm³)	Density Computed	Density (From Table III in Appendix)	Percent Error
Aluminum Cylinder							
Brass Cylinder							
Iron Cylinder							
Copper Wire							
Irregular Solid							

CALCULATIONS

1. From the data of Procedure 1, compute the length of the copper wire for each set of measurements. Calculate the average value of the length by finding the arithmetic mean of the lengths obtained, to the proper number of significant figures.

2. From the data of Procedure 2, compute the length of the copper wire for each set of measurements. Calculate the average value of the length by finding the arithmetic mean of the lengths obtained, to the proper number of significant figures.

3. Compute the number of centimeters in one inch from your measurements in Procedures 1 and 2. Use the calculated average values of the wire's length in making the computations. Compare your result with the accepted value by finding the percent error.

4. From the data of Procedure 3, compute the average value of the zero reading of the vernier caliper. From the data of Procedure 4, compute the average value of the setting of the vernier caliper for the length and diameter of each cylinder. Making sure to properly include its algebraic sign, use the average zero reading to obtain the measured length and diameter of each cylinder, and insert these values in the table of data for the density determinations.

5. From the data of Procedure 5, compute the average value of the zero reading of the micrometer caliper. From the data of Procedure 6, compute the average value of the setting of the micrometer caliper for the diameter of the copper wire. Calculate the diameter of the copper wire, making sure to properly include the algebraic sign of the zero correction.

6. From the data of Procedures 1 and 2, calculate the deviations of the lengths obtained from the average value of the length. Compute the average deviation from the mean (the a.d.), the average deviation of the mean (the A.D.), and the percent deviation of the mean (the percent A.D.) for both sets of measurements. (See the Theory of Errors in the Introduction.)

7. Compute the volume of each object measured.

8. Calculate the density of each substance used.

9. Compare your results with those given in Table III of the Appendix and find the percent error. *Note:* If the instructor has designated the substance of which the irregular solid is made as an unknown, try to identify it by comparing your computed value with the list in Table III.

QUESTIONS

1. Why are several observations taken for each measurement?

2. (a) What is the smallest part of a centimeter that can be read or estimated with a meter stick? (b) What is the smallest part of a centimeter that can be read or estimated with your vernier caliper? (c) Which readings are more reliable? Why?

3. What is the smallest part of a centimeter that can be read or estimated with your micrometer caliper? This represents the sensitivity of the micrometer, for the sensitivity of a measuring instrument is the value of the smallest quantity that can be read or estimated with it.

4. State the number of significant figures in the data of Procedures 1, 4, and 6.

5. (a) What is the significance of the average deviation from the mean (the a.d.)? What is the significance of the average deviation of the mean (the A.D.)?

6. In measuring the length and diameter of a cylinder, which dimension should be measured more carefully? Why? *Hint:* Remember that the length and diameter are being measured in order to determine the volume.

7. Why was the triple-beam balance used in determining the mass of the copper wire? Why was the micrometer used in preference to the vernier caliper in determining the wire's diameter?

8. (a) What is the volume in cubic millimeters of the largest cylinder you measured? (b) What is the volume in liters? (c) What is the mass in kilograms?

9. A thin circular sheet of copper has a diameter of 30.0 cm and a thickness of 1 mm. Find the weight of the sheet in grams.

10.   A certain graduated cylinder has an inside diameter of 4.00 cm. (a) The graduations on its side are labeled "cc" (cubic centimeters). How far apart are these graduations? (b) The cylinder is partially filled with water, and a solid sphere of radius 1.20 cm is then totally submerged therein. Through what distance does the water level rise when the sphere is so immersed?

Addition of Vectors. 2
Equilibrium of a Particle

When a system of forces, all of which pass through the same point, acts on a body, they may be replaced by a single force called the resultant. The purpose of this experiment is to show that the magnitude and direction of the resultant of several forces acting on a particle may be determined by drawing the proper vector diagram, and that the particle is in equilibrium when the resultant force is zero.

THEORY

A scalar is a physical quantity that possesses magnitude only; examples of scalar quantities are temperature, mass, and density. A vector is a quantity that possesses both magnitude and direction; examples of vector quantities are velocity, acceleration, and force. A vector may be represented by drawing a straight line in the direction of the vector, the length of the line being made proportional to the magnitude of the vector. The sense of the vector (for example, whether it is pointing toward the right or toward the left) is indicated by an arrowhead placed at the end of the line.

Vectors may be added either graphically or analytically. The sum or resultant of two or more vectors is the single vector which produces the same effect. For example, if two or more forces act at a certain point, their resultant is that force which, if applied at that point, has the same effect as the two separate forces acting together. Thus in Fig. 9 the resultant of two forces **A** and **B** is shown. The equilibrant, defined as the force equal and opposite to the resultant, is also shown. Note that if the equilibrant is regarded as a third force (in addition to **A** and **B**), the sum of **A**, **B**, and the equilibrant is zero.

The operation of adding vectors graphically consists of

constructing a figure in which a straight line is drawn from some point as origin to represent the first vector, the length of the line being proportional to the magnitude of the vector and the direction of the line being the same as the direction of the vector. From the arrowhead (nose) end of this line and at the proper angle with respect to the first vector, another line is drawn to represent the second vector, and so on with the remaining vectors. The resultant is the vector drawn from the origin or tail of the first vector to the nose of the last vector (see Fig. 10). If a closed polygon is formed, that is, if the nose of the last vector falls upon the tail of the first, then the resultant is zero. If the vectors represent forces, they are in equilibrium.

If no more than two vectors are to be added, the parallelogram method may be used. This consists of constructing both vectors with their tails at the origin and completing the parallelogram, as shown in Fig. 11. The diagonal of the parallelogram is then the resultant. Note, however, that this method is really identical with the previously described method of graphical addition by constructing the vectors nose-to-tail. Thus the side of the parallelogram opposite **A** is parallel to and has the same

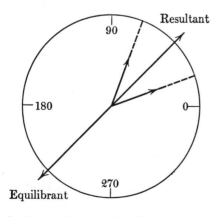

Figure 9   The Resultant and Equilibrant for Two Forces

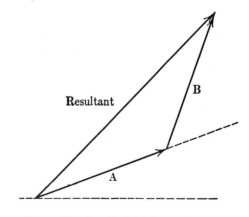

Figure 10   Graphical Addition of Vectors

13

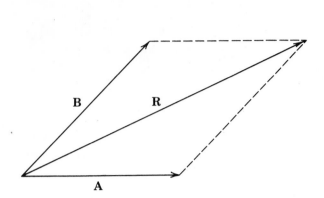

Figure 11   Parallelogram Method of Adding Vectors

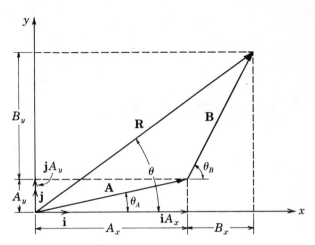

Figure 12   Addition of Vectors by Components

length as **A**. Hence it is really a reconstruction of **A** with its tail on the nose of **B**. The resultant **R** is then drawn from the tail of **B** to the nose of **A** as before. Similarly, the side of the parallelogram opposite **B** is equal to **B** and is therefore a reconstruction of **B** with its tail on the nose of **A**. Here the resultant may be considered as being drawn from the tail of **A** to the nose of **B**. The parallelogram method thus demonstrates the commutative property of vector addition: in vector addition, as in algebraic addition, the order in which the quantities are added doesn't matter. This property may be stated mathematically as

$$\mathbf{A} + \mathbf{B} = \mathbf{B} + \mathbf{A}$$

and may seem obvious, but the student should bear in mind that mathematical processes do exist for which it does not hold.

Vectors may also be added analytically, and this is in fact the preferred method since it does not require the making of precise drawings and does not involve the inaccuracies inherent in the measurements made on them. The method is illustrated in Fig. 12, in which the vectors of Fig. 10 have been placed on the usual cartesian (x–y) coordinate system. In Fig. 12, the vectors **A** and **B** have been resolved into their x and y components. For example, the x component A_x of the vector **A** is the projection of **A** on the x axis and is equal to $A \cos \theta_A$, where A is the magnitude of **A**. Similarly the y component of **A** is the projection of **A** on the y axis and is given by $A \sin \theta_A$.

It is easy to show how a vector is represented by its components. Fig. 12 introduces the two *unit* vectors **i** and **j**, which are defined as vectors of unit magnitude and with directions along the x and y axes respectively. Then $\mathbf{i}A_x$ is a vector along the x axis having magnitude A_x, and $\mathbf{j}A_y$ is a vector along the y axis having magnitude A_y. Fig. 12 shows that, by the parallelogram method of vector addition

$$\mathbf{A} = \mathbf{i}A_x + \mathbf{j}A_y \qquad (14)$$

By the same arguments, vector **B** is given by

$$\mathbf{B} = \mathbf{i}B_x + \mathbf{j}B_y \qquad (15)$$

where B_x and B_y, the x and y components of **B**, have the values $B \cos \theta_B$ and $B \sin \theta_B$ respectively.

Finally, Fig. 12 shows that the x component R_x of the resultant **R** is just the sum of the two x components A_x and B_x, while R_y is the sum of A_y and B_y. Thus the analytical method consists of finding the x and y components of all the vectors to be added, adding the x components algebraically (paying due attention to sign) to get the resultant's x component, and adding the y components to get the resultant's y component. In mathematical terms

$$R_x = A_x + B_x \qquad (16)$$

and

$$R_y = A_y + B_y \qquad (17)$$

With the x and y components of **R** known, the magnitude and direction of this vector are easily found if desired. Inspection of Fig. 12 shows that

$$R^2 = R_x{}^2 + R_y{}^2 \qquad (18)$$

and

$$\tan \theta = \frac{R_y}{R_x} \qquad (19)$$

However, since a vector is completely specified by its components, it is often left in this form, the magnitude and direction being calculated only if they are needed.

The conclusion that the components of a resultant are each the algebraic sum of the respective components of the vectors to be added may be arrived at analytically as well as by inspection of Fig. 12. Thus, in the addition of vectors **A** and **B**, these vectors may be given by Equations 14 and 15 so that

$$\mathbf{R} = \mathbf{A} + \mathbf{B} = \mathbf{i}A_x + \mathbf{j}A_y + \mathbf{i}B_x + \mathbf{j}B_y \qquad (20)$$

Writing **R** as $iR_x + jR_y$ and grouping terms appropriately in Equation 20 leads to

$$iR_x + jR_y = i(A_x + B_x) + j(A_y + B_y) \qquad (21)$$

This can be true only if $R_x = A_x + B_x$ and, separately, $R_y = A_y + B_y$ in agreement with Equations 16 and 17.

The apparatus used in this experiment (see Fig. 13) consists of a horizontal force table graduated in degrees and provided with pulleys which may be set at any desired angles. A string passing over each pulley supports a weight holder upon which weights may be placed. A pin holds a small ring to which the strings are attached and which acts as the particle. When a test for equilibrium is to be made the pin is removed; if the forces are in equilibrium, the particle will not be displaced.

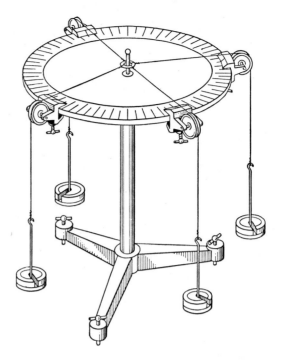

Figure 13   Force Table

APPARATUS

1. Force table
2. Four pulleys
3. Four weight hangers
4. Set of slotted weights, including:
    four 100-gm weights
    four 50-gm weights

two 20-gm weights
two 10-gm weights
5. Set of small weights (5-, 2-, 2-, 1-gm)
6. Protractor
7. Metric ruler

PROCEDURE

1. Mount a pulley on the 20° mark on the force table and suspend a total of 100 grams over it. Mount a second pulley on the 120° mark and suspend a total of 200 grams over it. Draw a vector diagram to scale, using a scale of 20 grams per centimeter, and determine graphically the direction and magnitude of the resultant by using the parallelogram method.

2. Check the result of Procedure 1 by setting up the equilibrant on the force table. This will be a force equal in magnitude to the resultant, but pulling in the opposite direction. Set up a third pulley 180° from the calculated direction of the resultant and suspend weights over it equal to the magnitude of the resultant. Cautiously remove the center pin to see if the ring remains in equilibrium. Before removing the pin, make sure that all the strings are pointing exactly at the center of the pin; otherwise the angles will not be correct.

3. Mount the first two pulleys as in Procedure 1, with the same weights as before. Mount a third pulley on the 220° mark and suspend a total of 150 grams over it. Draw a vector diagram to scale and determine graphically the direction and magnitude of the resultant. This may be done by adding the third vector to the sum of the first two, which was obtained in Procedure 1. Now set up the equilibrant on the force table and test it as in Procedure 2.

4. Clamp a pulley on the 30° mark on the force table and suspend a total of 200 grams over it. By means of a vector diagram drawn to scale, find the magnitude of the components along the 0° and the 90° directions. Set up these forces on the force table as they have been determined. These two forces are equivalent to the original force. Now replace the initial force by an equal force pulling in a direction 180° away from the original direction. Test the system for equilibrium.

CALCULATIONS

1. Calculate the resultant in Procedure 1 by solving for the third side of the force triangle algebraically. Find the magnitude of the resultant vector by using the law of cosines and its direction from the law of sines. Compare your result with that obtained in Procedure 1. How do these predicted results compare with your experimental finding in Procedure 2?

2. Calculate the resultant in Procedure 1 by using the analytical method of adding vectors. Compare this result with those already obtained in Procedures 1 and 2 and in Calculation 1.

3. Calculate the resultant in Procedure 3 by the analytical method. Note that you have already found the components of two of the force vectors in Calculation 2 and need only find and add the components of the third. Compare your result with that obtained in Procedure 3.

QUESTIONS

1. State how this experiment has demonstrated the vector addition of forces.

2. In Procedure 3 could all four pulleys be placed in the same quadrant or in two adjacent quadrants and still be in equilibrium? Explain.

3. State the condition for the equilibrium of a particle.

4. The x and y components of a certain force are measured and found to be 68 ± 3 and 42 ± 2 newtons, respectively. Calculate the direction and magnitude of this force, expressing your result to the proper number of significant figures and showing the error in both magnitude and angle. Note that the error in the latter is obtained from the error in its tangent. Devise a procedure for doing this and show your work.

5. The analytical method of adding vectors expressed in terms of their components may be applied to vectors in three dimensions, for which graphical work is inconvenient. Find the magnitude of the resultant of the vectors $\vec{A} = \vec{i}12 - \vec{j}37 + \vec{k}58$ and $\vec{B} = \vec{i}5 + \vec{j}30 - \vec{k}42$, where $\vec{i}, \vec{j}$, and $\vec{k}$ are unit vectors along the x, y, and z axes, respectively.

Equilibrium of a Rigid Body 3

When a rigid body is acted upon by a system of forces that do not all pass through the same point, a change may be produced in the angular (rotational) velocity of the body as well as in its linear (translational) velocity. Under certain conditions the body will be in equilibrium, that is, there will be no tendency for either its translational or rotational motion to change. If it is at rest, it will remain at rest. This experiment presents a study of the conditions for the equilibrium of a rigid body under the action of several forces.

THEORY

In the preceding experiment the equilibrium of a particle was investigated, and it was found that the particle was in equilibrium (had no tendency to change its translational velocity) when the resultant force acting on it was zero. In that case the particle was a body so small that it could be considered a point and therefore the question of where on the body the various forces were applied did not arise. Moreover, rotation was not considered, since rotation of a very small body is of little consequence.

More generally, bodies of appreciable extent must be considered. Then the points on a body at which forces are applied become important and rotation of the body as a unit is a new motion that must be considered in addition to its translational motion, that is, its displacement as a whole from one location to another. The bodies in question are assumed to be rigid, which means that their parts do not change their distances from one another. Internal forces (between atoms and molecules) hold a rigid body together. These forces maintain all parts in fixed relative positions and are assumed strong enough to do this no matter what external forces may be applied. The body then remains rigid and unbroken, and the internal forces need be considered no further.

A force's tendency to produce rotation of an extended rigid body is called the torque or the moment of the force. It is equal to the product of the force's magnitude and the perpendicular distance from the axis of rotation to the force's line of action. This definition is illustrated in Fig. 14. Here a force **F** is shown applied to a disc wheel at point P a distance r from the wheel's axis. The dashed line is the line of action of force **F** and is simply a line constructed through P along the direction of **F**. It therefore includes the arrow that represents **F** and is really no more than an extension of that arrow in both directions. The perpendicular distance from the axis of rotation to the line of action of **F** is shown as l in Fig. 14, and the magnitude of the torque is Fl. The distance l is called the lever arm or moment arm of the force **F** about the wheel's axis.

Since torque has a direction associated with it, it is a vector, and in general this vector lies along the axis of rotation, that is, perpendicular to the plane determined by the line of action of the force and the lever arm. In Fig. 14 this plane is the plane of the page, and the torque vector may be imagined as being perpendicular to the page and lying along the wheel's axis, which is shown in Fig. 14 as a point representing the axis's end view. This will be the situation in all cases encountered in elementary mechanics, so that all torque vectors will be parallel to each other and only their sense need be considered when they are added. The usual convention that torques tending to produce counterclockwise rotation are taken to be positive whereas those tending to produce clockwise rotation are taken to be negative will be adhered to throughout this book.

An important problem arises when the lever arm of the force of gravity on (the weight of) a rigid body is to be found about some given axis. The force of gravity acts on every little piece of the body and is therefore distributed over the extent of the body rather than being applied at one particular point. However, it is found that the body can be balanced, meaning that it can be sup-

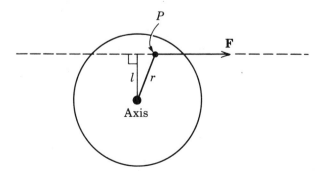

Figure 14   The Torque or Moment of a Force

ported at a single point about which (if the point is properly chosen) it will show no tendency to rotate. But if the body has no tendency to rotate, the total torque on it must be zero. Moreover, if the body is just being supported, the upward supporting force must be exactly equal to its weight. Fig. 15 now shows that if the entire weight W of the body is considered to act at the balance point, then the lever arms of the weight and the supporting force will be the same about any arbitrarily chosen axis, and their respective torques will be equal and opposite. The net torque will be zero and there will be no tendency for the body to rotate. Since in fact the body is observed not to rotate when supported at its balance point, the actual situation is equivalent to the hypothetical one in which the entire weight of the body is considered to be applied at the balance point. In fact, correct results will be obtained if the weight is taken to act at the balance point, and this point is therefore called the *center of gravity*. Thus the moment or lever arm of the weight of a rigid body may be calculated by assuming the body's entire weight to be concentrated at its center of gravity.

The foregoing shows that for a rigid body to be in equilibrium, it is not enough for the resultant force to be zero. That merely guarantees that the body's translational velocity will not change. If the body's rotational velocity is not to change either, the net torque due to all the applied forces must also be zero. There are thus two conditions for the equilibrium of a rigid body: (1) the vector sum of all the forces acting on the body must be zero; (2) the algebraic sum of all the torques about any axis must be zero. The first condition means that the sum of the forces in any direction must be equal to the sum of the forces in the opposite direction. The second

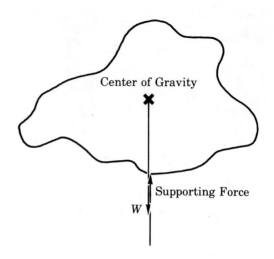

Figure 15   Center of Gravity

condition means that the sum of the torques producing clockwise rotation around any point must be equal to the sum of the torques producing counterclockwise rotation around the same point. The present experiment illustrates these principles using a meter stick as the rigid body. The arrangement is shown in Fig. 16. The meter stick is supported at an axis of rotation near its center and various forces are applied to it by weights and the supporting stand. The torque due to each force is calculated and the unknown force or distance computed by application of the two conditions of equilibrium. The weight and location of the center of gravity of the meter stick are also found and the torque due to this weight is calculated as if the weight were concentrated at this point.

Figure 16   A Meter Stick as a Rigid Body in Equilibrium

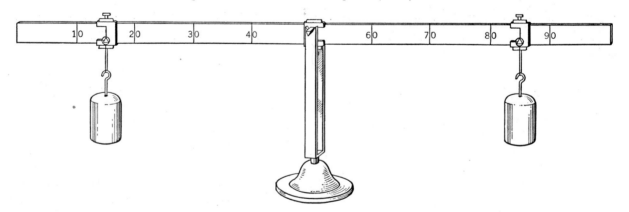

APPARATUS

1. Meter stick
2. Three meter stick knife-edge clamps
3. Stand for supporting the meter stick with a clamp
4. Set of hooked weights
5. Equal-arm balance and weights
6. Body of unknown mass (about 100 grams)

PROCEDURE

1. Weigh the meter stick and record the weight.

2. Weigh the three meter stick clamps together and compute their average weight.

3. Find the center of gravity of the meter stick to the nearest 0.5 millimeter by balancing it in one of the clamps.

4. Put another clamp near one end of the stick and hang a 100-gram weight from it. Slide the stick through the supporting clamp until the position of balance is found. Record the position of the axis of rotation, that is, the point of support, and the position of the weight.

5. Put another clamp near the other end of the stick and suspend a 200-gram weight from it. Leaving the other weight in place, find the new point of balance by again sliding the meter stick through the supporting clamp. Record the position of the 200-gram weight and of the new point of balance.

6. Remove the weights and clamps. Clamp the meter stick in the support clamp at its center of gravity. Put another clamp near one end of the stick and suspend the body of unknown weight from it. Slide another clamp, with a 200-gram weight attached, along the other end of the stick until the position of equilibrium is found. Record the positions of both weights.

7. Weigh the body of unknown mass used.

DATA

Weight of the meter stick	_____	New point of balance	_____
Weight of the three meter stick clamps	_____	Position of the unknown weight	_____
Average weight of the meter stick clamps	_____	New position of the 200-gram weight	_____
		Actual weight of the unknown mass	_____
Center of gravity of the meter stick	_____	Weight of the meter stick computed by the method of moments	_____
Position of the 100-gram weight	_____		
Position of the axis of rotation	_____	Weight of the unknown mass computed by the method of moments	_____
Position of the 200-gram weight	_____		

CALCULATIONS

1. Compute the weight of the meter stick from the data of Procedure 4 by the method of moments. Compare your result with that obtained by direct weighing of the stick. In particular, note whether the two measurements agree within the errors associated with each.

2. Using the point of support as the axis in Procedure 5, compute the moment of force of each of the weights and also of the meter stick, assuming its weight to be concentrated at its center of gravity. Add all these moments together, paying attention to their algebraic signs. Compare this net torque with zero, noting in particular whether zero lies within the error associated with your result.

3. Compute the weight of the body used in Procedure 6 by the method of moments. Compare the measured weight of this body with the computed weight, noting in particular whether the two weights agree within the experimental errors involved.

QUESTIONS

1. State the two conditions for the equilibrium of a rigid body and tell how this experiment demonstrates their validity.

2. In Calculation 2 why was the supporting force exerted on the meter stick by the stand not considered?

3. Why was the meter stick clamped at its center of gravity in Procedure 6? Was there any advantage in doing this?

4. Using the data of Procedure 5, assume the axis of rotation to be at one end of the stick and compute the moments of all the forces about this axis. The total upward force at the point of support is equal to the sum of all the downward forces. Add all the positive moments together and all the negative moments together, then find their algebraic sum. What conclusion do you draw from the result?

5. A meter stick is pivoted at its 50-cm mark but does not balance because of non-uniformities in its material that cause its center of gravity to be displaced from its geometrical center. However, when weights of 150 and 200 grams are placed at the 10-cm and 75-cm marks respectively, balance is obtained. The weights are then interchanged and balance is again obtained by shifting the pivot point to the 43-cm mark. Find the mass of the meter stick and the location of its center of gravity.

Uniformly Accelerated Motion. 4
The Atwood Machine

In accordance with Newton's first law of motion, when the resultant of all the forces acting on a body is zero, if the body is at rest it will remain at rest, and if it is in motion, it will continue to move with constant speed in a straight line. Newton's second law of motion describes what happens if the resultant is different from zero. This law states that if an unbalanced force is acting on a body, it will produce an acceleration in the direction of the force, the acceleration being directly proportional to the force and inversely proportional to the mass of the body. If the acceleration is constant, the body is said to be moving with uniformly accelerated motion. The purpose of this experiment is to measure the acceleration of a given mass produced by a given force and to compare it with that calculated from Newton's second law of motion.

THEORY

The Atwood machine consists of two weights connected by a light, flexible string which passes over a light pulley as shown in Fig. 17. The pulley should be as nearly frictionless as possible. The machine is used in measuring the acceleration produced by an arbitrarily chosen force acting upon a given mass. Once the mass and the force have been chosen, the acceleration produced is determined by Newton's second law of motion

$$F = ma \qquad (22)$$

where F is the net force in dynes acting on a body, m is the mass of the body in grams, and a is the acceleration in centimeters per second per second.

In the Atwood machine the total mass that is being accelerated is the sum of the two masses. The driving force, which is expressed in dynes, is the difference in the weights on the two ends of the string. The immediate object in performing this experiment is to bring out the dependence of acceleration on force when the mass is kept constant. The only way in which the force can be varied without varying the total mass is to transfer masses from one side of the moving system to the other. First of all, the force of friction must be overcome; this is done by transferring masses from the ascending side to the descending side until the mass on the descending side moves downward with constant velocity when given a very slight push. In this manner friction is compensated for. The value of the force of friction should be recorded.

For convenience the distance should be the same in all of the observations. The starting point is taken as the position of the moving system in which one of the masses rests on the floor. As this mass ascends, the other will descend an equal distance, and the stopping point is taken as the instant at which this mass strikes the floor. The distance traversed should be about one and a half meters. The time required for the mass to move through this distance is measured with a stop watch.

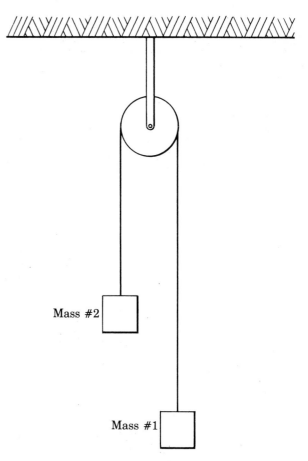

Figure 17 The Atwood Machine

25

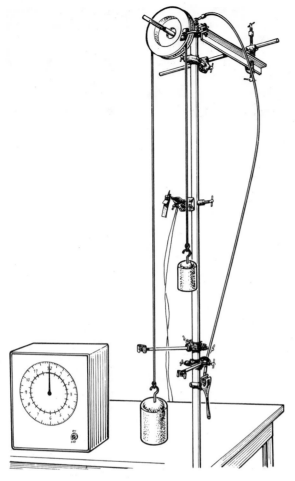

In uniformly accelerated motion the velocity is increased by the same amount in each succeeding second. The distance traveled is equal to the average velocity multiplied by the time. Since the system starts from rest and experiences a uniform acceleration, the final velocity will be twice the average velocity. From the time taken to acquire this final velocity, the corresponding acceleration can be computed. The equations of motion involved in uniformly accelerated motion are discussed below. The distance is given by

$$s = \bar{v}t \qquad (23)$$

where s is the distance in centimeters, $\bar{v}$ is the average velocity in centimeters per second, and t is the time in seconds. The average velocity is given by

$$\bar{v} = \frac{v_1 + v_2}{2} \qquad (24)$$

where $\bar{v}$ is the average velocity; v_1 is the initial velocity, which is zero in this case; and v_2 is the final velocity. The final velocity is given by

$$v_2 = v_1 + at \qquad (25)$$

where v_2 is the final velocity; v_1 is the initial velocity, which is zero in this case; a is the acceleration in centimeters per second per second; and t is the time in seconds.

Figure 18 Demonstration Atwood Machine with Electric Timing Device

APPARATUS

1. Pulley clamped to a vertical rod
2. Two weight holders
3. Set of slotted weights
4. Set of small weights, including:
     one 10-gm weight
     one 5-gm weight

      five 2-gm weights
      one 1-gm weight
5. String (strong fishing line)
6. Stop watch or stop clock
7. Two-meter stick

PROCEDURE

1. Using a total mass of about 2000 grams, determine the force of friction in the machine by transferring masses from the ascending side to the descending side until the mass on the descending side moves downward with constant velocity when given a very slight push. Record the mass on the descending side, the mass on the ascending side, and the force of friction. Be sure to include the mass of each weight hanger. Arrange the weights so as to have five 2-gram weights on the ascending side when the machine is in the balanced condition. CAUTION: Always stand clear of the suspended weights, for the string may break.

2. Transfer two more grams from the ascending side to the descending side, and thus determine the acceleration produced by a net force of four grams times 980 cm/sec², or 3920 dynes. Begin the observation when the ascending mass is on the floor, starting the stop watch at the instant when you let the weights go, and stopping it at the instant when the other mass strikes the floor. Make and record four independent observations.

3. Measure the distance traversed.

4. Repeat Procedure 2, using accelerating forces of 8 × 980, 12 × 980, 16 × 980, and 20 × 980 dynes, by transferring two additional grams each time. Make four independent observations for each accelerating force.

DATA

Distance traversed _____ Ascending mass, for constant velocity _____

Descending mass, for constant velocity _____ Force of friction _____

Descending Mass	Ascending Mass	Net Force	Time				
			1	2	3	4	Average

Net Force	Average Velocity	Final Velocity	Acceleration Experimental Value	Acceleration From $F = ma$	Percent Error

CALCULATIONS

1. Compute the average time taken with each accelerating force. From the known distance and the time taken in each case, compute the average velocity corresponding to each accelerating force.

2. Calculate the final velocity for each set of observations.

3. Calculate the acceleration produced by each accelerating force from the values of the final velocity and the time.

4. Compute the theoretical value for the acceleration from Newton's second law of motion. Assuming this to be the correct value, compute the percent error of the observed value for each case.

QUESTIONS

1. (a) State what this experiment tested. (b) State the relation between force and acceleration observed in this experiment.

2. What is the difference between uniform motion (constant velocity) and uniformly accelerated motion?

3. If you gave the system an initial velocity different from zero, how would this affect your results?

4. What is the advantage of transferring masses from one side to the other, instead of adding masses to one side?

5. Why should the pulley be as light as possible as well as nearly frictionless?

The Ballistic Pendulum. 5
Projectile Motion

The principle of conservation of momentum follows directly from Newton's laws of motion. According to this principle, if there are no external forces acting on a system containing several bodies, then the momentum of the system remains constant. In this experiment the principle is applied to the case of a collision, using a ballistic pendulum. A ball is fired by a gun into the pendulum's bob. The initial velocity of the ball is determined in terms of the masses of the ball and the bob and the height to which the bob rises after impact. This velocity can also be obtained by firing the ball horizontally and allowing it to fall freely toward the earth. The velocity is then determined in terms of the range and the vertical distance of fall. It is the object of this experiment to study the law of conservation of momentum and the elements of projectile motion; more precisely, the purpose is to determine the initial velocity of a projectile (1) by measurements of its range and vertical distance of fall and (2) by means of a ballistic pendulum.

THEORY

The momentum of a body is defined as the product of the mass of the body and its velocity. Newton's second law of motion states that the net force acting on a body is proportional to the time rate of change of momentum. Hence if the sum of the external forces acting on a body is zero, the linear momentum of the body is constant. This is essentially a statement of the principle of conservation of momentum. Applied to a system of bodies, the principle states that if no external forces act on a system containing two or more bodies, then the momentum of the system does not change.

In a collision between two bodies, each one exerts a force on the other. These forces are equal and opposite, and if no other forces are brought into play, the total momentum of the two bodies is not changed by the impact. Hence the total momentum of the system after collision is equal to the total momentum of the system before collision. During the collision the bodies become deformed and a certain amount of energy is used to change their shape. If the bodies are perfectly elastic, they will recover completely from the distortion and will return all of the energy that was expended in distorting them. In this case, the total kinetic energy of the system remains constant. If the bodies are not perfectly elastic, they will remain permanently distorted, and the energy used up in producing the distortion is not recovered.

Inelastic impact can be illustrated by a device known as the ballistic pendulum, which is sometimes employed in determining the speed of a bullet. If a bullet is fired into a pendulum bob and remains imbedded in it, the momentum of the bob and bullet just after the collision is equal to the momentum of the bullet just before the collision. This follows from the law of conservation of momentum. The velocity of the pendulum before collision is zero, while after the collision, the pendulum and the bullet move with the same velocity. Hence the momentum equation gives

$$mv = (M + m)V \qquad (26)$$

where m is the mass of the bullet in grams, v is the velocity in centimeters per second of the bullet just before the collision, M is the mass of the pendulum bob, and V is the common velocity of the bob and bullet just after the collision.

As a result of the collision, the pendulum with the imbedded bullet swings about its point of support, and the center of gravity of the system rises through a vertical distance h. From a measurement of this distance it is possible to calculate the velocity V. The kinetic energy of the system just after the collision must be equal to the increase in potential energy of the system as the pendulum reaches its highest point. This follows from the law of conservation of energy; here we assume that the loss of energy due to friction at the point of support is negligible. The energy equation gives

$$\tfrac{1}{2}(M + m)V^2 = (M + m)gh \qquad (27)$$

where M is the mass in grams of the pendulum bob, m is the mass of the bullet, V is the common velocity in centimeters per second of the pendulum bob and bullet just after the collision, g is the known value of the acceleration of gravity, and h is the vertical distance in centimeters through which the center of gravity of the system rises. The left hand side of Equation 27 represents

29

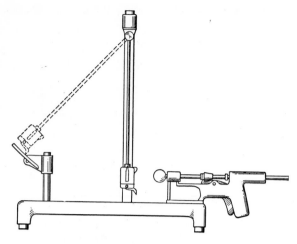

Figure 19   The Blackwood Ballistic Pendulum

the kinetic energy of the system just after the impact; the right hand side represents the change in the potential energy of the system. Solving Equation 27 for V, one obtains:

$$V = \sqrt{2gh} \qquad (28)$$

By substituting this value of V and the values of the masses M and m in Equation 26, it is possible to calculate the velocity of the bullet before the collision.

The velocity of the bullet can also be determined from measurements of the range and vertical distance of fall when the bullet is fired horizontally and allowed to fall to the floor without striking the pendulum bob.

The motion of a projectile is a special case of a freely falling body in which the initial velocity may be in any direction with respect to the vertical. The path of the projectile is a curved path produced by a combination of the uniform velocity of projection and the velocity due to the acceleration of gravity. This type of motion may be studied very effectively by considering it as made up of two independent motions, one of constant speed in the horizontal direction, and the other of constant accelera-

tion in the vertical direction. Here the problem is simplified by neglecting the effect of air friction.

For the special case where the projectile is fired horizontally with an initial velocity v, the path is determined by the two equations of motion to be discussed presently, one for the horizontal direction, and the other for the vertical direction.

Since the horizontal component of the velocity remains constant, the distance in the horizontal direction is given by

$$x = vt \qquad (29)$$

where x is the distance in centimeters that the projectile travels in the horizontal direction, v is the initial velocity in centimeters per second, and t is the time in seconds.

The initial velocity in the y direction is zero. Hence the motion in this direction is the same as that of a freely falling body with zero initial velocity. Therefore the distance traveled in the y direction is given by

$$y = \tfrac{1}{2} gt^2 \qquad (30)$$

where y is the distance in centimeters that the projectile travels in the vertical direction, g is the known value of the acceleration of gravity, and t is the time in seconds.

The apparatus used in this experiment (see Fig. 19) is a combination of a ballistic pendulum and a spring gun. The pendulum consists of a massive cylindrical bob, hollowed out to receive the projectile, and suspended by a strong, light rod pivoted at its upper end. The projectile is a brass ball which is fired into the pendulum bob and is held there by a spring in such a position that its center of gravity lies on the axis of the suspension rod. A brass index is attached to the pendulum bob to indicate the height of the center of gravity of the loaded pendulum. When the projectile is fired into the bob, the pendulum swings upward and is caught at its highest point by a pawl which engages a tooth on a curved rack.

APPARATUS

1. Blackwood ballistic pendulum
2. Equal-arm balance and set of weights
3. Metric steel scale
4. Meter stick
5. Plumb bob
6. Carbon paper
7. White paper and masking tape

PROCEDURE

1. In the first part of the experiment, the initial velocity of the projectile is obtained from measurements of the range and fall. The apparatus should be set near one edge of a level table. In this part the pendulum is not used and should be swung up onto the rack so that it will not interfere with the free flight of the ball.

2. Get the gun ready for firing by placing the ball on the end of the firing rod and pushing it back, compressing the spring until the trigger is engaged. The ball is fired horizontally so that it strikes a target placed on the floor. Fire the ball and determine approximately where it strikes the floor. Place a sheet of white paper on the floor so that the ball will hit it near its center and cover it with carbon paper. Secure the paper to

the floor with masking tape to keep it from moving around. In this way a record can be obtained of the exact spot where the ball strikes the floor. Fire the ball five more times.

3. Measure the range for each shot; this is the horizontal distance from the point of projection to the point of contact with the floor. Use a plumb bob to locate the point on the floor directly below the ball as it leaves the gun.

4. Measure the ball's vertical fall, *i.e.*, the vertical distance of the point of projection above the floor.

5. Get the gun ready for firing. Release the pendulum from the rack and allow it to hang freely. When the pendulum is at rest, pull the trigger, thereby firing the ball into the pendulum bob. This will cause the pendulum with the ball inside it to swing up along the rack where it will be caught at its highest point. Record the notch on the curved scale reached by the pawl when it catches the pendulum. To remove the ball from the pendulum, push it out with the finger or a rubber-tipped pencil while holding up the spring catch.

6. Repeat Procedure 5 four more times, recording the position of the pendulum on the rack each time.

7. From the data of Procedures 5 and 6, compute the average value of the position of the pendulum on the rack. Set the pendulum with the pawl engaged in the notch which corresponds most closely with the average reading. Measure the vertical distance h_1 from the base of the apparatus to the index point attached to the pendulum. This index point indicates the height of the center of gravity of the pendulum and ball. Use the metric steel scale and read the measurement to 0.1 millimeter.

8. With the pendulum hanging in its lowest position, measure the vertical distance h_2 from the base of the apparatus to the index point.

9. Loosen the thumbscrew holding the axis of rotation of the pendulum and carefully remove the pendulum from its support. Weigh the pendulum and the ball separately and record the values obtained.

DATA

	Trials

Range of the projectile   1. _____ 2. _____ 3. _____ 4. _____ 5. _____

Average range   _____

Vertical distance of fall   _____   Velocity of projection from measurements of range and fall

Time of flight   _____   _____

Highest position of ballistic pendulum: Trials
Reading of curved scale   1. _____ 2. _____ 3. _____ 4. _____ 5. _____

Average reading _____

Vertical distance h_1 or the average height of the center of gravity of the pendulum in its highest position   _____   Mass of the pendulum bob   _____

Mass of the ball   _____

Vertical distance h_2 or the height of the center of gravity of the pendulum in its lowest position   _____   Velocity of the pendulum and ball just after the collision   _____

Velocity of the ball before the collision   _____

Vertical distance h through which the center of gravity of the pendulum was raised   _____   Difference between this value of v and that found from the range measurement   _____

Percent difference   _____

CALCULATIONS

1. From the data of Procedure 3, compute the average range of the projectile.

2. From the measured vertical distance that the ball falls and the known value of g, calculate the time of flight of the ball by using Equation 30.

3. Compute the velocity of projection from the time of flight obtained in Calculation 2 and the range obtained in Calculation 1 by using Equation 29.

4. From the data of Procedures 7 and 8, calculate the vertical distance h through which the center of gravity of the loaded ballistic pendulum was raised as a result of the collision.

5. Compute the value of V, the common velocity of the pendulum bob and ball just after the collision, by using Equation 28.

6. Calculate the velocity of the ball before the collision by substituting the value of V found in Calculation 5 and the measured values of the masses of the pendulum bob and of the ball in Equation 26.

7. Compare the values of v obtained by your two methods by finding the difference between them and expressing it as a percent.

QUESTIONS

1. Using the data of your experiment, calculate the kinetic energy of the ball just before impact from the value of the velocity of the ball obtained with the ballistic pendulum and the mass of the ball.

2. Calculate the kinetic energy of the pendulum bob and ball just after impact from the value of their common velocity and of their masses.

3. (a) Using the results of Questions 1 and 2, calculate the fractional loss of energy during this inelastic impact. Express it in percent. (b) What became of the energy lost?

4. (a) Compute the ratio of the mass of the pendulum bob to the total mass of the bob and the ball. Express it in percent. (b) How does this ratio compare with the fraction of energy lost during the impact?

5. A bullet weighing 10 grams is fired horizontally into a block of wood weighing 2000 grams and suspended like a ballistic pendulum. The bullet sticks in the block and the impact causes the block to swing so that its center of gravity rises 10 centimeters. Find the velocity of the bullet just before the impact.

6. Compare the momentum and the kinetic energy of an automobile weighing 2000 pounds and moving with a velocity of 60 miles per hour with the momentum and the kinetic energy of a projectile weighing 70 pounds and moving with a velocity of 2500 feet per second.

Uniform Circular Motion **6**

Whenever a body is moving in a circular path, a force directed toward the center of the circle must act on the body to keep it moving in this path. This force is called centripetal force. The reaction, which is equal and opposite, is the pull of the body on the restraining medium and is called centrifugal force. The purpose of this experiment is to study uniform circular motion and to compare the observed value of the centripetal force with the calculated value.

THEORY

If a body is moving with constant speed in a circle, it is said to be moving with uniform circular motion. Even though the speed is constant, the velocity is continuously changing because the direction of the motion is continuously changing. Thus such a body has an acceleration. It can be shown that the direction of the acceleration is always toward the center of the circle and the magnitude of the acceleration is given by

$$a = \frac{v^2}{r}$$

where v is the speed of the body in centimeters per second and r is the radius of the path in centimeters.

A force is necessary to produce this acceleration, and this force is called centripetal force because it is always directed toward the center of rotation. By Newton's second law of motion, the magnitude of this force is given by the relation $F = ma$, or

$$F = m\frac{v^2}{r}$$

where F is the force in dynes and m is the mass in grams; v and r are the same as before. But by Newton's third law of motion, an equal and opposite force is exerted by the body on the restraining medium because of the inertia of the body. This reaction is called centrifugal force.

The centripetal force can also be expressed in terms of the angular speed, since

$$v = r\omega \quad \text{and} \quad \omega = 2\pi f$$

where v is the linear speed in centimeters per second, r is the radius of the path in centimeters, and ω is the angular speed in radians per second; f is the number of revolutions per second. Thus

$$F = mr\omega^2 \quad \text{or} \quad F = 4\pi^2 f^2 rm$$

where F is the centripetal force in dynes.

In the apparatus used for this experiment (see Fig. 20), the centripetal force of a stretched spring balances the centrifugal force due to the rotating mass. When the critical speed is reached, the sensitive indicator will stand opposite the index. This speed is kept constant and is measured by means of a revolution counter and a stop watch. The force required to stretch the spring the same amount is subsequently measured by applying weights.

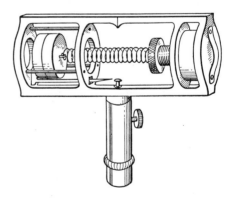

Figure 20 Centripetal Force Apparatus

APPARATUS

1. Centripetal force apparatus
2. Rotator with counter
3. Stop watch or stop clock
4. Supporting stand
5. Weight hanger and weights
6. Vernier caliper

35

PROCEDURE

1. Adjust the speed of the rotator and keep it constant at the critical speed for which the sensitive indicator of the centripetal force apparatus stands exactly opposite the index. Do not let this speed be exceeded. Begin by adjusting the variable speed rotator to a low speed and gradually increase the speed until the pointer of the indicator is exactly opposite the index. Keep on adjusting the speed very gradually during the experiment so that the pointer is kept opposite the index, vibrating about this position as little as possible. CAUTION: It is dangerous to let the apparatus rotate at excessive speeds.

2. Measure the speed of rotation with the revolution counter and a stop watch. First record the reading of the counter. At the proper instant engage the counter and set the stop watch going. To perform the experiment more conveniently, it is a good plan to let one observer pay strict attention to the proper adjustment of the speed. The other observer can obtain the value of the speed by manipulating the stop watch and the revolution counter. At the end of exactly one minute, disengage the counter and record the reading.

3. Repeat Procedure 2 and take three more readings of the speed.

4. Remove the apparatus, suspend it on a supporting stand, and determine the weight necessary to make the index and pointer coincide. Include the weight of the rotating mass and of the weight hanger in computing the total force needed to stretch the spring.

5. Measure the radius of rotation of the mass, using the vernier caliper, by measuring the distance from the axis of rotation to the center of the mass. Do this before removing the weight so that the mass will be in the same position it occupies during rotation when you make the measurement.

DATA

Weight of the rotating mass	_____	Average speed of rotation	_____
Weight applied to the spring	_____	Centripetal force calculated from the theory	_____
Total weight necessary to stretch the spring	_____	Measured value of the centripetal force	_____
Radius of rotation	_____	Percent discrepancy	_____

Time Interval	Readings of the Revolution Counter		Speed of Rotation
	At the Beginning	At the End	

CALCULATIONS

1. From the data of Procedures 2 and 3 calculate the speed of rotation for each set of readings. Compute the average speed of rotation.

2. Calculate the centripetal force from the theory, using the average speed of rotation obtained in Calculation 1.

3. Compare the calculated value of the centripetal force with the value measured directly, computing the percent discrepancy.

QUESTIONS

1. State what the experiment checks.

2. (a) How does the centripetal force vary with the speed of rotation for a constant radius of the path? (b) How does it vary with the radius of the path for a constant speed of rotation?

3. Distinguish between centripetal force and centrifugal force. Explain in what direction each force is acting and on what it is acting.

4. Calculate at what speed the earth would have to rotate in order that objects at the equator would have no weight. Assume the radius of the earth to be 6400 kilometers. What would be the linear speed of a point on the equator? What would be the length of a day (time from sunrise to sunset) under these conditions?

5. Engines for propeller-driven aircraft are limited in their maximum rotational speed by the fact that the tip speed of the propeller must not approach the speed of sound in air (Mach 1). Taking 6 feet as a typical diameter for a propeller of a light airplane and 1100 ft/sec as the speed of sound, find the upper limit on the rpm (revolutions per minute) of the propeller shaft.

Friction 7

Whenever a body slides along another body, a resisting force is called into play which is known as the force of friction. This is a very important force and serves many useful purposes, for a person could not walk without it, nor could a car propel itself along a highway without the friction between the tires and the road surface. On the other hand, friction is very wasteful; it reduces the efficiency of machines because work must be done to overcome it, and this energy is wasted as heat. The purpose of this experiment is to study the laws of friction and to determine the coefficient of friction between two surfaces.

THEORY

Friction is the resisting force encountered when one tries to slide one surface over another; this force acts along the tangent to the surfaces in contact. The force necessary to overcome friction depends on the nature of the materials in contact, their roughness or smoothness, and on the normal force, but not on the area of contact, within wide limits. It is found experimentally that the force of friction is directly proportional to the normal force. The constant of proportionality is called the coefficient of friction.

When the contacting surfaces are actually sliding one over the other, the force of friction is given by

$$F_r = \mu_k N \tag{31}$$

where F_r is the force of friction and is directed parallel to the surfaces and opposite to the direction of motion, N is the normal force, and μ_k the coefficient of friction. The subscript k stands for *kinetic*, meaning that μ_k is the coefficient that applies when the surfaces are moving one with respect to the other. It is therefore more precisely called the coefficient of *kinetic* or *sliding* friction.

A method of checking the proportionality of F_r and N and of determining the proportionality constant μ_k is to have one of the surfaces in the form of a plane placed horizontally with a pulley fastened at one end. The other surface is the bottom face of a block which rests on the plane and to which is attached a cord that passes over the pulley and carries weights. These are varied until the block moves at constant speed after having been started with a slight push. Since there is thus no acceleration, the net force on the block is zero, which means that the frictional force is equal to the tension in the cord. This tension, in turn, is equal to the total weight attached to the cord's end. The normal force between the two surfaces is equal to the weight of the block and can be increased by placing weights on top of the block. Thus

corresponding values of F_r and N can be found, and plotting them will show whether F_r and N are indeed proportional. The slope of this graph gives μ_k.

When a body lies at rest on a surface and an attempt is made to push it, the pushing force is opposed by a frictional force. As long as the pushing force is not strong enough to start the body moving, the body remains in equilibrium, which means that the frictional force automatically adjusts itself to be equal to the pushing force and thus to just balance it. There is, however, a threshold value of the pushing force beyond which larger forces will cause the body to break away and slide. We conclude that in the static case (body at rest) the frictional force automatically adjusts itself to keep the body at rest up to a certain maximum; but if static equilibrium demands a frictional force larger than this maximum, static equilibrium conditions will cease to exist because this force is not available, and the body will start to move. This situation may be expressed in equation form as

$$F_r \leq \mu_s N \qquad \text{or} \qquad F_{r_{max}} = \mu_s N \tag{32}$$

where F_r is the frictional force in the static case, $F_{r_{max}}$ is the maximum value this force can assume, and μ_s is the coefficient of *static* friction. It is found that μ_s is slightly larger than μ_k, which means that a somewhat larger force is needed to break a body away and start it sliding than is needed to keep it sliding at constant speed once it is in motion. This is why a slight push is necessary to get the block started for the measurement of μ_k.

One way of investigating the case of static friction is to observe the so-called *limiting angle of repose*, defined as the maximum angle to which an inclined plane may be tipped before a block placed on the plane just starts to slide. The arrangement is illustrated in Fig. 21. The block has weight W whose component $W \cos \theta$ (where θ is the plane angle) is perpendicular to the plane and is

39

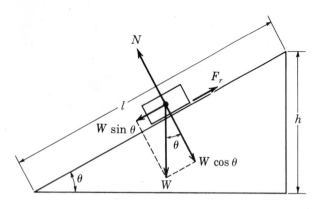

Figure 21 The Inclined Plane

thus equal to the normal force N. The component $W \sin \theta$ is parallel to the plane and constitutes the force urging the block to slide down the plane. It is opposed by the frictional force F_r, and as long as the block remains at rest F_r must be equal to $W \sin \theta$. If the plane is tipped up until at some value θ_{max} the block just starts to slide, then $F_{r_{max}} = W \sin \theta_{max}$. But $F_{r_{max}} = \mu_s N = \mu_s W \cos \theta_{max}$. Hence $W \sin \theta_{max} = \mu_s W \cos \theta_{max}$ or

$$\mu_s = \frac{\sin \theta_{max}}{\cos \theta_{max}} = \tan \theta_{max} \qquad (33)$$

Thus if the plane is gradually tipped up until the block just breaks away and the plane angle is then measured, the coefficient of static friction is just equal to the tangent of this angle. It is interesting to note that W cancelled out in the derivation of Equation 33, so that the weight of the block doesn't matter.

APPARATUS

1. Board with pulley at one end
2. Supporting rod and clamp for adjusting the angle of the inclined plane
3. Wood block with a cord attached to it
4. Glass block
5. Set of known weights
6. Pan to support the known weights
7. Equal-arm balance and weights
8. Protractor
9. Lintless dust cloth or paper wipers

PROCEDURE

1. Weigh the wood block and record the weight.

2. Place the board in a horizontal position on the laboratory table with its pulley projecting beyond the table's edge. Be sure that the surfaces of both the board and the wood block are clean, dry, and free of any dust or grit. Wipe them off if necessary with a clean, dry, lintless cloth or paper wiper. *After this has been done, do not touch these surfaces with your hands.* Handle the block with the cloth or a wiper and set it down *only* on the clean board. Begin the experiment by setting the block on the board with its largest surface in contact with the board's surface. Run the cord attached to the block over the pulley and attach it to the weight pan. Place some weights in the pan and slowly increase the load until it is just sufficient to keep the block sliding slowly with constant speed after it has been started with a very small push. Record this load.

3. Repeat Procedure 2 placing weights of 200, 400, 600, 800, and 1000 grams successively on top of the wood block. Record the load needed in each case.

4. Turn the wood block on its side and repeat Procedure 2 with a mass of 400 grams on top of the block. Record the load needed.

5. Again turn the wood block with the largest surface in contact with the plane and place 400 grams on top of the block. Gradually increase the load on the pan until the block just starts to move, without any initial push. Be careful to place the weights on the pan gently so as not to jerk the cord. Notice whether this time the block moves with uniform speed or whether it is being accelerated. Record the load needed under these conditions.

6. Adjust the board as an inclined plane. Place the wood block on the plane with its largest surface in contact, and gradually tip the plane up until the block just breaks away and starts to slide down. Be very careful to tip the plane slowly and smoothly so as to get a precise value of the angle with the horizontal at which the block just breaks away. This is the limiting angle of repose. Measure it by means of a protractor and record the result obtained in three separate trials. These trials should be independent, meaning that in each case the plane should be returned to the horizontal, the block placed on it, and the plane carefully tipped up until the limiting angle of repose is reached.

7. Repeat Procedure 6 using the glass block. Record the limiting angle of repose obtained in three independent trials.

DATA

Weight of wood block   _____

Position of Block	Weight Placed on the Block	Total Normal Force	Force to Keep Block Moving Uniformly
Flat	0 gm		
Flat	200 gm		
Flat	400 gm		
Flat	600 gm		
Flat	800 gm		
Flat	1000 gm		
On Side	400 gm		

Coefficient of kinetic friction μ_k from graph   _____

Coefficient of kinetic friction μ_k from Procedure 4   _____

Position of Block	Weight Placed on the Block	Total Normal Force	Force to Start Block Moving	Coefficient of Static Friction μ_s
Flat	400 gm			

Block Used	Trial 1		Trial 2		Trial 3		Average Value of μ_s
	Angle	μ_s	Angle	μ_s	Angle	μ_s	
Wood							
Glass							

Percent error in agreement of the two values of μ_s _____

CALCULATIONS

1. From the data of Procedures 2 and 3 plot a curve using the values of the total normal force as abscissas and the values of the force of friction as ordinates. See if your curve is a straight line and obtain the coefficient of kinetic friction μ_k for wood on wood by finding its slope.

2. Use the data of Procedure 4 to calculate the coefficient of kinetic friction for the case of the wood block sliding on its side. Record your result and see how it compares with the value of μ_k obtained from your graph.

3. From the data of Procedure 5 compute the coefficient of static friction μ_s for wood on wood.

4. From the data of Procedure 6 calculate μ_s for wood on wood from each of your three trials. Calculate an average value of μ_s and the a.d. Record your result on the data sheet. Compare this value of μ_s with that obtained in Procedure 5 by finding the percent error in the agreement of the two values.

5. From the data of Procedure 7 calculate μ_s for glass on wood for each of your three trials and obtain the average value and the a.d. Record your result on the data sheet.

QUESTIONS

1. Explain in your own words why it is necessary that the block move at constant velocity in Procedures 2–4.

2. (a) How does the coefficient of friction depend upon the normal force between the surfaces in contact? (b) How does it depend upon the area of the surfaces in contact?

3. How does the coefficient of static friction compare with the coefficient of kinetic friction for the same surfaces, areas, and normal forces? In this connection explain what happened in Procedure 5.

4. Calculate the force needed to pull a mass of 20 kilograms at a uniform slow speed up a plane inclined at an angle of 30° with the horizontal if the coefficient of kinetic friction is 0.20.

5. Using an average value of the limiting angle of repose obtained from your data in Procedure 6 and the value of μ_k from your graph, find the acceleration of the block after it just breaks away in Procedure 6.

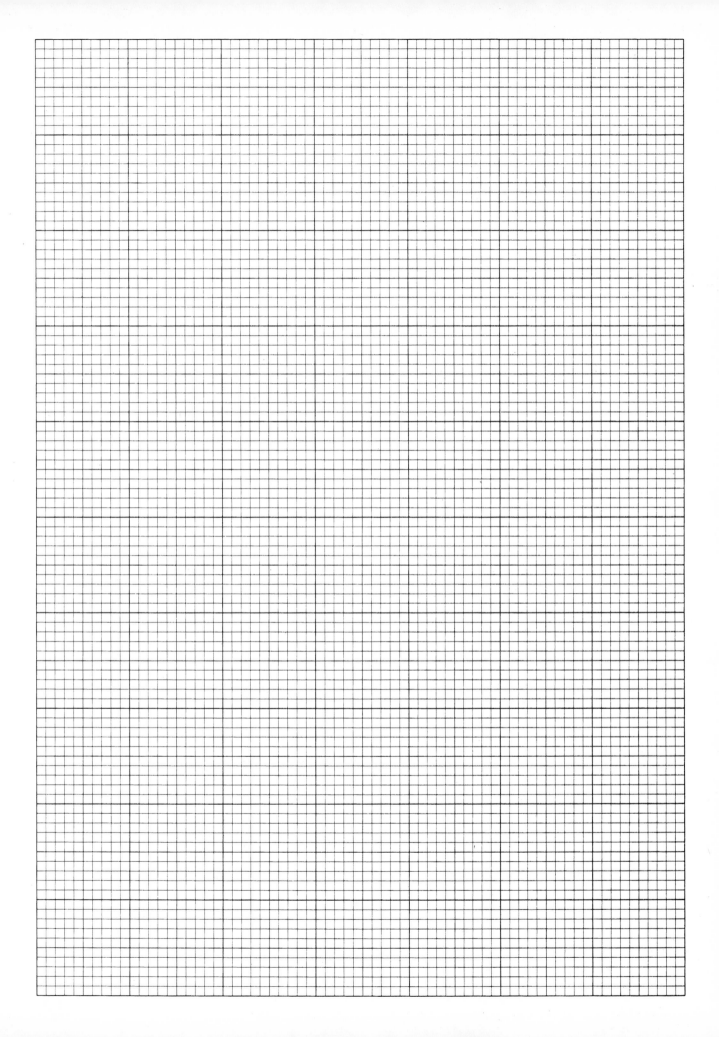

Simple Machines 8
and the Principle of Work

A simple machine is a device by which a force applied at one point is changed into another force applied at a different point for some useful advantage. The pulley system, the jack, the lever, and the inclined plane are well-known examples of such machines. The purpose of this experiment is to study an inclined plane, a wheel and axle, and a pulley system and to determine their characteristics.

THEORY

A simple machine is a device by means of which a force applied at one point (called the input) gives rise to another force acting at some other point (the output). The purpose of the machine is to make the new force different in magnitude and/or direction from the applied force in such a way as to make some particular piece of work possible or at least easier. We must recall that in physics the word "work" means a force acting through a distance and is given by the product of that force and the distance through which it acts. The units of work are thus foot-pounds or newton-meters (joules) or dyne-centimeters (ergs). Clearly a given amount of work may be accomplished by a small force moving through a large distance or a large force moving through a small distance; in either case the product of force and distance will be the same. A simple machine may be thought of as a device for changing the "mix" of force and distance. Ideally (that is, in the absence of losses due to friction) the work put into a simple machine will be the same as that received from it, the purpose of the machine being to increase the force at the expense of distance or *vice versa*. The machine may also change the direction of the force in the interests of convenience.

A common example is that of the 2000-pound car which must be raised 6 inches off the road to permit the changing of a tire. The work required is 1000 foot-pounds, but the average person cannot do this directly. To lift the car at all requires a 2000-pound effort, which is beyond most people's capacity. The fact that the car need be raised only 6 inches is no help — it might as well be 6 miles!

However, by the use of a jack, the required 2000-pound force is easily developed, the car gets raised the necessary 6 inches, and the 1000 foot-pounds of work gets done. The operator accomplishes this by applying a force of, say, 50 pounds at the jack handle, the jack multiplying this force by 40. But the operator has to pump the handle up and down many times. By the time the car has been raised 6 inches, the handle will have been pushed down through a total distance of $6 \times 40 = 240$ inches or 20 feet. The work of $20 \times 50 = 1000$ foot-pounds at the input is the same as that of $(\frac{1}{2}) \times 2000 = 1000$ foot-pounds at the output, but the proportion of force and distance involved has been drastically altered. The operator's hand has had to move through a much larger distance than that through which the car has been lifted, but this is usually considered a small price to pay for the ability to do the job at all.

The ratio of the force delivered by a simple machine at its output to the force applied at its input is called its *mechanical advantage*. In the above example of an ideal machine in which there are no frictional losses, this ratio is the same as the ratio of the distance moved at the input to the distance moved at the output. In any practical machine, however, frictional losses do exist and must be supplied by additional work at the input. Thus, in the jack, frictional forces in its mechanism might require that a force of 60 pounds rather than 50 be exerted on the handle in order to develop the 2000-pound lifting force at the output. The *actual* mechanical advantage (A.M.A.) is then only 2000/60, or 33.3 instead of 40. The handle must nevertheless be pumped through the same total of 20 feet to raise the car 6 inches as before. Thus the mechanical advantage the jack would have under the ideal conditions of no friction remains the ratio of the total distance moved by the handle to the distance the car is raised: $20 \div \frac{1}{2}$, or 40. This is called the *ideal mechanical advantage* (I.M.A.) and is given for any simple machine by the ratio of the distance moved at the input to the distance moved at the output. In equation form

$$\text{I.M.A.} = \frac{s}{h} \tag{34}$$

where s is the distance through which the force applied at the machine's input moves and h is the distance through which the force delivered by the machine moves. The actual mechanical advantage is given by

$$\text{A.M.A.} = \frac{P}{F} \qquad (35)$$

where P is the output force exerted by the machine (such as the weight of the car being lifted) and F is the applied force (such as the force applied to the jack handle). Since in most cases the primary purpose of the simple machine is to multiply force, the actual mechanical advantage will be the machine's most important characteristic.

Ideally the same amount of work is delivered at the output of a simple machine as is put in at the input, but in all actual cases extra work must be put in to make up for frictional losses in the machine. We are therefore interested in the machine's efficiency, which is a measure of how much of the work put in is actually delivered at the output. Specifically, the efficiency of a machine is defined as the ratio of the work delivered at the machine's output to the work done at the input. Thus

$$\text{Efficiency} = \frac{\text{output work}}{\text{input work}} = \frac{hP}{sF} \qquad (36)$$

Note that for the ideal machine the efficiency is 1 or 100% since all of the work put in is delivered at the output. For practical machines the efficiency will be a fraction less than 1 and may be expressed as a percent by multiplying by 100. Note also that since the I.M.A. is s/h and the A.M.A. is P/F, the efficiency can be expressed as

$$\text{Efficiency} = \frac{\text{A.M.A.}}{\text{I.M.A.}} \qquad (37)$$

In the example of the jack, in which frictional losses required the input force to be 60 pounds rather than 50 pounds, the above considerations show that the efficiency is $2000(\frac{1}{2})/60(20) = 33.3/40 = .83$ or 83%.

In this experiment an inclined plane, a wheel and axle, and a pulley system will be studied. In all of these cases a known load is raised against the gravitational force, and the applied force is measured. The ideal mechanical advantage is easily derived in each case as follows.

● **Inclined Plane**   When a block rests on an inclined plane as shown in Fig. 21 (Experiment 7), its weight W, which acts vertically downward, may be resolved into two components, one perpendicular to the plane and one parallel to it. The component of the weight perpendicular to the plane is $W \cos \theta$, and the component parallel to the plane is $W \sin \theta$, where θ is the angle made by the plane with the horizontal. The plane also exerts a force on the block. If the contacting surfaces are frictionless, this force (designated by N in Fig. 21) is perpendicular to the plane and is equal to the component of the block's weight perpendicular to the plane. In the absence of friction, a force $W \sin \theta$ directed up the plane will then keep the block in equilibrium, either at rest or moving up or down the plane with uniform velocity. Thus the plane enables us to raise the block to height h with a force $W \sin \theta$ instead of a force W, which would be required to lift the block directly. The plane allows the required force to be reduced by a factor of the sine of the plane angle. However, the distance through which the block must be moved to get it to the top of the plane is l, so that the work done is $(W \sin \theta)l$. Since, as Fig. 21 shows, $l \sin \theta = h$, this work is the same as Wh, the work required to lift the block directly. Notice that for the inclined plane

$$\text{I.M.A.} = \frac{l}{h} = \frac{l}{l \sin \theta} = \frac{1}{\sin \theta}$$

Thus the ideal mechanical advantage of an inclined plane is the reciprocal of the sine of the plane angle (the cosecant of the plane angle).

If there is friction between the block and the plane, the force of friction will be directed up the plane when the block is moving downward and down the plane when the block is moving upward. In Fig. 21 this force is designated F_r and is drawn in the direction it will assume if the block is moving down the plane. If a force F parallel to the plane is pushing the block up the plane, the frictional force will be reversed and the value of F needed to keep the block moving up the plane at constant speed will be $W \sin \theta + F_r$. Here F_r is the force of kinetic or sliding friction and is given by

$$F_r = \mu_k N = \mu_k W \cos \theta$$

Then
$$F = W \sin \theta + \mu_k W \cos \theta$$

and
$$\text{A.M.A.} = \frac{W}{W \sin \theta + \mu_k W \cos \theta}$$

$$= \frac{1}{\sin \theta + \mu_k \cos \theta} \qquad (38)$$

Moreover, the input work is $Fl = Wl(\sin \theta + \mu_k \cos \theta)$, so that the efficiency is

$$\text{Efficiency} = \frac{Wh}{Fl} = \frac{Wh}{Wl(\sin \theta + \mu_k \cos \theta)}$$

$$= \frac{\sin \theta}{\sin \theta + \mu_k \cos \theta}$$

$$= \frac{1}{1 + \mu_k \cot \theta} \qquad (39)$$

Thus the efficiency is reduced below 100% by the addition of the μ_k term in the denominator. Only in the case of $\mu_k = 0$ (no friction) do we get a 100% efficient inclined plane.

• **Wheel and Axle** Figure 22 is an end view of a wheel and axle. The load W is supported by a cord wrapped around an axle whose radius is r, while the operator applies force F to another cord wrapped around a wheel of radius R fixed on the axle. The load W exerts a clockwise torque Wr about the axis of the system, and the operator exerts a counterclockwise torque FR about this axis by pulling with force F on the cord wrapped around the wheel. If there is no friction in the bearings, equilibrium will exist with W at rest or ascending or descending at constant speed if the two torques are equal. Thus $W = (R/r)F$ and the ideal mechanical advantage is the ratio R/r of the wheel radius to the axle radius. That this is the I.M.A. as defined in Equation 34 may be seen from the fact that when the wheel and axle turns through one revolution, a length of the load-supporting cord equal to one axle circumference gets wrapped around the axle, so that the load is raised by a distance $2\pi r$. Similarly the operator must draw a length $2\pi R$ of cord off of the wheel. Therefore the input force F moves through this distance. Hence

$$\text{I.M.A.} = \frac{s}{h} = \frac{2\pi R}{2\pi r} = \frac{R}{r}$$

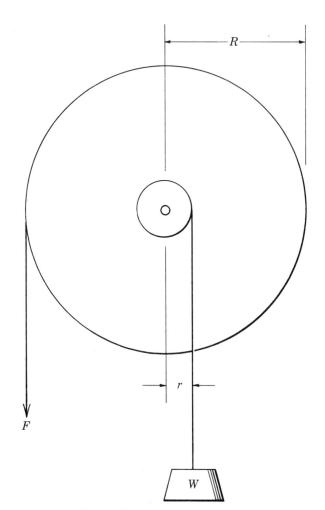

Figure 22 The Wheel and Axle

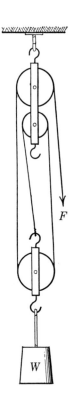

Figure 23 Pulley System

If there is friction in the bearings, the operator must exert an additional torque beyond FR to overcome this friction and must therefore pull with a force greater than F. The A.M.A. and the efficiency are thus reduced accordingly.

• **Pulley System** A typical pulley system is shown in Fig. 23. In the absence of friction in the pulleys, the tension in the cord is the force F with which the operator pulls on the end. But in effect, three strands of the cord pull on the load W, two via the pulley that is attached to W and one that is attached to W directly. Thus the upward pull on W is $3F$. At equilibrium (W at rest or moving up or down at constant speed), $3F$ must be equal to W, and the ideal mechanical advantage is 3. If there is friction in the pulleys, a force larger than $W/3$ must be exerted on the cord by the operator and the actual mechanical advantage becomes less than 3. However, in either case, lifting the weight W through 1 foot requires that each strand attached to it be shortened by 1 foot, which means that 3 feet of cord must be pulled out of the system. Thus force F must move through a distance three times that through which the load is raised, confirming that I.M.A. = 3. An interesting point to note here is that in any pulley system consisting of a set of pulleys attached to a fixed support and another set attached to and moving with a load, the ideal mechanical advantage is equal to the number of strands of the cord attached to the load either directly or via the load-end pulleys. A pulley system using just one cord is, of course, assumed.

APPARATUS

1. Board with pulley at one end
2. Supporting rod and clamp for adjusting the angle of the inclined plane
3. Wood block with a cord attached to it
4. Set of known weights
5. Pan to support the known weights
6. Equal-arm balance and weights
7. Meter stick
8. Protractor
9. Wheel and axle
10. Pulley system
11. Spring balance
12. Lintless dust cloth or paper wipers

PROCEDURE

● Inclined Plane

1. Weigh the wood block and record the weight.

2. Measure the length of the board and record this value.

3. Adjust the board as an inclined plane and set it at an angle of 10° to the horizontal with the pulley at the upper end. Be careful about measuring the angle so that precise values of the sine and cosine may be obtained.

4. Clean the surface of the plane and the bottom of the block with the lintless dust cloth or paper wipers. *After this has been done, do not touch these surfaces with your hands.*

5. Lay the block on the plane, run the cord over the pulley, and attach the weight pan to the cord's end. Place some weights in the pan and slowly increase this load until it is just sufficient to keep the block sliding slowly up the plane with constant speed after it has been started with a very small push. Record the load so determined.

6. Measure the height *h* of the inclined plane. Be careful in doing this not to include the thickness of the board. Thus if you measure *h* from the laboratory bench surface, measure it to the edge where the *bottom* surface of the board meets the board's upper end. Record your value of *h*.

7. Set the plane at an angle of 20° and repeat Procedures 3, 5, and 6. Again use care in measuring the plane angle and the height.

8. Set the plane at an angle of 30° and repeat Procedures 3, 5, and 6, once more using care to obtain precise measurements of your new values of θ and *h*.

● Wheel and Axle

9. Wind one string several times around the axle of the wheel-and-axle assembly. Wind the other string several times in the reverse direction around the wheel.

10. Hang a 500-gram mass on the string coming from the axle. This is the load to be lifted. Attach the pan to the string coming from the wheel and place weights in the pan until the load moves upward at a small, constant speed. Record the total weight needed to do this.

11. Measure the diameter of the wheel up to the surface on which the string is wound. Similarly, measure the diameter of the axle. Record these values.

● Pulley System

12. With a spring balance, measure the force necessary to lift a mass of one kilogram with the pulley system set up in the laboratory. Observe the force while the mass is being lifted at a slow, uniform speed and record the value of this force.

13. Count the number of supporting strands in the pulley system and record it.

DATA

● Inclined Plane

Weight of wood block _____   Length of board used as plane _____

| | Angle of the Inclined Plane | | |
	10°	20°	30°
Force needed to pull block up plane			
Height h of the inclined plane			
Sine of the plane angle θ			
Cosine of the plane angle θ			

• Wheel and Axle

Load used in the wheel-and-axle experiment _____

Force needed to lift this load _____

Diameter of the wheel _____
Diameter of the axle _____

• Pulley System

Load on the pulley system _____
Force needed to lift this load _____

Number of supporting strands _____

• Mechanical Advantage and Efficiency

Machine	Actual Mechanical Advantage	Ideal Mechanical Advantage	Efficiency	Coefficient of Kinetic Friction
Inclined Plane at 10°				
Inclined Plane at 20°				
Inclined Plane at 30°				
Wheel and Axle				
Pulley System				

CALCULATIONS

1. From the data of Procedures 2, 3, and 6 calculate the sine of the plane angle by dividing the height of the plane by its length. Compare this result with the value given in a table of trigonometric functions for the sine of 10° (see Appendix). Your two values should agree to three significant figures. Record your calculated value in the space provided on the data sheet.

2. Using the formula $\sin^2\theta + \cos^2\theta + 1$, calculate the value of cos 10° from your value of sin 10°. Check your result with the tabulated value of cos 10° and enter it on the data sheet.

3. Repeat Calculations 1 and 2 for the 20° and 30° plane angles using the data obtained in Procedures 7 and 8.

4. From the data of Procedures 1, 3, and 5 determine the actual mechanical advantage of the inclined plane set at an angle of 10°. The ideal mechanical advantage is the reciprocal of the sine of the plane angle already found. Record this I.M.A. and calculate and record the 10° inclined plane's efficiency.

5. Use either Equation 38 or 39 to compute the coefficient of kinetic friction between the block and the plane.

6. Repeat Calculations 4 and 5 for the inclined plane set at 20° using the data from Procedure 7.

7. Repeat Calculations 4 and 5 for the inclined plane set at 30° using the data from Procedure 8.

8. From the data of Procedures 9–11 determine the actual mechanical advantage, the ideal mechanical advantage, and the efficiency of the wheel and axle used.

9. From the data of Procedures 12 and 13 determine the actual mechanical advantage, the ideal mechanical advantage, and the efficiency of the pulley system used.

QUESTIONS

1. Should the coefficients of kinetic friction obtained for the three angular settings of the inclined plane be the same? Are they?

2. The instructions for Calculation 1 suggest that your calculated value of the sine of the plane angle is better than the value obtained by looking up the sine of this angle as measured by the protractor. Why should this be true?

3. (a) State how the mechanical advantage of the inclined plane varies with the inclination of the plane. (b) Does this apply to both the ideal and the actual mechanical advantage?

4. (a) How does the efficiency of the inclined plane vary with the inclination of the plane? (b) Explain the reason for this.

5. A body weighing 400 pounds is pulled up an inclined plane 15 feet long and 5 feet high. The force required to pull the body up the plane at a slow uniform speed is 200 pounds. Compute the ideal mechanical advantage, the actual mechanical advantage, and the efficiency of the plane under these conditions. Also compute the coefficient of kinetic friction between the body and the plane.

6. Calculate the amount of work done in drawing a mass of 20 kilograms at a uniform slow speed up a plane 2 meters long, making an angle of 30° with the horizontal, if the coefficient of kinetic friction is 0.20.

7. The block and tackle of Fig. 23 is used to lift a stone weighing 300 pounds. The force required is 125 pounds. Calculate the actual mechanical advantage and the efficiency of this simple machine.

8. In the pulley system of Fig. 23, does the top pulley (the large one attached to the supporting beam) contribute in any way to the mechanical advantage of the system? Could it be omitted without changing the system's I.M.A.? How does it affect the system's A.M.A.? If it does not improve the mechanical advantage, what is its purpose?

Young's Modulus 9

All bodies are deformed in some way by the application of a force. An elastic body is a body that changes in size or shape upon the application of a distorting force, but returns to its original condition when that force is removed. But there is a limit to the magnitude of the force which may be applied, if the body is to return to its original condition. This is called the elastic limit. A greater force will cause a permanent distortion or even a break. In this experiment a steel wire is stretched by the application of a force, and measurements are made of the changes in length produced. From these measurements, Young's modulus for a steel wire is calculated.

THEORY

When a wire is stretched by a mass hanging from it, the wire is said to experience a tensile strain, which is defined as the elongation per unit length. The force acting on the wire is a measure of the tensile stress, which is defined as the force per unit area over which it is acting.

Hooke's law states that as long as a body is not strained beyond its elastic limit, the stress is proportional to the strain. Thus the ratio of stress to strain is a constant and is called a modulus of elasticity. For the case of a wire stretched by a force, this constant is called Young's modulus. Thus

$$Y = \frac{\text{force per unit area}}{\text{elongation per unit length}} = \frac{F/A}{e/L} = \frac{FL}{eA}$$

where Y is Young's modulus in dynes per square centimeter, F is the force in dynes, L is the length of the wire in centimeters, e is the elongation in centimeters, and A is the area of cross section of the wire in square centimeters.

The arrangement of the apparatus used in this experiment is shown in Fig. 24. The elongations produced in this experiment are very minute, hence a method of magnifying them is used: the optical lever. This consists of a vertical mirror mounted on a horizontal T-shaped base having pointed legs at the three extremities. The wire whose elongation is to be measured is clamped at its upper end and has a small cylinder attached to it near the lower end. This cylinder is free to move up and down through a hole in a shelf supporting the mirror and its base. The two front feet of the base rest on the shelf and form an axis of rotation for the mirror. The rear foot rests on the cylinder and moves up and down with it due to the elongation of the wire. At a distance of about one meter and a half from the mirror, there is placed a reading telescope to which is attached a vertical scale. The image of the illuminated scale as seen in the mirror is viewed

through the telescope. The position of the cross hairs upon the image of the scale, for the unstretched wire, is the zero reading. As the wire is stretched by applying a load, the angular movement of the mirror causes a different portion of the scale to appear in the telescope. The

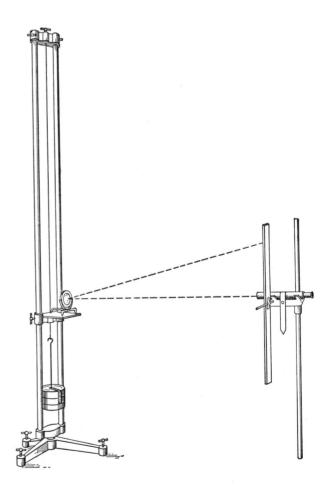

Figure 24   Young's Modulus Apparatus

elongation can be readily computed from geometrical considerations leading to the relation

$$e = \frac{Sd}{2D} \qquad (40)$$

where e is the elongation in centimeters, D is the distance in centimeters from the mirror to the scale, d is the perpendicular distance from the moving foot of the optical lever to the line joining the other two feet, and S is the difference in the scale readings before and after the application of the load. This relation is derived by using the principle of optics that the reflection of a beam of light falling on a plane mirror is turned through an angle twice as large as the angle through which the mirror is rotated, the incident beam remaining fixed.

APPARATUS

1. Young's modulus apparatus
2. Telescope and scale, with support stand
3. Mirror mounted on T-base (optical lever)
4. Desk lamp or scale illuminator
5. Slotted weights (ten 1-kg weights)
6. Micrometer caliper
7. Vernier caliper
8. Two-meter stick

PROCEDURE

1. Measure the length of the steel wire used, that is, the distance from the lower end of the upper chuck to the top of the cylinder at the lower end of the wire.

2. Measure the diameter of the wire in five places, using the micrometer caliper.

3. Measure the distance of the moving foot of the optical lever from the axis of rotation. This may be done by pressing the pointed legs on a piece of white paper and constructing the required distance. Use the vernier caliper for this measurement.

4. Carefully place the optical lever with its two front feet on the rigid support and its rear foot on the cylinder fixed to the wire.

5. Carefully measure the distance from the mirror to the scale, using the two-meter stick.

6. Apply a load of one kilogram to the wire to take up any slack, using great care not to jar the apparatus. Adjust the apparatus so that the image of the scale illuminated with the desk lamp, as seen in the mirror, is viewed through the telescope. Record the reading of the scale to the nearest tenth of a millimeter. This will be the zero reading.

7. Add another kilogram to the load on the wire and again record the scale reading.

8. Add one kilogram at a time up to a total of 10 kilograms, reading the scale after each added load. Record each reading.

9. Remove the load, one kilogram at a time, recording the scale reading for each load. *Note:* The values of the load to be used apply only to a steel wire. If a different kind of wire is used, the proper values of the load will be suggested by the instructor.

DATA

Length of wire used _____   Diameter of wire _____

Micrometer readings:

Trials

1. _____   2. _____   3. _____   4. _____   5. _____

Average reading _____   Zero reading _____

Cross-sectional area of the wire _____   Young's modulus (computed value) _____

Distance from front legs to the rear leg of the optical lever _____   Young's modulus (from Table IV, Appendix) _____

Distance from mirror to scale _____   Percent error _____

Load in Kilograms	Scale Reading (Increasing Load)	Scale Reading (Decreasing Load)	Average Scale Reading	Zero Reading / Elongation
1				
2				
3				
4				
5				
6				
7				
8				
9				
10				

CALCULATIONS

1. Compute the average of the readings for the diameter of the wire. This will be the value of the diameter when properly corrected for the zero reading of the micrometer.

2. Calculate the cross-sectional area of the wire in square centimeters.

3. Compute the total elongation for each total load, using the average of the scale readings obtained. Note that the value of the scale reading obtained when the load is one kilogram is taken as the zero reading. Hence the elongation produced by a load of one kilogram will be obtained by using the difference in the scale readings when the load is changed from 1 to 2 kilograms. The elongation produced by a load of 2 kilograms will be the difference in the scale readings for loads of 1 and 3 kilograms. A similar procedure is to be used for the other loads.

4. Plot a curve on the graph paper using loads in kilograms as abscissas and the resulting elongations in centimeters as ordinates. Note that the load producing a given elongation is the number of kilograms on the weight hanger minus the one kilogram used for the zero reading. Thus the maximum load appearing as an abscissa in this graph is 9 kilograms.

5. Compute Young's modulus for the wire from the slope of your graph. Compare your result with the value given in Table IV, Appendix, and find the percent error.

QUESTIONS

1. State what the curve shows and hence what the experiment checked.

2. Why was not the whole length of the wire measured?

3. Define elasticity, elastic limit, stress, and strain.

4. (a) State Hooke's law. (b) Does your curve agree with it? (c) How would the curve look if the elastic limit were exceeded?

5. Derive Equation 40 from a consideration of the apparatus geometry and the principle of optics stated in the theory section.

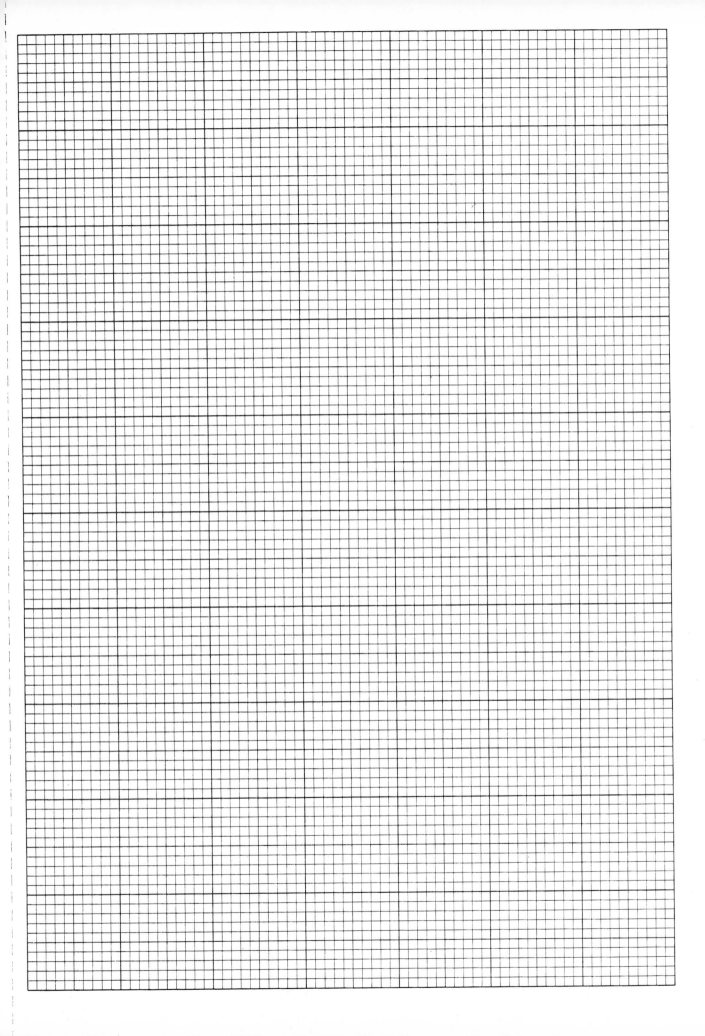

Rotational Motion 10

The motion of a flywheel of a steam engine, that of a propeller of an airplane engine, and that of any rotating wheel are examples of a very important type of motion called rotational motion. If a rigid body is acted upon by a system of torques, the body will be in equilibrium, as far as rotational motion is concerned, if the sum of the torques about any axis is zero. This means that if the body is at rest, it will remain at rest; if it is rotating about a fixed axis, it will continue to rotate about the same axis with uniform angular speed. If an unbalanced torque is acting on the body, it will produce an angular acceleration in the direction of the torque, the acceleration being proportional to the torque and inversely proportional to the moment of inertia of the body about its axis of rotation. The purpose of this experiment is to study rotational motion, to observe the effect of a constant torque upon a disk free to rotate, and to determine the resulting angular acceleration and the moment of inertia of the disk.

THEORY

Rotation refers to motion about some axis in space. In describing this type of motion, the angular displacement, the angular velocity, and the angular acceleration must be given. The unit of angular measurement usually used is the radian. This is the angle subtended at the center of a circle by an arc equal in length to the radius. The angular velocity is the time rate of change of angular displacement. It is equal to the angle through which the body rotates divided by the time. The angular acceleration is the time rate of change of angular velocity.

The equations of motion involved in uniformly accelerated angular motion are discussed below. The angular displacement is given by

$$\theta = \bar{\omega} t$$

where θ is the angular distance in radians, $\bar{\omega}$ is the average angular velocity in radians per second, and t is the time in seconds. The average angular velocity is given by

$$\bar{\omega} = \frac{\omega_1 + \omega_2}{2}$$

where $\bar{\omega}$ is the average angular velocity, ω_1 is the initial angular velocity, and ω_2 is the final angular velocity. The final angular velocity is given by

$$\omega_2 = \omega_1 + \alpha t$$

where ω_2 is the final angular velocity, ω_1 is the initial angular velocity, α is the angular acceleration in radians per second per second, and t is the time in seconds.

The corresponding equations of motion involved in uniformly accelerated linear motion are

$$s = \bar{v} t$$

$$\bar{v} = \frac{v_1 + v_2}{2}$$

and

$$v_2 = v_1 + at$$

where s is the distance, $\bar{v}$ is the average velocity, v_1 is the initial velocity, v_2 is the final velocity, a is the acceleration, and t is the time.

Using the radian as the unit of angular measurement gives rise to some very simple relations between the linear quantities and the angular quantities. Suppose that s represents the distance a particle moves around the circumference of a circle of radius r; let v be the linear speed, and let a be the linear or tangential acceleration of the particle. Then the angular distance θ, the angular velocity ω, and the angular acceleration α are given by the relations

$$s = r\theta \qquad (41)$$
$$v = r\omega \qquad (42)$$
$$a = r\alpha \qquad (43)$$

When a torque is applied to a body free to rotate about a fixed axis, then the body will acquire an angular acceleration given by the relation

$$L = I\alpha \qquad (44)$$

where L is the sum in centimeter-dynes of all the torques about the fixed axis of rotation, I is the moment of inertia in gm-cm^2 of the body about the same axis, and α is the angular acceleration in radians per second per second.

59

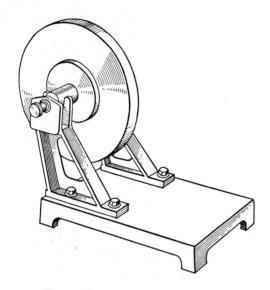

Figure 25 Moment of Inertia Apparatus

The moment of inertia is the inertia of a body as regards rotation about an axis. The moment of inertia of a body with respect to an axis is the sum of the products obtained by multiplying the mass of every particle of the body by the square of its distance from the axis. For a simple geometrical solid it can be readily determined by the use of integral calculus. For example, the moment of

inertia of a uniform cylinder, or disk, with respect to its longitudinal axis is given by

$$I = \tfrac{1}{2}MR^2$$

where M is the mass in grams of the cylinder, or disk; R is the radius in centimeters; and I is the moment of inertia in gm-cm.^{2}

The apparatus used in this experiment (see Fig. 25) consists of a massive disk mounted on an axle that can rotate freely in pivot bearings. The disk is set in motion by means of a weight attached to a string which is wound around the axle. As the weight descends, it will move with a constant linear acceleration, while the wheel and axle will rotate with a constant angular acceleration.

The angular acceleration of the wheel and axle depends on the applied torque, the frictional torque, and the moment of inertia. Assuming the last two to remain constant during the experiment, then the angular acceleration can be varied by changing the applied torque. This can be done by using different values of the suspended weight, or of the axle radius. The disk used in this experiment has two different axle radii. By varying the applied torque and thereby getting different values of the angular acceleration, it is possible to compute the moment of inertia and the frictional torque from the observations taken.

APPARATUS

1. Moment of inertia apparatus, consisting of an aluminum wheel mounted on a fixed axle
2. Two hooked weights (50 and 100 grams)
3. Stop watch or stop clock
4. Meter stick
5. Vernier calipers
6. Large calipers
7. String

PROCEDURE

1. Wind the string on the smaller axle in a single layer. Choose a length of string so that the weight will be disconnected when it strikes the floor. Do this by tying a loop in the string at the end opposite the one attached to the weight, placing this loop over a little pin set in the axle, and adjusting the string's length so that about half a turn remains on the axle surface when the weight reaches the floor.

2. For the first set of measurements use a 50-gram weight. Measure the distance that the weight has to fall.

3. Find the time required for the 50-gram weight to reach the floor. With the string wound around the smaller axle and the weight at the top of its path, release the wheel and start the stop watch at the same instant; make sure that no impulse is given to the wheel as it is released. Stop the watch at the instant when the weight strikes the floor.

4. Repeat Procedure 3, making two more observations of the time.

5. Repeat Procedures 2 and 3, using a 100-gram weight with the string wound on the smaller axle. Make three observations of the time of descent.

6. Repeat Procedures 1, 2, and 3, using a 50-gram weight with the string wound on the larger axle. Make three observations of the time of descent.

7. Repeat Procedures 1, 2, and 3, using a 100-gram weight with the string wound on the larger axle. Make three observations of the time of descent.

8. Measure the diameter of each axle with the vernier calipers.

9. Measure the diameter of the wheel with the large calipers.

10. Record the total mass of the wheel and axle. This is stamped on the wheel.

DATA

Diameter of the smaller axle _____

Radius of the smaller axle _____

Diameter of the larger axle _____

Radius of the larger axle _____

Diameter of the wheel _____

Radius of the wheel _____

Total mass of the wheel and axle _____

Moment of inertia of the wheel and axle from the slope of the line on the graph _____

Frictional torque from the y intercept on the graph _____

Radius of gyration of the wheel and axle _____

		Using Smaller Axle		Using Larger Axle	
		50-gm Mass	100-gm Mass	50-gm Mass	100-gm Mass
Distance that the Mass Falls					
Time of Fall:	1.				
	2.				
	3.				
	Average				
Average Velocity of Falling Mass					
Final Velocity of Falling Mass					
Acceleration of Falling Mass					
Final Angular Velocity					
Angular Acceleration					
Tension in the String					
Torque Acting on Wheel and Axle					

CALCULATIONS

1. From the data of Procedures 3 and 4, calculate the average time of descent.

2. Similarly, calculate the average time of descent for each set of data in Procedures 5, 6, and 7.

3. From the data of Procedures 8 and 9, compute the radius of each axle and the radius of the wheel.

4. Using the average time of fall, compute the average velocity of the falling mass for each set of observations.

5. From the value of the average velocity, calculate the final velocity of the mass as it strikes the floor, for each set of observations. Notice that since the system starts from rest and experiences a uniform acceleration, the final velocity will be twice the average velocity.

6. The value of the final velocity found in Calculation 5 is also the final tangential velocity of a point on the circumference of the axle. Using Equation 42, find the final angular velocity ω for each set of observations.

7. From the average time of fall and the final velocity, calculate the acceleration of the falling mass for each set of observations.

8. The value of the acceleration found in Calculation 7 is also the tangential acceleration of a point on the circumference of the axle. Using Equation 43, find the angular acceleration α for each set of observations.

9. There are two forces acting on the falling mass, the downward force of gravity, which is the weight, and the upward pull of the tension in the string.

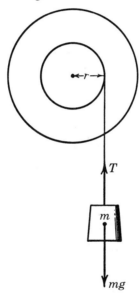

Figure 26   Forces Acting on a Falling Mass

Hence the resultant force acting on the falling mass is

$$F = mg - T$$

where F is the net force in dynes, m is the mass in grams, g is the acceleration of gravity, and T is the tension in the string in dynes. (See Fig. 26.) But by Newton's second law of motion, $F = ma$. Therefore

$$mg - T = ma$$

where m is again the mass in grams and a is the acceleration of the falling mass in centimeters per second per second.

From the value of the acceleration found in Calculation 7, calculate the tension in the string for each set of observations.

10. The torque acting on the wheel and axle is given by

$$L_a = Tr$$

where L_a is the applied torque in centimeter-dynes, T is the tension in the string in dynes, and r is the radius of the axle in centimeters.

Calculate the applied torque for each set of observations.

11. There are two torques acting on the rotating wheel and axle, the applied torque, calculated in (10), and the frictional torque. Hence the resultant torque is the difference between these two, since the frictional torque is acting in a direction opposite to the direction of rotation. But, from the laws of angular motion (Equation 44)

$$L \equiv \text{resultant torque} = I\alpha$$

we get the relation

$$L_a - L_f = I\alpha \qquad (45)$$

where L_f is the frictional torque, I is the moment of inertia in gm-cm^2 of the wheel and axle, and α is the angular acceleration in radians per second per second.

Equation 45 may be written

$$L_a = I\alpha + L_f$$

Here the quantities I and L_f are constant in the sets of observations taken, while L_a and α are variable. This is the equation of a straight line, since it is of the form

$$y = mx + b$$

Hence I is the slope of the line, and L_f is the y intercept.

Plot the values of the angular acceleration α along the x axis and the corresponding values of the applied torque L_a along the y axis. Draw the best straight line through the four points.

12. Determine the value of the moment of inertia I of the wheel and axle by computing the slope of the straight line drawn in Calculation 11.

13. Determine the value of the frictional torque L_f by reading the value of the y intercept of the straight line drawn in Calculation 11.

14. The radius of gyration of a rotating body is the distance from the axis of rotation to the point at which the entire mass of the body may be considered concentrated without altering the moment of inertia. It is defined by the expression

$$I = MK^2$$

where I is the moment of inertia in gm-cm^2 of the body, M is the total mass of the body in grams, and K is the radius of gyration in centimeters.

Using the value of the moment of inertia computed in Calculation 12 and the value of the total mass of the wheel and axle stamped on the wheel, calculate the radius of gyration of the wheel and axle.

QUESTIONS

1. By referring to your graph, determine if it is possible for the y intercept to be negative. What would this mean physically?

2. How does your calculated value of the radius of gyration compare with the radius of the wheel?

3. Could the radius of gyration be greater than the radius of the wheel? Explain.

4. What is the theoretical relation between the radius of gyration K and the actual radius R of a uniform disk? Do your results in the present experiment agree with this relation? If not, why not?

5. A uniform disk 30 centimeters in diameter and having a mass of 2000 grams is free to rotate about its horizontal axle. An object with a mass of 50 grams is attached to a string wound around the rim of the disk. The object is released from rest and descends with constant linear acceleration. Calculate: (a) the moment of inertia of the disk; (b) the linear acceleration of the descending object; (c) the angular acceleration of the disk; and (d) the tension in the string. Neglect the moment of inertia of the axle and the friction in the bearings.

6. Find the moment of inertia of a flywheel consisting of a disk-like rim of mass M, outside radius R_2, and inside radius R_1, mounted on a hub by means of light spokes. *Hint:* Neglect the mass of the hub and spokes and consider the rim to be a disk of radius R_2 with a hole of radius R_1 cut out of it.

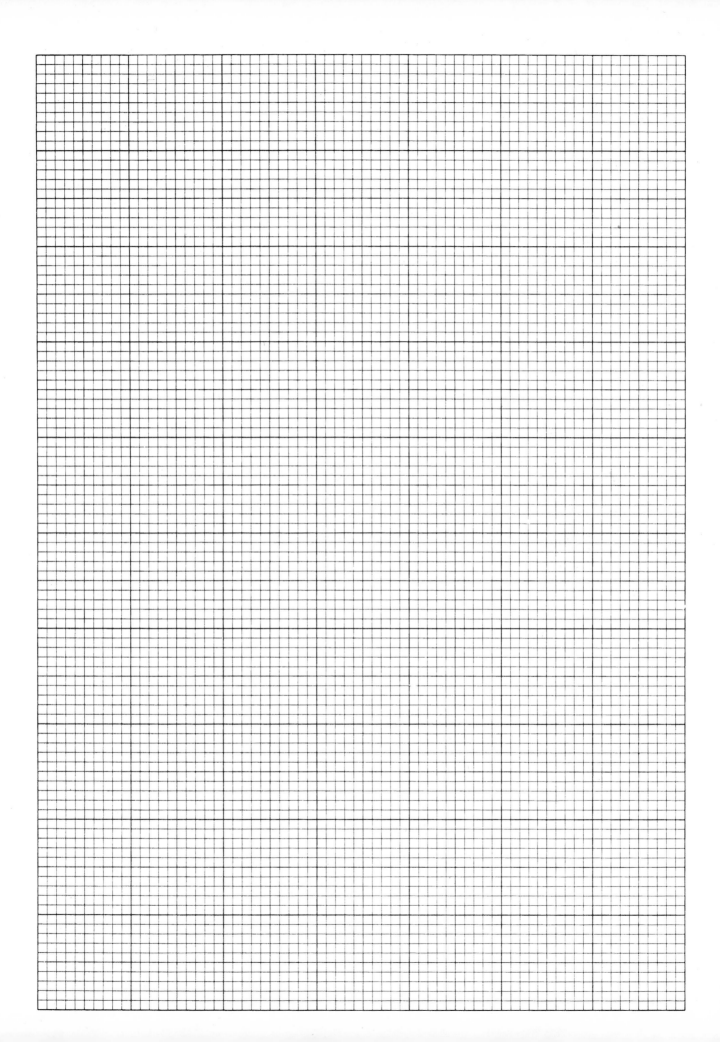

Conservation of Angular Momentum **11**

The principle of conservation of angular momentum is the rotational counterpart of the principle of conservation of linear momentum. According to this principle, the angular or rotational momentum of a rotating system remains constant provided no external torques act on it. In this experiment the system will consist of a rotating bar with a catcher mounted on it into which a ball is fired. The system consists of the rotating bar and the ball, so that the collision forces between the ball and the catcher are internal to the system. Thus, except for the friction in the bearing on which the bar is pivoted (which must be taken into account), no external torques act on the system and its angular momentum should be conserved. The purpose of this experiment is to investigate the law of conservation of angular momentum in this arrangement.

THEORY

Just as the quantity mv is the linear momentum of a moving object of mass m and velocity v, so $I\omega$ is the angular momentum of a rotating body having moment of inertia I and angular velocity ω. And just as it follows from Newton's second law of motion that the linear momentum of a moving system remains constant in the absence of forces external to the system, so also do we find that the angular momentum of a rotating system is constant in the absence of external torques.

In this experiment the rotating system consists of a heavy bar mounted on a very good bearing (so that it is free to rotate about its center with very little friction), a catcher cup mounted on one end and balanced by a coun-terweight mounted on the other, and a metal ball which gets fired into the catcher by a spring gun, as shown in Fig. 27. After the ball has been fired but before it enters the catcher, it has a linear momentum mv, where m is its mass and v its velocity. However, any object moving in a straight line with a linear momentum can be considered to have an angular momentum about some selected pivot point, as illustrated in Fig. 28. This angular momentum is found by multiplying the linear momentum mv by the lever or moment arm l defined as the perpendicular distance from the pivot point in question to the line of action of the linear momentum vector. As a consequence of this definition, angular momentum is sometimes called the *moment of momentum*. Like linear momentum, it is a vector, its direction lying along the supposed axis of rotation, which is a line through the pivot point perpendicular to the plane defined by that point and the line of action of mv. This axis is a line perpendicular to the page in Fig. 28. The sense of the angular momentum vector along this axis is given by the right-hand rule: if the axis is gripped by the right hand so that the fingers curl in the direction of rotation, the thumb points along the axis in

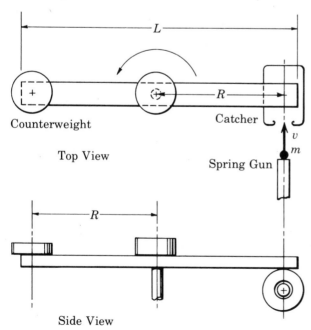

Counterweight

Top View

Catcher

Spring Gun

Side View

Figure 27 Angular Momentum Apparatus

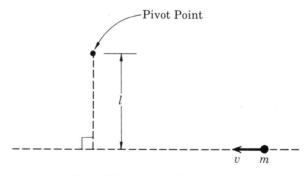

Figure 28 Moment of Momentum

the direction of the vector. This is into the paper in the situation shown in Fig. 28.

After the ball is in the catcher, the system (consisting of the rotating bar, the catcher with the ball in it, and the counterweight) rotates with angular velocity ω. The moment of inertia of the system can be thought of as made up of three parts. The first is the moment of inertia of the bar alone and is given by

$$I_b = \tfrac{1}{12}ML^2$$

where I_b is the moment of inertia of the bar in gram-centimeters^2, M is the mass of the bar in grams, and L is the bar's overall length in centimeters. The second is the moment of inertia of the catcher with the ball in it rotating in a circle of radius R, where R is the distance from the pivot to the center of mass of the catcher assembly (see Fig. 27). This moment of inertia is given by

$$I_{cb} = m_{cb}R^2$$

where m_{cb} is the total mass of the catcher, its mounting nut, and the ball in grams. Similarly the third part of the total moment of inertia is that of the counterweight. It moves in a circle of radius R just as the catcher does and has mass m_{cw} which is close to but not necessarily exactly the same as m_{cb}. Its moment of inertia is thus given by

$$I_{cw} = m_{cw}R^2$$

so that the total moment of inertia of the system is

$$I = I_b + I_{cb} + I_{cw}$$
$$= \tfrac{1}{12}ML^2 + (m_{cb} + m_{cw})R^2 \qquad (46)$$

The angular momentum of the system right after the ball has been caught in the catcher is then $I\omega$, where ω is the angular velocity of the bar right after the ball has been caught. Just before this, the bar was at rest but the ball in flight from the spring gun had angular momentum mvl, where l is the perpendicular distance from the ball's line of flight to the bar's pivot point (the supporting bearing shaft). In Fig. 27, where the axis of the catcher cup is set perpendicular to the bar, $l = R$, but the catcher axis can be set at some smaller angle θ, in which case l will be equal to $R \sin \theta$ and the effective angular momentum reduced accordingly. Whatever the value of l, the angular momentum before the collision will be mvl and afterwards $I\omega$, so that

$$mvl = [\tfrac{1}{12}ML^2 + (m_{cb} + m_{cw})R^2]\omega \qquad (47)$$

All the quantities in Equation 47 can be measured, hence its validity and thus the principle of conservation of angular momentum can be checked.

Most of these quantities are easily found by direct measurement; but two, the speed v given the ball by the spring gun and the angular velocity ω of the rotating system just after the ball has been caught, have to be determined by special procedures. The initial velocity of the ball is found from measurements of its range and vertical distance of fall when it is fired horizontally and allowed to fall onto the laboratory bench without striking the catcher. The method is the same as that discussed in Experiment 5. The ball is fired horizontally, hence its horizontal velocity component is equal to v. This component remains constant throughout the motion, so that the horizontal distance covered in time t is

$$x = vt \qquad (29)$$

where x is in centimeters, t is in seconds, and v is in centimeters per second.

The initial velocity in the vertical direction is zero, hence the vertical motion is that of a freely falling body dropped from rest. Therefore, the distance traveled in the vertical direction in time t is

$$y = \tfrac{1}{2}gt^2 \qquad (30)$$

where y is the distance of fall in centimeters, g is the known value of the acceleration of gravity in centimeters per second^2, and t is the same as in Equation 29. If the height of the gun's firing rod above the bench surface is substituted for y, the corresponding value of t is the ball's time of flight. When this is put into Equation 29, x becomes the range, the distance along the bench surface from a point directly under the end of the firing rod to the impact point. This is easily measured, after which v can be found.

The value needed for ω is the one in effect immediately after the ball has been caught. If there were no friction in the bearing supporting the rotating system, rotation would continue at this angular velocity indefinitely. However, although the bearing is a very good one, it does exert a small frictional torque which will eventually bring the system to a stop. If the system rotates through a total angle ϕ before stopping and takes t_r seconds to do this, then the average value $\bar{\omega}$ of ω is ϕ/t_r and can be measured by simply timing the rotational motion. But to a very good approximation the decelerating torque due to friction in the bearing can be considered constant, so that the rules of constant negative angular acceleration apply. One of these is that the average angular velocity is half the sum of the initial and final velocities. Since the final angular velocity is zero, $\bar{\omega} = \tfrac{1}{2}\omega$ or $\omega = 2\bar{\omega}$. Thus the desired value of ω is determined by dividing the total angle turned through by the time required for this motion and doubling the result.

APPARATUS

1. Rotating bar with catcher and counterweight mounted on force table
2. Spring gun from Blackwood ballistic pendulum
3. Mounting rod and clamps for mounting the spring gun
4. Meter stick
5. 45° right triangle
6. Stop watch or stop clock
7. Equal-arm balance and set of weights
8. Plumb bob
9. Level
10. Carbon paper
11. Sheet of white paper and masking tape

PROCEDURE

1. Weigh the bar, the catcher and its retaining nut, the counterweight, and the ball separately. Notice that the catcher can be removed from the bar by unscrewing the retaining nut, and the bar can be removed from the bearing shaft in a similar manner.

2. Measure the bar's total length L with the meter stick. Also measure the distance from the midpoint of its central hole to the midpoint of either the catcher mounting hole or the counterweight mounting hole. This is the radius R in which the catcher and counterweight revolve.

3. Set the spring gun up for the range measurements. To do this, mount it using the mounting rod and bench clamps, with the firing rod axis exactly 50 centimeters above the bench surface. With the meter stick, measure from the surface to the middle of the firing rod's tip. In this part the rotating system is not used and should be set aside, out of the way of the ball's trajectory.

4. Get the gun ready for firing by placing the ball on the end of the firing rod and pushing it back, compressing the spring until the trigger is engaged. The ball is fired horizontally along the bench so that it will strike the bench surface some distance from the gun. Fire the ball and determine approximately where it lands. Place a sheet of white paper there so that the ball will hit near its center. Secure the paper in position with masking tape. Put a sheet of carbon paper on top of the white paper with the carbon side down so that when the ball strikes, a mark will be left on the white paper at the impact point.

5. Fire the ball five times to obtain five range measurements. These measurements are made from a point on the bench surface directly below the end of the firing rod to each impact point. To determine the point directly below the rod end, hang the plumb bob from this end and mark the indicated point on the bench surface. Note that you want the point directly below the center of the ball as it leaves the gun, and so you should hang the plumb bob from the very end of the firing rod with the rod in its released (fired rather than cocked) position.

6. You are now ready to start the conservation of angular momentum trials. Mount the rotating bar on its bearing support, put the counterweight in place on one end, and mount the catcher cup on the other. Before tightening down the cup, adjust it so that its axis is perpendicular to that of the bar. This is done accurately by placing one short edge of the 45° right triangle against the flat provided on the side of the cup and lining the perpendicular edge up with the edge of the bar. Tighten the retaining nut when the cup is properly aligned.

7. Place the rotating assembly in front of the gun, as shown in Fig. 29. The gun position will have to be changed so that the gun is aimed directly into the catcher cup. The force table supporting the rotating assembly should be placed so that when the gun's firing rod points directly into the catcher cup, the firing rod axis and the rotating bar are at right angles to each other and the indicator on the bar behind the catcher points to zero degrees on the force table scale. Note that aiming the gun at the catcher is facilitated if the two are quite close, although enough space should be left so that the ball is well clear of the firing rod before entering the catcher. However, even with enough clearance in this respect, the end of the firing rod may be close enough to the catcher after firing to interfere with the free rotation of the bar. Be on the lookout for this, and if you see that the back of the catcher will hit the gun as it comes around, release the gun mounting clamp during the first revolution and turn the gun out of the way. The need for doing this can be avoided by placing the gun far enough back from the catcher to begin with, but this requires extra care in aiming and probably a few trial runs before an adjustment is found that lets the ball enter the catcher properly on every shot. Finally, level the table by placing the level on the bar and adjusting the screws in the table feet until the bar is seen to be level regardless of its orientation.

8. When the apparatus is working correctly, you can start taking data. Load the gun and set the rotating bar at rest in the proper position. Fire the gun, start the stop watch at the instant the ball enters the catcher,

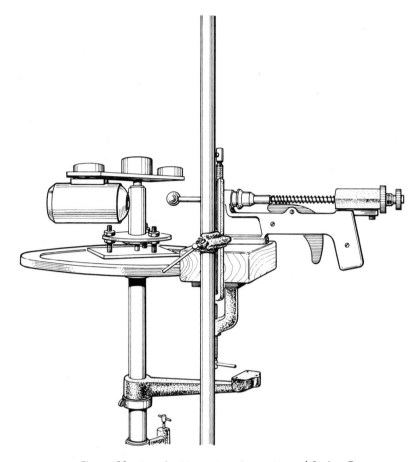

Figure 29   Angular Momentum Apparatus and Spring Gun

and stop the stop watch at the instant the bar just stops turning. While the bar is in motion, count the number of turns, and note the reading of the indicating pointer on the force table scale after rotation stops. Record both the number of turns and the pointer reading, this latter giving the number of degrees through which the bar turned in the uncompleted final revolution.

9. Repeat Procedure 8 twice to obtain a total of three sets of data on the angle turned through and the time required.

10. Reset the catcher so that its axis is at 45° to the bar. Use the 45° right triangle as in Procedure 6, but this time align the hypotenuse with the bar edge. The rotating system will then have to be moved and the gun position changed so that the gun is once more properly aimed into the catcher. Repeat Procedures 7–9 with the new catcher position.

DATA

Mass of the bar	_____	Length L of the bar		_____
Mass of the catcher and its retaining nut	_____	Radius R of the counterweight and catcher orbits		_____
Mass of the counterweight	_____	Moment of inertia I of the rotating system		_____
Mass of the ball	_____			

Range of the ball   1. _____   2. _____   3. _____   4. _____   5. _____

     Average range   _____

Time of flight   _____   Velocity of projection from range measurements   _____

Catcher Angle, Degrees	Rotation Time, Seconds	Number of Turns Completed	Angle of Incomplete Turn, Degrees	Total Angle Turned, Radians	Average Angular Velocity $\bar{\omega}$, Radians/sec	Initial Angular Velocity ω, Radians/sec
90°						
45°						

Catcher at 90°

Average value of the initial angular velocity ω _____

Angular momentum after the ball is caught _____

Moment arm l of the ball in flight _____

Angular momentum before the ball is caught _____

Percent discrepancy _____

Catcher at 45°

Average value of the initial angular velocity ω _____

Angular momentum after the ball is caught _____

Moment arm l of the ball in flight _____

Angular momentum before the ball is caught _____

Percent discrepancy _____

CALCULATIONS

1. Calculate the moment of inertia of the rotating system using Equation 46 and the data of Procedures 1 and 2. Remember that m_{cb} is the mass of the catcher complete with its mounting nut and with the ball inside it.

2. From the data of Procedure 5, compute the average range of the ball.

3. Knowing that the ball falls a vertical distance of 50 centimeters and that $g = 980$ centimeters per second^2, calculate the time of flight of the ball from Equation 30.

4. Compute the velocity of projection from the time of flight obtained in Calculation 3 and the range obtained in Calculation 2 by using Equation 29.

5. Calculate the total angle through which the rotating arm turned following each firing of the ball into the catcher. This is done by multiplying the number of complete turns by 360 and adding the number of degrees turned through in the incomplete final turn. The result is converted from degrees to radians by multiplying by $\pi/180$.

6. Calculate the average angular velocity $\bar{\omega}$ in each case by dividing the total angle turned through *in radians* by the corresponding time in seconds. Then double each value of $\bar{\omega}$ to obtain the corresponding initial angular velocity ω.

7. Find the average of your three values of ω for the case of the catcher set at 90° to the bar and multiply this result by the moment of inertia found in Calculation 1 to obtain the angular momentum of the system after the ball is caught.

8. For the case of the catcher set at 90° to the bar, the moment arm l of the ball in flight is equal to the radius R of the circle in which the catcher moves. Multiply this radius, the mass of the ball, and the ball's velocity from Calculation 4 together to obtain the angular momentum mvl due to the ball before it is caught. Since before the ball is caught the rotating system is at rest, mvl is the total angular momentum of the complete system at that time and should be equal to the system angular momentum afterwards as found in Calculation 7. Compare your before and after values of the angular momentum by calculating the difference between them and the percent discrepancy.

9. Repeat Calculations 7 and 8 for the case of the catcher set at 45° to the rotating bar. Note that in this case, $l = R \sin 45° = 0.707R$.

QUESTIONS

1. Explain in your own words how this experiment demonstrates the principle of conservation of angular momentum.

2. Where do you think errors came into your experiment? What could be done to reduce them?

3. The counterweight was considered to be a point mass traveling in a circle of radius R, but actually it is a cylinder with a diameter of about 5 centimeters. Since it is mounted on the bar, it turns about its own center of mass with the same angular velocity as that of the rotating system. In other words, its total motion has two components: the motion of its center of mass in a circle of radius R and the rotational motion about its own center of mass. Both contribute to the moment of inertia of the complete rotating system. Why could you neglect the contribution due to the counterweight's own rotation?

4. Calculate the energy $\frac{1}{2}mv^2$ of the system before the ball was caught in the catcher and the energy $\frac{1}{2}I\omega^2$ afterwards. Which is greater? What happened to the difference? Was the difference greater when the catcher was at 45° to the bar or when it was at 90°? Why?

5. A wheel having a moment of inertia of 10^5 gram-centimeters^2 is mounted in a horizontal plane on a frictionless shaft. The wheel has a diameter of 50 centimeters and is initially at rest. A 40-gram ball is fired in a direction tangent to the rim of the wheel with a velocity of 1000 centimeters per second and is caught in a cup mounted on the rim. What is the angular velocity of the wheel after the ball settles down in the cup?

6. (a) Calculate the conversion factor between angular velocity in radians per second and in revolutions per minute. (b) Express your answer to Question 5 in revolutions per minute.

Archimedes' Principle 12

Buoyancy is the ability of a fluid to sustain a body floating in it or to diminish the apparent weight of a body submerged in it. This apparent reduction in weight being equal to the weight of the fluid displaced is called Archimedes' principle. It is the purpose of this experiment to study Archimedes' principle and its application to the determination of density and specific gravity. In particular, the specific gravities of a solid heavier than water, a solid lighter than water, and a liquid other than water will be measured.

THEORY

The density of a body is defined as its mass per unit volume. It is usually expressed in grams per cubic centimeter. The specific gravity of a body is the ratio of its density to the density of water at the same temperature. Since for a given location the weight of a body is taken as a measure of its mass, the specific gravity may be taken as the ratio of the weight of a given volume of a substance to the weight of an equal volume of water. Because the mass of one cubic centimeter of water at 4°C is 1 gram, the specific gravity of a body at this temperature is also numerically equal to its density in grams per cubic centimeter.

Archimedes' principle states that the apparent loss of weight of a body immersed in a fluid is equal to the weight of the fluid displaced. The specific gravity of a solid heavier than water may be easily determined by the application of this principle. The body is weighed in air; then it is weighed in water, that is, suspended by a thread from the arm of a balance so as to be completely submerged (see Fig. 30). The loss of weight in water is $W - W_1$, where W is the weight in air and W_1 is the weight in water. But the loss of weight in water is equal to the weight of the water displaced, or the weight of an equal volume of water. Thus the specific gravity S will be

$$S = \frac{W}{W - W_1}$$

The specific gravity of a liquid may be found by measuring the loss of weight of a convenient solid body when immersed in that liquid and the loss of weight when immersed in water. The procedure is as follows : A heavy body is weighed in air; this weight is called W. Then it is weighed in water; this weight is called W_1. Finally it is weighed in the liquid whose specific gravity is to be determined; this weight is called W_2. The specific gravity of the liquid will be

$$S = \frac{W - W_2}{W - W_1}$$

since this expression represents the weight of a certain volume of the liquid divided by the weight of an equal volume of water. Here $W - W_1$ is the loss of weight in water, and $W - W_2$ is the loss of weight in the given liquid.

In order to find the specific gravity of a solid lighter than water, it is necessary to employ an auxiliary body, or sinker, of sufficient weight and density to hold the other body completely submerged. The specific gravity of a solid lighter than water, as obtained by the sinker method, is given by

$$S = \frac{W}{W_1 - W_2} \qquad (48)$$

where W is the weight of the solid in air; W_1 is the weight of the solid and the sinker, with the sinker alone immersed; and W_2 is the weight when both solids are immersed in water.

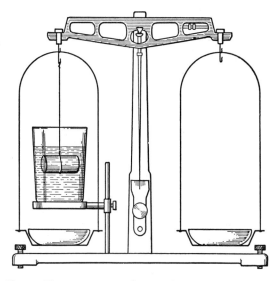

Figure 30 Arrangement for Weighing a Body in Water

The hydrometer is an instrument designed to indicate the specific gravity of a liquid by the depth to which it sinks in the liquid. To measure the specific gravity of a liquid by means of a hydrometer, it is only necessary to let the hydrometer float in the liquid and to read the specific gravity directly on the calibrated scale. The reading is taken, if possible, by placing the eye below the liquid surface and seeing where this surface cuts the hydrometer scale.

APPARATUS

1. Equal-arm balance with platform and set of weights
2. Metal cylinder
3. Wooden cylinder
4. Lead sinker
5. Distilled water
6. Alcohol
7. Hydrometer
8. Hydrometer jar
9. 1000-cc Pyrex beaker
10. Fine thread

PROCEDURE

1. Weigh the metal cylinder in air.

2. Weigh the metal cylinder in water. Suspend the cylinder by a fine thread from the lower hook over the scale pan, without removing the pan. Place a beaker of distilled water on the adjustable circular platform over the scale pan and immerse the cylinder completely. Weigh the cylinder while it is immersed but not touching the beaker. Remove the beaker and dry the metal cylinder.

3. Empty the beaker, dry it thoroughly, and refill it with alcohol. Then repeat Procedure 2. When you are finished, pour the alcohol into the hydrometer jar and rinse out the beaker with distilled water.

4. Weigh the wooden cylinder in air.

5. Refill the beaker with distilled water. Attach the sinker to the wooden cylinder and weigh the combination with the sinker alone immersed in water.

6. Now weigh the two solids when they are both immersed in water.

7. Measure the specific gravity of the alcohol with the hydrometer. See that there is enough alcohol in the hydrometer jar and let the hydrometer float in the alcohol. Read the specific gravity directly.

DATA

Weight of metal cylinder in air _____
Weight of metal cylinder in water _____
Weight of metal cylinder in alcohol _____
Specific gravity of metal cylinder _____
Specific gravity of metal cylinder from Table III, Appendix _____
Percent error _____
Specific gravity of alcohol _____
Specific gravity of alcohol from Table III, Appendix _____
Percent error _____
Weight of wooden cylinder in air _____
Weight of cylinder in air and sinker immersed in water _____
Weight of cylinder and sinker, both immersed in water _____
Specific gravity of wooden cylinder _____
Specific gravity of alcohol by using the hydrometer _____

CALCULATIONS

1. From the data of Procedures 1 and 2, calculate the specific gravity of the metal cylinder.

2. From the data of Procedures 1–3, calculate the specific gravity of the alcohol.

3. From the data of Procedures 4–6, compute the specific gravity of the wooden cylinder.

4. Compute the percent error of your measurements by comparing your results for the specific gravity of the metal cylinder and of the alcohol with the accepted values.

5. Compare your measurement of the specific gravity of alcohol using the hydrometer with the accepted value and note whether this latter value falls within the limits of precision on the hydrometer reading.

QUESTIONS

1. (a) Explain how you can obtain the volume of an irregular solid insoluble in water. (b) How can you obtain the weight of an equal volume of water?

2. (a) Why does a block of wood apparently lose more than its entire weight in air, when completely submerged in water? (b) How is this made possible?

3. (a) What becomes of the lost weight of an object when immersed in a liquid? (b) How is this apparent loss of weight explained?

4. Suppose there were a bubble of air on the bottom of the metal cylinder immersed in water. How would this affect the calculations of the density of the metal?

5. A piece of cork which weighs 25 grams in air and has a specific gravity of 0.25 is attached to a lead sinker which weighs 226 grams in air. What will be the apparent weight of the two solids when they are both immersed in water?

6. The cork of Question 5 is allowed to float in water. What fraction of its volume is above the surface?

7. Derive Equation 48.

8. Consider a solid cylinder just immersed in a liquid with its axis vertical so that the top of the cylinder is just level with the liquid's surface. Knowing that the gauge pressure at depth h in the liquid is $h\rho g$, where ρ is the liquid's density and g is the acceleration of gravity, derive Archimedes' principle.

9. Suggest a modification of the apparatus arrangement shown in Fig. 30 that will demonstrate your answer to Question 3.

Simple Harmonic Motion **13**

Simple harmonic motion is one of the most common types of motion found in nature, and its study is therefore very important. Examples of this type of motion are found in all kinds of vibrating systems, such as water waves, sound waves, the rolling of ships, the vibrations produced by musical instruments, and many others. In fact, any time a *linear restoring force* exists, simple harmonic motion results. By a linear restoring force is meant a force that is proportional to the displacement of the body on which the force acts from an equilibrium position (where the force is zero) and is always directed back towards that equilibrium position. The object of this experiment is to study two important examples of such a force, the simple pendulum and the vibrating spring, and to determine the time of vibration (called the period) in each case.

THEORY

A linear restoring force as defined above may be expressed mathematically by the equation

$$F = -kx \qquad (49)$$

where F is the force, x represents the displacement from the equilibrium point $x = 0$, k is the proportionality constant, and the minus sign expresses the restoring nature of the force by indicating that the direction of F is always opposite to the displacement x from equilibrium.

If a body to which a linear restoring force is applied is displaced from equilibrium and released, the restoring force will bring the body back to its equilibrium position. But the body's inertia will then carry it beyond this position, and so a restoring force will build up in the opposite direction. This force first brings the body to rest and then accelerates it back towards the equilibrium position again. The action is repeated and the body executes a type of vibration known as simple harmonic motion. In this motion the maximum displacement from the equilibrium point achieved by the body on either side is called the amplitude, one complete vibration is called a cycle, the time required for one cycle is called the period, and the number of cycles completed per second (the reciprocal of the period) is called the frequency. Clearly the period is measured in seconds and the frequency in cycles per second (sec^{-1}). The unit of cycles per second is now commonly called hertz in honor of Heinrich Hertz (1857–1894), who investigated electrical oscillations and first showed (1887) that high frequency electrical oscillations could produce electromagnetic radiation. Further analysis of the vibrations resulting from a linear restoring force shows that the period is given by

$$T = 2\pi \sqrt{\frac{m}{k}} \qquad (50)$$

where T is the period, m is the mass of the vibrating body, and k is the proportionality constant from Equation 49. If m is expressed in grams and k in dynes per centimeter, T will come out in seconds.

Two physical examples of this type of force are considered in the present experiment. The first of these is the simple pendulum, which consists of a concentrated mass suspended at the end of a cord of negligible weight. A close approximation to this is a small metal sphere on a long, thin thread. Such an arrangement is diagrammed in Fig. 31, which shows a sphere of mass m hung on the end of a thread whose length is such that the distance

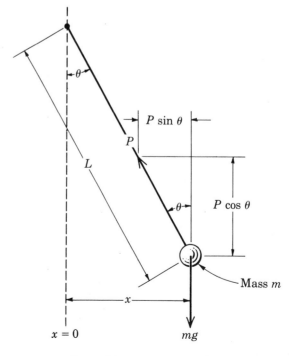

Figure 31 The Simple Pendulum

81

from the suspension point to the center of the sphere is L. The sphere has been drawn aside to a horizontal distance x, the thread consequently making an angle θ with the vertical. Note that the vertical line through the suspension point is at $x = 0$, for this is the equilibrium position, mass m being in equilibrium when it hangs straight down.

Inspection of Fig. 31 shows that the sum of the x components of all the forces acting on mass m is simply $-P \sin \theta$ and the sum of all the y components is $P \cos \theta - mg$, where P is the tension in the thread. If the pendulum is never swung by more than a very small amount so that θ is always a small angle even at the extremes of the swing, then two simplifying assumptions may be made. The first of these is that for small angles the cosine remains approximately equal to 1, so that $P \cos \theta - mg$ may be approximated by $P - mg$. Secondly, in a small swing there is very little vertical motion of mass m, so that zero vertical acceleration may be assumed. Hence

$$P \cos \theta - mg \approx P - mg \approx 0$$

or $$P \approx mg$$

Further inspection of Fig. 31 shows that, by definition, $\sin \theta = x/L$, and therefore by substitution the x-directed force $-P \sin \theta$ becomes $-mgx/L$. This qualifies as a linear restoring force with the proportionality constant k equal to mg/L, and Equation 50 shows that the period of the simple pendulum should then be

$$T = 2\pi \sqrt{\frac{mL}{mg}} = 2\pi \sqrt{\frac{L}{g}} \qquad (51)$$

Again the period will be in seconds if the length L is in centimeters and g, the acceleration of gravity, is in centimeters per second per second (980 cm/sec²).

The second example of a linear restoring force presented in this experiment is that of a stretched spring. Ideally the elongation of a spring is proportional to the stretching force. Therefore, if a mass is hung on a spring, the spring will be stretched by an amount proportional to the weight of the mass. Furthermore, an equilibrium position will be reached at which the spring pulls up on the mass with a force just equal to the mass's weight. The situation is illustrated in Fig. 32, where initially the spring hangs relaxed, with no mass attached, and has a length y_1. When a weight of mass m is attached, the spring elongates to length y_0, and if the elongation is proportional to the stretching force (the weight of the mass), $mg = k(y_0 - y_1)$.

When the weight is at any arbitrary distance y below the spring support, the elongation of the spring is $y - y_1$, the upward pull of the spring on the weight is $k(y - y_1)$ and the downward force of gravity is mg. Thus the net force on the weight is

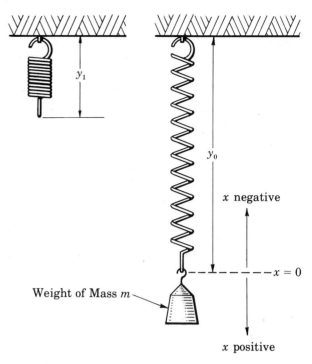

Figure 32   Weight Suspended on an Ideal Spring

$$F = mg - k(y - y_1) \qquad (52)$$

If a new coordinate x, measured in the same direction as y but having the equilibrium point as its origin, is introduced as shown on the right in Fig. 32, then $x + y_0 = y$, for the value of y will always be y_0 greater than the value of x. Substitution in Equation 52 yields

$$F = mg - k(x + y_0 - y_1) \qquad (53)$$

But since $mg = k(y_0 - y_1)$, Equation 53 becomes

$$F = k(y_0 - y_1) - k(x + y_0 - y_1) = -kx$$

Thus again we have a linear restoring force, the constant k here being a characteristic of the spring called its force constant. The force constant can be determined by measuring the elongation in centimeters suffered by the spring when a stretching force of a given number of dynes is applied. A part of this experiment will be devoted to making such a determination.

If the weight in Fig. 32 is now pulled down below its equilibrium position and released, it will execute simple harmonic motion with a period given by Equation 50. Note, however, that m in this equation will be the mass of the weight only if the spring has negligible mass. In the general case where the spring's mass cannot be neglected, m is the effective mass of the vibrating system, which is made up of the mass of the suspended weight plus a part of the mass of the spring since the spring itself is also vibrating. It is found by analysis that one third of the mass of the spring must be added to the mass of the suspended weight in calculating the value of m to be used in Equation 50.

APPARATUS

1. Metal sphere on a long string (the simple pendulum)
2. Supporting rod with pendulum clamp and hook for suspending spring
3. Two-meter stick with caliper jaws
4. Vernier caliper
5. Spring (cylindrical type, coils not touching when relaxed)
6. Set of hooked weights (100-, 200-, 200-, 500-grams)
7. Stop watch or stop clock
8. Equal-arm balance and weights

PROCEDURE

1. Measure the diameter of the metal sphere with the vernier caliper.

2. Clamp the string so that the center of the sphere is about 60 centimeters below the point of support. Carefully measure the distance from the lower edge of the support to the upper surface of the sphere. Record this distance as the length of string used. The length of the pendulum will be this distance plus the radius of the metal sphere.

3. Displace the sphere to one side through an angle of not more than about 5° and let the pendulum oscillate. Record the time it takes the pendulum to make 50 vibrations. In counting vibrations, be sure to start the stop watch on the count of zero and not on the count of one.

4. Repeat Procedures 2 and 3, making the length of the pendulum successively about 80, 100, 120, 140, and 160 centimeters. In the last three cases record the time for 25 vibrations instead of 50.

5. Using a length of about 50 centimeters, determine the time for 50 vibrations when the sphere is displaced about 5°, 30°, and 45° from the equilibrium position. Record the length of the pendulum and the time for 50 vibrations at each displacement.

6. Weigh the spring and record its mass.

7. Remove the pendulum from the support and suspend the spring in its place. Using the two-meter stick and a caliper jaw, observe the position of the lower end of the spring and record the reading.

8. Suspend 100 grams from the spring and again record the position of its lower end.

9. Repeat Procedure 8 with loads of 200, 300, 400, and 500 grams suspended from the spring. Record the position of the spring's lower end in each case.

10. Suspend a mass of 200 grams from the spring, displace it about 5 centimeters downward from its position of equilibrium, and set it into vibration. Using a stop watch, measure the time for 50 complete vibrations of the mass and record it.

11. Repeat Procedure 10, again using a mass of 200 grams but displacing it about 10 centimeters from its position of equilibrium and setting it into vibration. Again record the time for 50 complete vibrations.

12. Suspend a mass of 500 grams from the spring, displace it about 5 centimeters from its position of equilibrium, and set it into vibration. Record the time for 50 complete vibrations.

DATA

Diameter of sphere _____   Radius of sphere _____

Length of String Used	Length of Pendulum	Number of Vibrations	Time	Period	Square of Period

Value of g from slope _____   Percent error _____

Length of string used _____   Length of pendulum _____

Initial Displacement of Sphere	Number of Vibrations	Time	Period	Square of Period
5°				
30°				
45°				

Mass Suspended From the Spring	Scale Reading	Elongation
0 gm		
100 gm		
200 gm		
300 gm		
400 gm		
500 gm		

Mass of the spring _____   Force constant of the spring _____

| Mass Suspended From the Spring | Mass of the Vibrating System | Amplitude of Vibration | Time For 50 Vibrations | Period | | Percent Discrepancy |
				Experimental Value	Calculated Value	
200 gm		5 cm				
200 gm		10 cm				
500 gm		5 cm				

CALCULATIONS

1. Calculate the period of the pendulum for each observation.

2. Calculate the square of the pendulum's period for each observation. Use only three significant figures in recording the square of the period, rounding off the third figure.

3. Plot a curve using the values of the square of the period as ordinates and the lengths of the pendulum as abscissas. Use the entire sheet of graph paper. Read the instructions on plotting graphs in the Introduction.

4. Determine your measured value of g from the slope of the curve just plotted and compare it with the known value, $g = 980$ cm/sec^2, by finding the percent error. The acceleration of gravity varies slightly with latitude and elevation. The value given is approximately correct for sea level and 45° latitude.

5. From the data of Procedures 7–9, determine the elongation of the spring produced by each load by subtracting the zero reading from the reading corresponding to each load.

6. Plot a curve using the values of the elongation as ordinates and the forces due to the corresponding loads as abscissas.

7. Obtain the force constant of the spring from the slope of the curve plotted in Calculation 6. See the instructions in the Introduction.

8. From the data of Procedures 10–12 calculate the experimental value of the period of the oscillating weight for each set of observations.

9. Using Equation 50, calculate the value of the period for each of the two masses used. The equivalent mass of the vibrating system to be used in 50 is equal to the mass suspended from the spring plus one third of the mass of the spring.

10. Calculate the percent discrepancy between the experimental values of the period and the calculated values.

QUESTIONS

1. State what you conclude from your work with the simple pendulum.

2. Explain in your own words why the angle through which the pendulum swings must be no more than about 10°.

3. Why was it unnecessary to measure the mass of the metal sphere?

4. (a) At which point on the path of its vibration does the weight suspended on the spring have the greatest acceleration? (b) Where does it have the greatest velocity? (c) Where does it have its least acceleration? (d) Where does it have its least velocity?

5. Is the motion of the piston in a one-cylinder steam engine simple harmonic? Explain.

6. What does your curve show about the dependence of your spring's elongation upon the applied force?

7. (a) If it takes a force of 500 dynes to stretch a spring 2 centimeters, what force is needed to stretch the spring 5 centimeters? (b) How much work is done in stretching the spring 5 centimeters?

8. What percentage error is introduced in the calculated value of the period of vibration, for the 200-gram load and the 500-gram load respectively, if the mass of the spring is neglected?

9. If adding a certain weight in Procedure 10 would double the period, what would the mass of the added weight be?

10. Rotational simple harmonic motion occurs when there is a linear restoring *torque* given by $L_r = -K\theta$, where θ is the *angular* displacement from equilibrium. The period of the resulting rotational oscillation is given in analogy with Equation 50 by $T = 2\pi\sqrt{I/K}$. Derive Equation 51 by treating the simple pendulum as a rotating system with the suspension point as the pivot. Note that for small angles, $\sin\theta \approx \theta$.

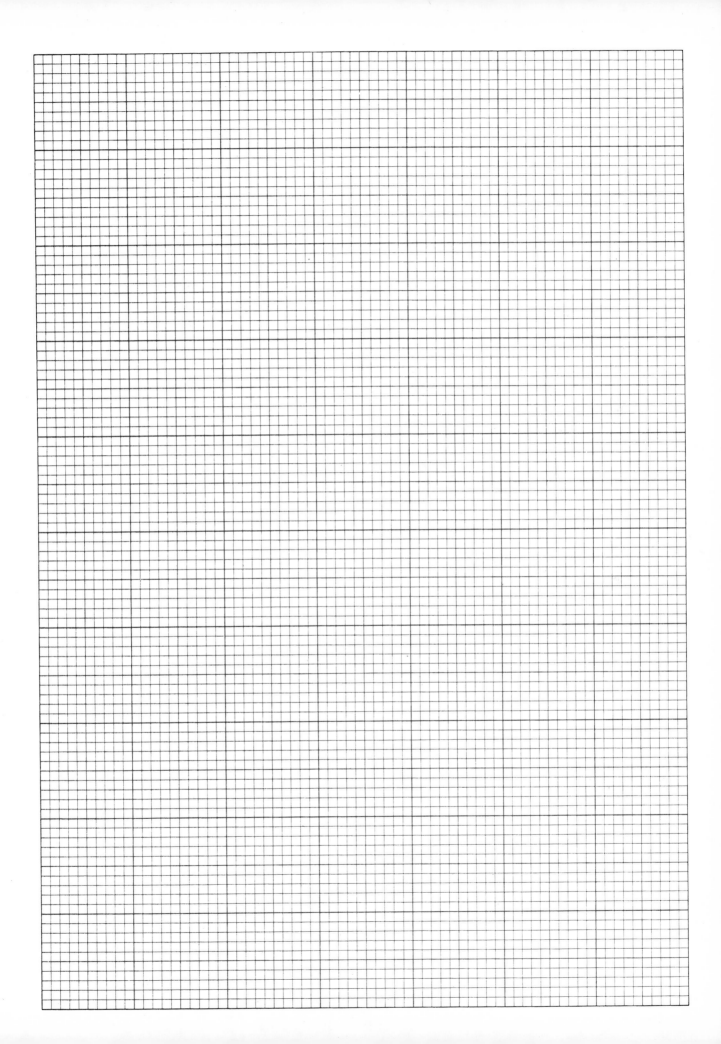

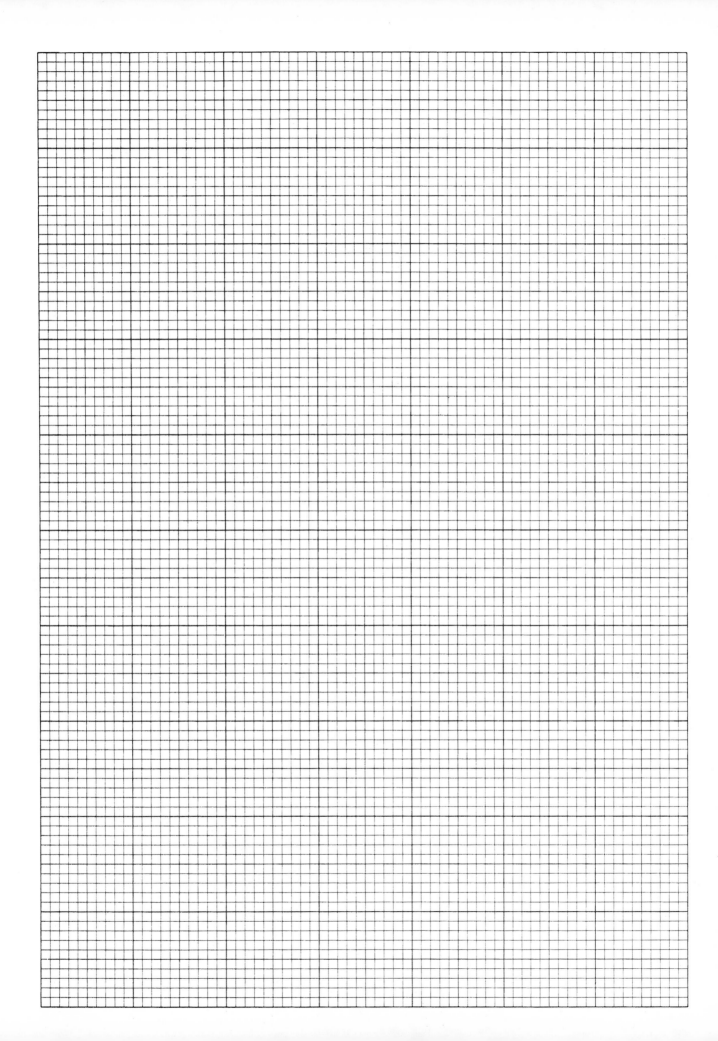

Standing Waves in Strings 14

The general appearance of waves can be shown by means of standing waves on a string. This type of wave is very important because most of the vibrations of bodies, such as the prongs of a tuning fork or the strings of a piano, are standing waves. The purpose of this experiment is to study the relation between stretching force and wavelength in a vibrating string and to determine the natural vibrational frequencies of a stretched string by means of standing waves.

THEORY

Standing waves, or stationary waves, are produced by the interference of two wave trains of the same wavelength, velocity, and amplitude traveling in opposite directions through the same medium. The necessary conditions for the production of standing waves can be met in the case of a stretched string by having a train of waves, set up by some vibrating body, reflected at the end of the string. These will then interfere with the oncoming waves.

A stretched string has many modes of vibration. It may vibrate as a single segment; its length is then equal to one half the wavelength of the vibrations produced. It may also vibrate in two segments, with a node at each end and one in the middle; the wavelength of the vibrations produced is then equal to the length of the string. It may also vibrate in a larger number of segments. In every case the length of the string is some whole number of half wavelengths.

When standing waves are produced, a condition of resonance exists between the vibrating body and the string. That is, the frequency of vibration of the body is the same as the frequency of that particular mode of vibration of the string. Corresponding to this frequency, there is a particular wavelength λ such that

$$V = f\lambda \qquad (54)$$

where f is the frequency and V is the velocity of the wave on the string. This velocity is given by

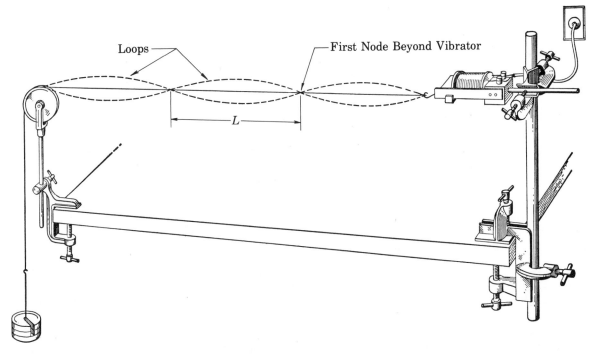

Loops — First Node Beyond Vibrator

Figure 33 Standing Waves in a String

$$V = \sqrt{\frac{F}{d}} \qquad (55)$$

where F is the tension in dynes and d is the mass per unit length of the string in grams per centimeter. Changing the tension thus changes the velocity of the wave and hence (if the frequency remains constant) the wavelength. Therefore, as the tension is varied, the conditions for the vibration of the string in different numbers of parts (the conditions for different vibrational modes) are fulfilled as each is brought into resonance at the fixed driving frequency.

In this experiment standing waves are set up in a stretched string by the vibrations of an electrically driven string vibrator operated by a 60-cycle alternating current. The arrangement of the apparatus is shown in Fig. 33. Since the blade of the vibrator is attracted towards the pole face once during each half cycle, its frequency will be double that of the supply current, or 120 vibrations per second. The tension in the string is measured by the masses suspended over the pulley by means of a weight hanger and is altered by changing these masses. Since the distance L between consecutive nodes is equal to half a wavelength λ, Equation 54 becomes

$$f = \frac{V}{\lambda} = \frac{1}{2L} \sqrt{\frac{F}{d}} \qquad (56)$$

The relation between the wavelength and the tension is

$$\lambda = \frac{1}{f} \sqrt{\frac{F}{d}} \qquad (57)$$

These relations between the wavelength, the tension, and the length of the vibrating segment apply to each mode of vibration of the string, where L in each case is the length of one vibrating segment corresponding to a particular tension.

APPARATUS

1. Electrically driven string vibrator
2. Support rod and clamps for vibrator
3. String
4. Triple-beam balance
5. Two-meter stick
6. Weight hanger
7. Set of slotted weights
8. Rod-mounted pulley

PROCEDURE

1. Measure and cut off a piece of string exactly 150 centimeters long. Weigh this length of string.

2. Fasten one end of the string to the vibrator armature, pass the string over the pulley, and attach the weight hanger at the other end. The pulley should be mounted as shown in Fig. 33, about a meter away from the vibrator, which is supported from the laboratory table by means of the rod and clamps as shown.

3. Connect the vibrator directly to a 110-volt, 60-cycle outlet and switch the current on. Increase the tension by adding weights to the weight hanger until the string vibrates in two segments. Adjust the tension by adding or removing small weights until the loops formed are of maximum width. Obtaining this maximum with the weights available may prove difficult, therefore a fine adjustment of the length of string between the vibrator and the pulley should be made to assure maximum amplitude of vibration of the vibrating parts of the string. This may be done by carefully loosening the clamp holding the vibrator and sliding this latter back and forth to "tune" for maximum amplitude.

4. Measure the distance from the point where the string contacts the pulley to the first node beyond the vibrator (see Fig. 33). With the string vibrating in only two segments, this node is the one in the middle. The measurement may be made by simply holding the meter stick alongside the vibrating string, but a more accurate result is usually obtainable by grasping the string between your thumb and forefinger with your thumb nail right at the node, shutting off the vibrator, and then using the meter stick to measure the distance from your nail to the pulley. In doing this, be sure not to pull on the string in such a direction as to raise the weights and thus increase the length being measured. If you pull at all, make sure you pull against the vibrator. Divide the length so obtained by the number of vibrating segments included in this length to obtain the length of one segment. Note that in the case of a total of two vibrating segments, only one is included between the node beyond the vibrator and the pulley, so that the measured length is itself the length of one segment. Record this length and the tension.

5. Repeat Procedure 4, decreasing the tension by removing weights until the string vibrates in a total of 3, 4, 5, 6, 7, and 8 segments in turn. In each case adjust the tension to produce loops of approximately maximum width and fine tune for an exact maximum by adjusting the distance from the vibrator to the pulley as described in Procedure 3. Measure the length of one vibrating segment as described in Procedure 4, and record this length and the corresponding tension in each case.

DATA

Length of string _____   Average value of the frequency _____

Mass of string _____   Percent error _____

Mass per unit length _____   Wave velocity for vibration in 3 segments _____

Number of Segments	Tension		Length of One Segment	Wavelength	Square Root of Tension
	Grams	Dynes			
2					
3					
4					
5					
6					
7					
8					

CALCULATIONS

1. Determine the mass per unit length of the string. Express it in grams per centimeter.

2. Calculate the tension in dynes and the square root of this tension for each observation.

3. Calculate the wavelength in centimeters by multiplying the length of a single vibrating segment by 2 in each case.

4. Plot the square root of the tension against the wavelength using your values of $\sqrt{F}$ as abscissas and those of λ as ordinates. Use an entire sheet of graph paper for the curve.

5. According to Equation 57, λ and $\sqrt{F}$ are proportional with proportionality constant $1/(f\sqrt{d})$. This constant should therefore be the slope of the curve drawn in Calculation 4. Obtain this slope and determine from it the frequency f. This represents an average value of f for all your observations.

6. Compare your average value of the frequency found in Calculation 5 with the known value, 120 vibrations per second, by calculating the percent error.

7. Compute the velocity of the wave in the string when it vibrates in a total of three segments.

QUESTIONS

1. Why was the length of the string between the first node beyond the vibrator and the pulley considered in this experiment, rather than the entire length between the vibrator and the pulley? What error would have been introduced if the entire length and total number of vibrating segments had been used? With your experimental setup, is this error likely to have been large?

2. Why should string that doesn't stretch easily be used in this experiment?

3. Should the curve drawn in Procedure 4 pass through the origin? What does this mean physically?

4. What is meant by resonance?

5. Upon what physical properties do (a) the loudness, (b) the pitch, and (c) the quality of a musical note depend?

6. A copper wire 1 meter long and weighing 0.61 gram per centimeter vibrates in two segments when under a tension produced by a load of 250 grams. What is the frequency of this mode of vibration?

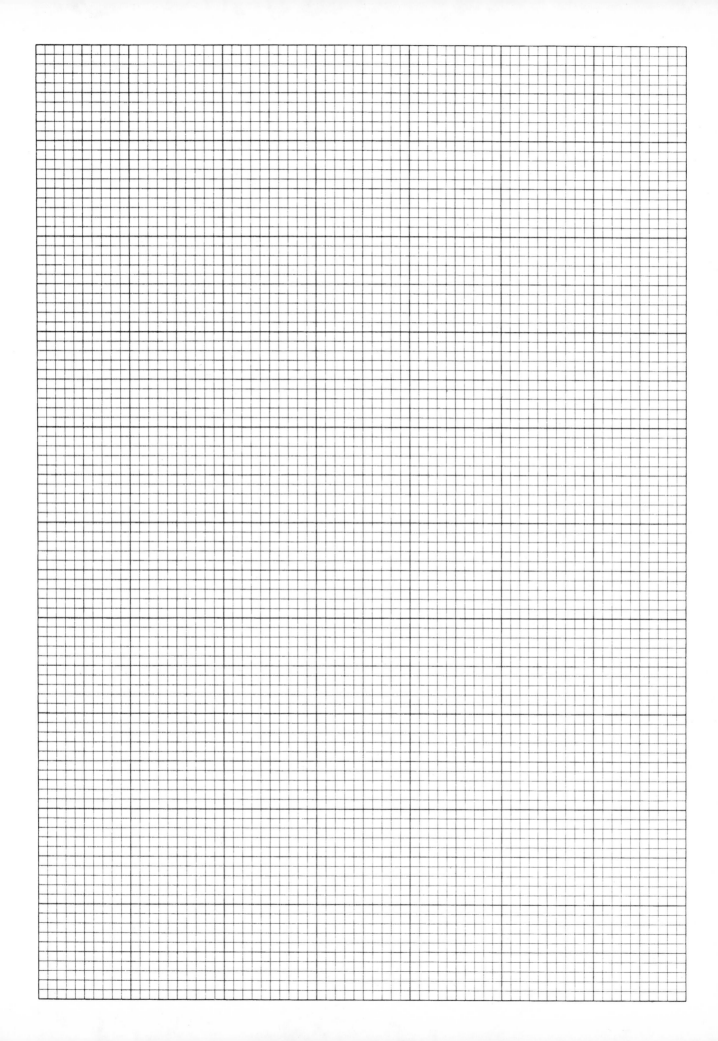

The Sonometer 15

A stretched string that is set into vibration will emit a musical note, the frequency of which is determined by the length of the string, the mass per unit length, and the tension. By varying any one of these quantities, the frequency may be changed. The purpose of this experiment is to determine (1) how the frequency of a vibrating string varies with its length when the tension is kept constant, (2) how the frequency of a vibrating string varies with the tension in the string when its length is kept constant, and (3) how the frequencies of two strings of the same length and material and under the same tension depend on their cross-sectional areas.

THEORY

The frequency, or number of vibrations per second, of the fundamental of a stretched string depends upon the length of the string, the tension, and the mass per unit length. The relation between these quantities is

$$f = \frac{1}{2L} \sqrt{\frac{F}{d}} \qquad (56)$$

where f is the frequency of the note emitted in vibrations per second, L is the length of the vibrating segment in centimeters, F is the tension in dynes, and d is the mass per unit length of the string expressed in grams per centimeter.

In this experiment, the above relation will be tested by the use of a sonometer. This device consists of a resonant case provided with wires which may be stretched. The tension on the wires, or the stretching force, can be easily measured. Adjustable bridges are provided so that the lengths of the vibrating strings may be varied and measured. The general appearance of a sonometer is shown in Fig. 34.

The experimental work consists of tuning the sonometer to several tuning forks of known frequencies so that the law expressed by the relation given above may be tested. Thus the variation of the frequency of a vibrating string with stretching force may be studied, as well as the variation of frequency with length. Finally, the effect of the mass per unit length may be shown by using two wires of different diameters.

To detect when the wire and the fork are in unison, the two may be sounded alternately, making adjustments on the length of the wire until beats are observed. Then the two are sounded simultaneously and the adjustments are continued until the beats disappear. Unison may also be tested by placing the tip of the vibrating tuning fork on the top of the resonance board and noting whether the string is set into sympathetic vibration. Another method is to place a very small paper rider on the wire at the center of the vibrating segment. When the tip of the vibrating tuning fork is placed on the top of the resonance board, the rider will jump off if the string is in resonance with the tuning fork.

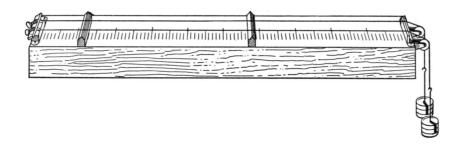

Figure 34 The Sonometer

APPARATUS

1. Sonometer with two steel wires of different diameters
2. Four tuning forks of different pitches (such as 256, 320, 384, and 512 vibrations per second)
3. Meter stick
4. Rubber hammer
5. Micrometer caliper
6. Card giving the mass per unit length of the wires

PROCEDURE

1. Adjust the tension in the string of smaller diameter to 4 kilograms. Vary the length of the string by means of the sliding bridge until its frequency is the same as that of the tuning fork of lowest pitch. Record the length of the string and the frequency of the tuning fork.

2. Repeat Procedure 1 with the other three tuning forks. Record the length of the string and the frequency of the tuning fork in each case.

3. Adjust the tension in the string of smaller diameter to 5 kilograms. Vary the length of the string by means of the sliding bridge until its frequency is the same as that of the tuning fork of highest pitch. Keeping the length fixed, reduce the tension gradually until the frequency of the string is the same as that of the tuning fork of the next lower pitch. Record the length of the string, the tension, and the frequency of the tuning fork.

4. Repeat Procedure 3 with the other two tuning forks. Record the value of the tension and the frequency of the tuning fork in each case.

5. Adjust the tension in both strings to 5 kilograms. Vary the length of one of the strings until its frequency is the same as that of the tuning fork of lowest pitch. Repeat this adjustment for the second string. Record the lengths of both strings and the frequency of the tuning fork.

6. Using the micrometer calipers, measure the diameter of each string and record the values.

DATA

Diameter of smaller wire _____   Mass per unit length of smaller wire _____

Diameter of larger wire _____   Mass per unit length of larger wire _____

Pitch of Tuning Fork	Length of Vibrating String	Tension	Diameter of Wire	Square Root of Tension	Reciprocal of Length of the String

CALCULATIONS

1. Using the values of the tension and length observed in Procedure 1 and the given value of the mass per unit length of the wire, calculate the frequency of the fundamental of the string by means of Equation 56. Compare the calculated value of the frequency with that of the tuning fork used by finding the percent error.

2. From the data of Procedures 1 and 2 calculate the reciprocals of the length of the string. Plot a curve using these reciprocals as abscissas and the corresponding frequencies as ordinates. What does this curve show about the way the frequency of a string under constant tension depends on the length of the string?

3. From the data of Procedures 3 and 4 calculate the square roots of the tension. Plot a curve using the square roots of the tension as abscissas and the corresponding frequencies as ordinates. What does this curve show about the dependence of the frequency of a string of constant length on the tension in the string?

4. From the data of Procedures 5 and 6 show how the lengths of two strings compare with the cross-sectional areas, if the strings vibrate at the same frequency, if they are under the same tension, and are made of the same material.

QUESTIONS

1. Upon what physical properties do (a) the loudness, (b) the pitch, and (c) the quality of a musical note depend?

2. What is meant by beats?

3. A wire weighing 0.003 gm/cm and 50 cm long is vibrating 200 times per second. What must be the tension in grams?

4. The first overtone of a stretched string 75 cm long has a frequency of 200 vibrations per second when the string is stretched by a weight of 4 kilograms. What is the mass per unit length of the vibrating string?

5. Two wires equal in length and made of the same material are subjected to tensions in the proportion of 1 to 4, the one under the greater tension being the thicker wire. If they vibrate at the same frequency, how do their diameters compare?

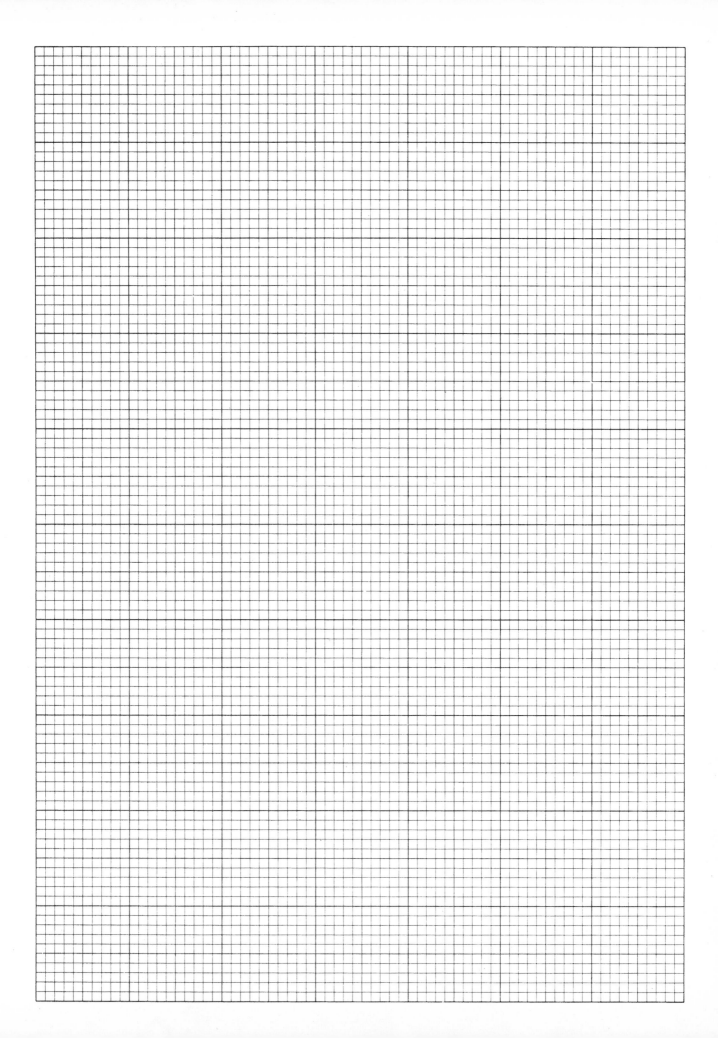

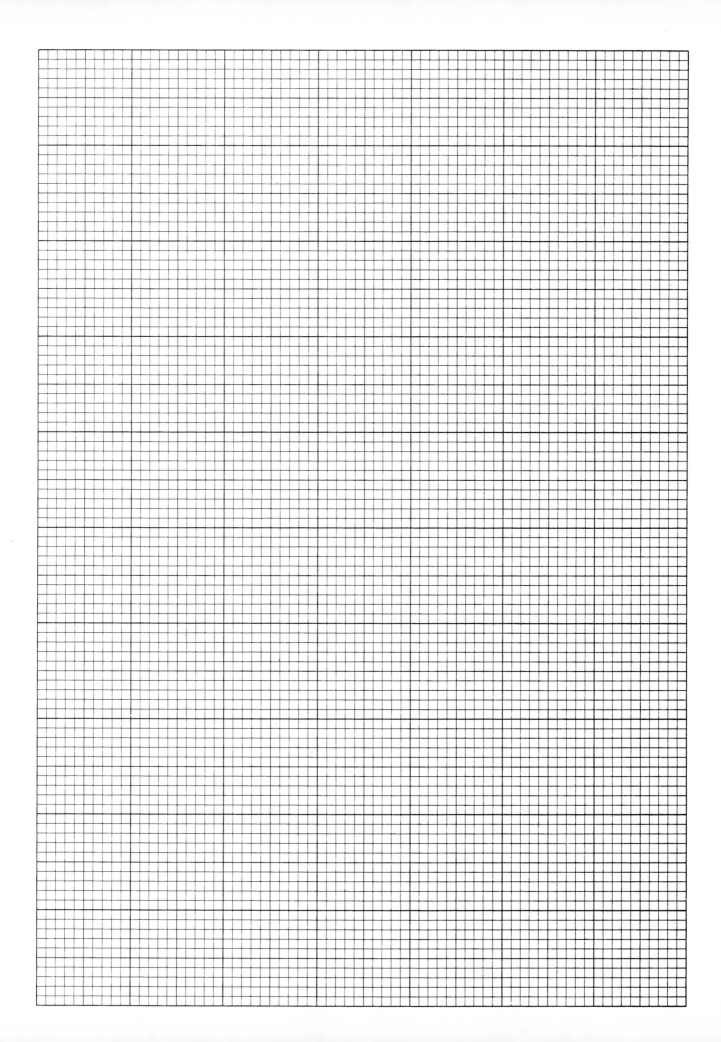

Resonance of Air Columns 16

The velocity with which a sound wave travels in a substance may be determined if the frequency of the vibration and the length of the wave are known. In this experiment the velocity of sound in air will be found by using a turning fork of known frequency to produce a sound wave whose wavelength can be measured by means of a resonating air column. The purpose of this experiment is thus to determine the wavelengths in air of sound waves of different frequencies by the method of resonance in closed pipes and to calculate the velocity of sound in air using these measurements.

THEORY

If a vibrating tuning fork is held over a tube open at the top and closed at the bottom, it will send disturbances made up of alternate compressions and rarefactions, *i.e.,* a sound wave, down the tube. These disturbances will be reflected at the tube's closed end, thus creating a situation in which identical waves are propagating in opposite directions in the same region. This is the condition for a standing wave, and so we have the possibility of a standing wave being set up in the air column in the tube. However, closing one end of the tube requires that there be a zero-amplitude point or node at that end, while at the other end, which is open, we must have a loop, or maximum-amplitude point. These requirements will be satisfied if the length of the tube is one-quarter of the sound wave's wavelength in air, for this is the distance from a node to an adjacent maximum, or loop. The requirements will also be satisfied if the tube's length is a quarter wavelength plus any integer number of half wavelengths, because half a wavelength is the distance from one node to the next. Whenever the tube has one of the lengths that meet the requirements for a standing wave at the frequency of the tuning fork, the tube and the fork are said to be in resonance, or put another way, the tube resonates at the frequency of the fork. The condition of resonance is indicated by an increase in the loudness of the sound heard when the air column has the resonant length.

When a tuning fork is held over a tube closed at one end, resonance will occur whenever the tube length is an odd number of quarter wavelengths, *i.e.,* when $L = \frac{1}{4}\lambda$, $\frac{3}{4}\lambda$, $\frac{5}{4}\lambda$, etc., where L is the length of the tube and λ is the wavelength of the sound waves in air. The relation between this wavelength and the frequency of the sound (*i.e.,* the pitch of the tuning fork) is

$$V = f\lambda \qquad (54)$$

where V is the velocity of sound in air in centimeters per second, f is the frequency (the number of vibrations per second), and λ is the wavelength in centimeters. From this relation, the velocity can be calculated from the known tuning-fork frequency and the wavelength obtained from measurements of L at resonance. It should be noted, however, that the center of the loop at the tube's open end does not fall right at the end but outside it by a small distance that depends on the wavelength and the tube diameter. However, the distance between successive tube lengths at which resonance is obtained gives the exact value of a half wavelength.

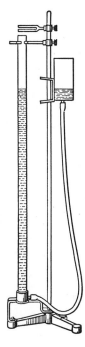

Figure 35 Resonance Tube Apparatus

In this experiment a closed pipe of variable length is obtained by changing the level of the water contained in a glass tube. The length of the tube above the water level is the length of the air column in use. The apparatus is shown in Fig. 35 and consists of a glass tube 110 cm long closed at the bottom with a stopper or brass cup and mounted on a heavy tripod base. A supply tank, which is connected with the bottom of the tube by a rubber hose, is attached to the support rod by means of a clamp, which permits rapid adjustment. A clamp for the tuning fork is also attached to the support rod and positioned directly above the tube.

The velocity of sound in air is 331.4 meters per second at 0°C. At higher temperatures the velocity is slightly greater than this and is given by

$$V = 331.4 + 0.6t \qquad (54a)$$

where V is the velocity in meters per second (not cm/sec in this relation) and t is the temperature in degrees Celsius. Equation 54a can be used to get a more accurate value for the speed of sound at the laboratory room temperature.

APPARATUS

1. Resonance tube apparatus
2. Two tuning forks of different pitches (such as 512 and 640 vibrations per second)
3. Meter stick
4. Rubber hammer
5. Celsius thermometer

PROCEDURE

1. Adjust the level of the water in the glass tube by raising the supply tank until the tube is nearly full of water. Clamp the tuning fork of higher frequency about 2 centimeters above the tube and in such a manner that the prongs will vibrate vertically.

2. Determine the shortest tube length for which resonance is heard. Start the tuning fork vibrating by striking it gently with the rubber hammer. Slowly lower the water level while listening for resonance to occur. At resonance, there is a sudden increase in the intensity of the sound when the air column is adjusted to the proper length. When the approximate length for resonance has been found, run the water level up and down near this point, until the position for maximum sound is found. Measure and record the length of the resonating air column to the nearest millimeter. Make two additional determinations of this length by changing the water level and locating the position for maximum sound again. Record these two readings also.

3. Repeat Procedure 2 with the same tuning fork to locate the second position at which resonance occurs. Make three independent determinations of this length and record the readings to the nearest millimeter.

4. Repeat Procedures 2 and 3 for the tuning fork of lower frequency.

5. Record the temperature of the room and the frequency of each tuning fork.

DATA

Room temperature _____

Calculated value of the velocity of sound in air (average) _____

Known value of the velocity of sound at room temperature _____

Percent error _____

Frequency of Tuning Fork	First Position of Resonance				Second Position of Resonance				Wavelength	Velocity of Sound in Air
	1	2	3	Average	1	2	3	Average		

CALCULATIONS

1. For each length of resonating air column calculate the average of the three readings taken.

2. For each tuning fork determine the value of the wavelength of its sound wave in air from the lengths of the resonating air column.

3. For each tuning fork calculate the velocity of sound in air from the frequency stamped on the fork and the wavelength found in Calculation 2. Find the average of these two values.

4. Calculate the velocity of sound in air at room temperature from Equation 54a.

5. Compare the average value of the velocity of sound as found in Calculation 3 with the known value as found in Calculation 4 by finding the percent error.

QUESTIONS

1. Suppose that in this experiment the temperature of the room had been lower. What effect would this have had on the length of the resonating air column for each reading? Explain.

2. (a) How would an atmosphere of hydrogen affect the pitch of an organ pipe? (b) How would it affect the pitch of a tuning fork? Explain.

3. An observer measured an interval of 10 seconds between seeing a lightning flash and hearing the thunder. If the temperature of the air was 20°C, how far away was the source of sound?

4. A tuning fork rated at 128 vibrations per second is held over a resonance tube. What are the two shortest distances at which resonance will occur at a temperature of 20°C?

5. A vibrating tuning fork is held over a resonance tube, and resonance occurs when the surface of the water in the tube is 10 cm below the fork. Resonance occurs again when the water is 26 cm below the fork. If the temperature of the air is 20°C, calculate the frequency of the tuning fork.

Coefficient of Linear Expansion **17**

Most substances expand with an increase of temperature. It is found that the change in length of a solid is proportional to the original length and to the change in temperature. The factor of proportionality, which is called the coefficient of linear expansion, depends on the material of which the solid is made. The purpose of this experiment is to determine the coefficient of linear expansion of several metals.

THEORY

The coefficient of linear expansion of a substance is the change in length of each unit of length when the temperature is changed one degree. The coefficient α may be expressed by the relation

$$\alpha = \frac{L_2 - L_1}{L_1(t_2 - t_1)} \tag{58}$$

where L_1 is the original length in centimeters at the temperature t_1 degrees Celsius, and L_2 is the length at the temperature t_2.

The apparatus consists of a metal rod which is placed in a brass jacket through which steam may be passed. The temperature of the rod is measured by a thermometer inserted in the jacket. One end of the rod is fixed and the other end is allowed to move. A micrometer screw, which may be adjusted to make contact with the movable end, is used to measure the change in length.

In determining the coefficient of linear expansion of a metal rod, the length of the rod is measured at room temperature. The rod is then placed in the brass tube of the apparatus and the initial reading of the attached micrometer screw is taken. Steam is then run through the tube, and the increase in length is measured by means of the micrometer screw. The supports of the apparatus carry binding posts for connecting an electric bell or a buzzer so as to indicate the exact point of contact of the screw. The coefficient of linear expansion is finally calculated from Equation 58.

APPARATUS

1. Coefficient of linear expansion apparatus
2. Boiler and tripod stand
3. Bunsen burner
4. 0–100°C thermometer
5. Rubber tubing
6. Meter stick
7. Dry cell
8. Electric bell or buzzer
9. Three metal rods: aluminum, copper, and iron
10. Paper towels or wipers

PROCEDURE

1. Measure the length of the aluminum rod with the meter stick to 0.1 mm at room temperature. Record the length and the temperature. Avoid handling the rod with your bare hands so as not to raise its temperature above that of the room. Use dry paper towels or wipers.

2. Adjust the rod in the brass tube and record the setting of the micrometer screw for the rod at room temperature. Make sure that one end of the rod is touching the fixed end of the apparatus and the other end is facing the micrometer screw, but not touching it. Carefully turn the micrometer screw until the bell begins to ring, showing that the screw has made contact with the rod. Record the setting of the micrometer screw.

3. Turn the micrometer screw back several millimeters to allow room for the expansion of the rod. Pass steam through the brass tube. When the rod has ceased to expand, carefully turn the micrometer screw until the bell begins to ring, showing that the screw has made contact with the rod. Record the setting of the micrometer screw. The temperature of the steam is 100°C at normal pressure.

4. Repeat Procedures 1–3 for the copper rod.

5. Repeat Procedures 1–3 for the iron rod.

DATA

	Aluminum	Copper	Iron
Room Temperature			
Length of Rod at Room Temperature			
Setting of Micrometer Screw at Room Temperature			
Setting of Micrometer Screw at 100°C			
Change of Length of the Rod			
Coefficient of Linear Expansion			
Value of the Coefficient of Linear Expansion from Table V, Appendix			
Percent Error			

CALCULATIONS

1. Compute the change in length of each rod in centimeters from the two readings of the micrometer screw.

2. Calculate the coefficients of linear expansion of aluminum, copper, and iron from your data.

3. Compare your results with those given in Table V, Appendix, and find the percent error.

QUESTIONS

1. Why must the increase in length of the rod be measured so carefully although the length itself can be determined by using an ordinary meter stick?

2. A compound bar made of an aluminum strip and an iron strip fastened firmly together is heated. Explain what happens to the shape of the bar.

3. When an ordinary mercury thermometer is placed in hot water, its reading drops at first, but quickly rises again. Explain why.

4. Why is the numerical value of the coefficient of linear expansion different when expressed in Fahrenheit degrees from what it is when expressed in Celsius degrees?

5. The Golden Gate Bridge in San Francisco Bay is 4200 feet long. Compute the total expansion of the bridge as the temperature changes from −20°C to +40°C. The coefficient of linear expansion for steel is given in Table V, Appendix.

6. An aluminum hoop whose inside diameter is 40 cm at room temperature (20°C) is to be ''shrunk'' on to a disc to form a wheel with an aluminum tire or rim. The process involves heating the hoop to make it expand, placing it on the disc, and letting it shrink down on the disc as it cools. The available furnace can heat the hoop to 300°C. To what diameter should the disc be machined so that the hoop will just fit on when so heated?

7. If a solid material is in the form of a block rather than a rod, its volume will grow larger when it is heated, and a *coefficient of volume expansion* β defined by

$$\beta = \frac{V_2 - V_1}{V_1(t_2 - t_1)}$$

may be quoted. Here V_1 and V_2 are the initial and final volumes of the block in cubic centimeters, respectively, and t_1 and t_2 are the initial and final temperatures as before. Find the relation between the coefficients α and β. *Hint:* Suppose the block is a rectangular parallelepiped whose volume is its length times its width times its height.

Specific Heat and Calorimetry 18

To raise the temperature of a body from a given initial temperature t_1 to a final temperature t_2 requires a total quantity of heat that depends on the mass of the body, the specific heat of the body's material, and the temperature difference. The process of measuring quantities of heat is called calorimetry. The purpose of this experiment is to determine the specific heat of one or more metals by the method of mixtures.

THEORY

The measurement of heat quantities by the method of mixtures makes use of the principle that when a heat interchange takes place between two bodies initially at different temperatures, the quantity of heat lost by the warmer body is equal to that gained by the cooler body, and some intermediate equilibrium temperature is finally reached. This is true provided no heat is gained from or lost to the surroundings.

Any given body is characterized by a *heat capacity*, which is the number of calories required to raise the temperature of that particular body one degree Celsius. Clearly the heat capacity depends both on the mass of the body and the nature of its material. To eliminate the dependence on mass and obtain a characteristic depending only on the material involved, the *specific heat* is defined as the heat capacity of one gram of the material in question. Notice that the units of heat capacity are calories per degree Celsius, whereas the units of specific heat are calories per degree Celsius per gram. Note also that the calorie is defined as the heat required to raise 1 gram of water (at 4°C) 1 degree Celsius, Thus by definition the specific heat of water is 1 cal/gm-°C. The heat capacity of m grams of water is thus numerically equal to m, whereas the heat capacity of the same mass of a substance whose specific heat is c cals/gm-°C is mc. Since $mc/m = c$, the specific heat of any substance may be thought of as the ratio of the heat capacity of a body made of that substance to the heat capacity of an equal mass of water.

In this experiment the specific heat of two different metals will be measured by studying the heat interchange between a sample of each metal and a mass of water. The vessel in which this interchange takes place is called a calorimeter. Any cup may be used as a calorimeter provided account is taken of the heat given up or absorbed by the cup and care is exercised to prevent heat exchange with the cup's surroundings. A normal way to do this is to use a thin metal cup whose heat capacity is small and easily measured and to mount it in an insulat-

ing jacket that prevents external heat exchange. A simpler and quite satisfactory procedure is to use a styrofoam cup, which both provides the needed insulation and has a negligibly small heat capacity.

To insure that an equilibrium temperature is reached in a short time after the metal sample is placed in the water in the calorimeter, the water should be stirred. The thermometer used to measure the equilibrium temperature may also serve this purpose. The thermometer itself will absorb some heat in coming to the equilibrium temperature, but its heat capacity is usually so small that negligible error is made by ignoring it.

In the experimental determination of the specific heat of a metal by the method of mixtures, a metal sample of known mass that has been heated to a known high temperature is dropped into a known mass of water at a known low temperature. After equilibrium has been established, the new temperature of the water-metal-calorimeter combination is measured. In arriving at this equilibrium temperature, the water and calorimeter must have gained the heat lost by the metal sample. This situation is expressed by the relation

$$Mc(t_1 - t_2) = (m + m_1 c_1)(t_2 - t_3) \qquad (59)$$

where M is the mass of the metal sample in grams, c is the specific heat of the metal, t_1 is the metal sample's initial temperature, t_2 is the final equilibrium temperature, t_3 is the initial temperature of the water and calorimeter, m is the mass of water, m_1 is the mass of the calorimeter cup, and c_1 is the specific heat of the material of which this cup is made. Note that $m_1 c_1$ is the heat capacity of the calorimeter cup. This quantity is sometimes called the *water equivalent* of the calorimeter because it is numerically equal to the mass of water having that heat capacity, the specific heat of water being 1 cal/gm-°C. Equation 59 may be used to determine c if all the other quantities in it have been measured.

APPARATUS

1. Calorimeter
2. Boiler and tripod stand
3. Bunsen burner
4. Equal-arm balance and set of weights
5. 0–100°C thermometer
6. 0–50°C thermometer
7. Sample cylinder no. 1 (metal to be specified by the instructor)
8. Sample cylinder no. 2 (metal to be specified by the instructor)

PROCEDURE

1. Fill the boiler with enough water to cover either sample cylinder with a couple of inches to spare and place it on the tripod. Put the bunsen burner under it and bring the water to a boil.

2. Weigh the empty calorimeter cup.

3. Weigh sample cylinder no. 1 and then lower it into the boiling water by means of a thread. Be sure the cylinder is completely immersed in the water but do not allow it to touch the bottom or sides of the boiler.

4. Pour cold water (about 3 degrees below room temperature) into the calorimeter cup until the cup is about half full. Then weigh the cup with the water in it. Note that the temperature of the cold water should be about as much below room temperature as the equilibrium temperature will be above it so as to balance out errors due to transfer of heat by radiation to or from the surroundings.

5. Replace the calorimeter in its insulating jacket and measure the temperature of the water with the 0–50°C thermometer. Record this temperature.

6. Quickly transfer the sample cylinder from the boiler to the calorimeter without splashing any water. Stir the water and record the equilibrium temperature as indicated on the 0–50°C thermometer.

7. Measure and record the temperature of the boiling water using the 0–100°C thermometer.

8. Repeat Procedures 3–7 for sample cylinder no. 2.

DATA

Weight of calorimeter _____   Specific heat of calorimeter cup material _____

	Data for Cylinder No. 1	Data for Cylinder No. 2
Weight of Cylinder		
Weight of Calorimeter and Cold Water		
Weight of Cold Water		
Initial Temperature of Cold Water		
Temperature of Boiling Water		
Equilibrium Temperature		
Calculated Specific Heat		
Specific Heat from Table VI, Appendix		
Percent Error		

CALCULATIONS

1. Calculate the specific heat of the sample no. 1 material from the data of Procedures 2–7.

2. Calculate the specific heat of the sample no. 2 material from the data of Procedure 8.

3. Compare your results for the specific heats with those given in Table VI, Appendix, and find the percent error.

QUESTIONS

1. What is the purpose of starting with the temperature of the water lower than room temperature and ending about the same amount above room temperature?

2. How would the computed value of the specific heat be affected if some boiling water were carried over with the metal?

3. What will be the biggest source of error if too much water is used?

4. What is meant by the water equivalent of a body?

5. A platinum ball weighing 100 grams is removed from a furnace and dropped into 400 grams of water at 0°C. If the equilibrium temperature is 10°C and the specific heat of platinum is 0.04 cal/gm-°C, what must have been the temperature of the furnace? Neglect the effect of the mass of the calorimeter.

6. Equation 59 appears to be dimensionally wrong in that the term $m(t_2 - t_3)$ on the right has the dimension gm-°C whereas all other terms have the dimension of heat (calories). Explain.

Heat of Fusion and Heat of Vaporization **19**

The present experiment deals with *changes of phase,* that is, the changing of a substance from the solid to the liquid state and from the liquid to the gaseous state. These three states, or phases, are normally exhibited by all substances, but the temperatures at which changes from one state to another take place, and the heat involved, vary widely. It is the purpose of this experiment to study the heats involved in the change of water from a solid to a liquid and from a liquid to a gas by the method of mixtures. This method has been described in detail in Experiment 18 on calorimetry, and a review of the Theory section of that experiment will be helpful preparation for the present work.

THEORY

When a solid body is heated, its temperature rises at a rate determined by the rate at which heat is being supplied and the heat capacity of the body. However, when a certain temperature called the melting point is reached, the body starts to melt, and as melting proceeds the temperature remains constant even though heat is being continually supplied. Increasing the supply rate merely increases the rate at which the solid changes to a liquid. No change in temperature occurs until the body has completely melted, after which further heating will raise the temperature of the resulting liquid at a rate determined by the liquid's heat capacity (its mass multiplied by its specific heat) and the rate at which heat is being added. Eventually a temperature called the boiling point will be reached. Here the liquid changes rapidly to a vapor, and again there is no change in temperature while this is going on even though the addition of heat continues. When all the liquid has been evaporated, continued addition of heat to the resulting gas (which must now be contained in a suitably sealed vessel) will once more raise the substance's temperature, this time at a rate determined by the heat supply rate and the heat capacity (mass times specific heat) of the gas. Upon cooling, the gas begins to condense when the boiling point is reached, and again there is no change in temperature while condensation takes place. The same amount of heat is removed during this condensation process as was added during vaporization, this heat representing the difference in the energy of the molecules in the gaseous state and their energy when bound (although loosely) in the liquid. Similarly, once the temperature has been reduced to the freezing point (which is the same as the melting point), further removal of heat causes the liquid to solidify. No change of temperature occurs during this process, and the heat removed (which is the same as the heat added during melting) represents the difference in the molecular energy when the molecules are loosely bound in the liquid and when they are tightly bound in the solid.

The student should note that vaporization takes place at all temperatures, because there are always a few molecules in the liquid with enough energy to break the bonds at the surface and emerge into the space above. One can thus think of a vapor pressure due to these escaped molecules above any liquid surface. Naturally, the higher the temperature, the more molecules will have enough energy to break through the surface and the higher the vapor pressure will be. If the liquid surface is exposed to the atmosphere, atmospheric pressure will prevent molecules entering the gaseous state near this surface from forming bubbles until the vapor pressure and atmospheric pressure become equal. At this point the vapor can form bubbles in opposition to the pressure of the atmosphere and the action we call boiling occurs. Molecules can then break from the liquid in large numbers, energy in the form of heat must be added to the liquid to replace the energy of the departing molecules, and a high rate of heat supply merely permits more molecules to evaporate per second. Note, however, that the boiling point is now seen to be that temperature at which the liquid's vapor pressure becomes equal to atmospheric pressure and is therefore highly dependent on the atmospheric pressure's value. This is why water boils at a lower temperature on top of a mountain, where the atmospheric pressure is less than at sea level. Thus, when the boiling point of a liquid is given without further information, normal sea-level atmospheric pressure is assumed. Because there is usually a small volume change when a solid changes to a liquid or vice versa, the melting point may also be affected by the external pressure, but this effect is usually so small that it can be ignored.

The amount of heat required to change a unit mass of a substance from the solid to the liquid state without a change in temperature is called the *latent heat of fusion*

113

of the substance, or simply its heat of fusion. Similarly the amount of heat required to change a unit mass of a substance from the liquid to the vapor state without a change in temperature is called the *latent heat of vaporization* of the substance, or simply the heat of vaporization. Since evaporation can go on at any temperature and the heat of vaporization is different at different temperatures, the temperature should be quoted when a heat of vaporization is given. However, heats of vaporization are usually given without this information, it being assumed that the temperature of the boiling point is meant. Since this temperature is dependent on the atmospheric pressure, normal sea-level pressure is also assumed.

In doing heat experiments where the quantity of heat added to or taken from a substance is to be determined, the range of temperature should extend equally above and below room temperature so that the amount of heat absorbed from the surroundings during the course of the experiment will be approximately equal to the amount of heat radiated to the surroundings. Best results will thus be obtained by starting with a calorimeter of warm water when the heat of fusion of ice is to be measured and starting with a calorimeter of cold water for the measurement of the heat of vaporization of steam. This is easily accomplished by measuring first one and then the other of the two latent heats of water. It is the purpose of this experiment to carry out these two measurements.

The experimental determination of the heat of fusion of ice and of the heat of vaporization of water is made by the method of mixtures. This makes use of the principle that when a heat interchange takes place between two bodies initially at different temperatures, the quantity of heat lost by the warmer body is equal to that gained by the cooler body, and some intermediate equilibrium temperature is finally reached. This is true provided no heat is gained from or lost to the surroundings.

In determining the heat of fusion of ice, a few small pieces of ice are placed, one by one, into a calorimeter containing water. As the ice melts, heat is absorbed from the water and calorimeter until the mixture comes to a final equilibrium temperature. The heat absorbed by the ice in melting plus the heat absorbed by the ice water thus produced is equal to the heat lost by the warm water and calorimeter. The working equation is

$$ML_f + M(t_2 - 0) = (m + m_1 c_1)(t_1 - t_2)$$

where M is the mass of ice, in grams, L_f is the latent heat of fusion of ice, t_2 is the equilibrium temperature, t_1 is the initial temperature of the water and calorimeter, m is the mass of warm water, $m_1 c_1$ is the water equivalent of the calorimeter, and m_1 is the mass and c_1 the specific heat of the calorimeter.

The heat of vaporization of water is determined in a similar manner by the method of mixtures. A quantity of steam is passed into a known mass of cold water in a calorimeter, where it condenses and raises the temperature of the water. The heat lost by the steam in condensing plus the heat lost by the condensed steam is equal to the heat absorbed by the cold water and calorimeter. The working equation is

$$ML_v + M(100 - t_2) = (m + m_1 c_1)(t_2 - t_1)$$

where M is the mass of steam in grams, L_v is the latent heat of vaporization of water, t_2 is the equilibrium temperature, t_1 is the initial temperature of the water and calorimeter, m is the mass of cold water, and m_1 and c_1 are the mass and the specific heat of the calorimeter.

APPARATUS

1. Calorimeter
2. Boiler and tripod stand
3. Water trap
4. Rubber tubing
5. Bunsen burner

6. Equal-arm balance and set of weights
7. One 0–100°C thermometer
8. One 0–50°C thermometer
9. Ice
10. Paper towels

PROCEDURE

1. Weigh the empty calorimeter.

2. Fill the calorimeter half full of water about 10° above room temperature and weigh it. Replace it in the outer calorimeter jacket and record the temperature of the water.

3. Dry some small pieces of ice on a paper towel and add them to the water without touching the ice with the fingers, so as not to melt it. Add the ice until the temperature is about 10° below room temperature, keeping the mixture well stirred with the 0–50°C thermometer. Record the equilibrium temperature when the ice is entirely melted. Weigh the calorimeter with its contents again, but without the thermometer.

4. Fill the calorimeter three quarters full of water about 15° below room temperature and again weigh it. Replace it in the outer calorimeter jacket.

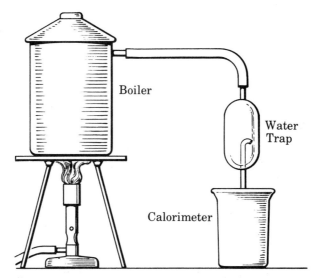

Figure 36   Heat of Vaporization Apparatus

5. Set up the steam generator as shown in Fig. 36 and adjust it so that the steam passes through the water trap and flows freely from the steam tube. Record the temperature of the cold water and quickly immerse the steam tube. Allow the steam to pass in and condense until the temperature of the water is about 15° above room temperature, stirring continuously with the 0–100°C thermometer. Remove the steam tube and record the equilibrium temperature. Weigh the calorimeter with its contents again, but without the thermometer.

DATA

Weight of calorimeter _____   Specific heat of calorimeter _____

● Heat of Fusion

Weight of calorimeter and water _____   Weight of ice _____

Weight of water _____   Calculated heat of fusion of ice _____

Initial temperature of water _____   Accepted value of heat of fusion
of ice _____

Equilibrium temperature _____   Percent error _____

Weight of calorimeter, water,
and melted ice _____

● Heat of Vaporization

Weight of calorimeter and water _____   Weight of steam _____

Weight of water _____   Calculated heat of vaporization of water _____

Initial temperature of water _____   Accepted value of heat of
vaporization of water _____

Equilibrium temperature _____   Percent error _____

Weight of calorimeter, water, and
condensed steam _____

CALCULATIONS

1. From the data of Procedures 1–3 compute the heat of fusion of ice.

2. From the data of Procedures 4 and 5 compute the heat of vaporization of water.

3. Compare your results for the heat of fusion and the heat of vaporization with the known values by finding the percent error.

QUESTIONS

1. Discuss the principal sources of error in this experiment.

2. Why is it necessary to stir the mixtures?

3. What error would have been introduced if the ice were not dry?

4. (a) What is the purpose of the water trap in determining the heat of vaporization? (b) How would its absence have affected the results?

5. Suppose the steam generator produced steam so fast that some of it bubbled up through the water and escaped from the calorimeter without condensing. Would this cause an error in your results? Explain.

6. The Theory section of this experiment suggests that when a liquid starts to boil, the formation of bubbles should begin near the surface. Why do the bubbles actually form first at the bottom of the containing vessel?

7. A glass tumbler of 200 grams mass and 5 cm inside diameter is filled to a depth of 10 cm with tap water whose temperature is 10°C. 50 grams of ice at 0°C are then added. At the end of an hour it is noted that the ice has just finished melting. What was the average rate, in calories per second, at which heat flowed into the glass from its surroundings?

The Gas Laws 20

When a substance is in the gaseous state, its molecules fly about at random, there being only very small forces of interaction between them (sometimes called Van der Waals forces), and because the molecules are small compared to their average separations, there is only a very small chance of any two of them colliding. A so-called "ideal" gas is defined as the idealization of the situation just described, that is, a gas in which the molecules are mass points that have *no* chance of colliding with each other and that also have *no* interaction forces, not even weak ones. Actual gases like nitrogen, oxygen, and hydrogen satisfy this definition to a good approximation at ordinary temperatures and pressures, so that the study of the ideal gas has a practical application. Only at extreme temperatures and pressures do the molecules of real gases get close enough together for their interaction forces to have an appreciable effect and

their finite size to become significant.

In order to specify fully the condition of any substance, its pressure, volume, and temperature must be known. These quantities are always interrelated through a relationship characteristic of the substance called its *equation of state,* so that if two of them are known, the equation of state gives the third one. The equation of state for an ideal gas (often called the ideal gas law) is a particularly simple one, derivable theoretically on the basis of the assumptions noted above with regard to no collisions and no interaction forces among the ideal gas molecules. It applies at ordinary temperatures and pressures to real gases and can thus be checked experimentally. The purpose of this experiment is to carry out such a check and incidentally to determine the so-called *absolute zero* of temperature.

THEORY

In studying the behavior of a gas under different conditions of pressure, temperature, and volume, it is convenient to keep one of these constant while varying the other two. Thus, if the temperature is kept constant, a relation is obtained between the pressure and the volume; if the volume is kept constant, a relation between the pressure and temperature can be found. Such studies were carried out independently by three different investigators whose names are now associated with the results they obtained. The three possible relations (pressure and volume with temperature constant, volume and temperature with pressure constant, and pressure and temperature with volume constant) are respectively called the laws of Boyle, Charles, and Gay-Lussac and are collectively referred to as the gas laws.

● **Boyle's Law** If a gas is kept at constant temperature, we would expect an increase in pressure to "squeeze the gas together," that is, to cause a decrease in volume. Conversely, reducing the pressure allows the gas to expand, that is to occupy a larger volume. This relation was studied quantitatively by Robert Boyle (1627–1691), who found that for a fixed amount of gas at constant temperature the pressure and volume are inversely

proportional over any ordinary pressure range. This result may be stated mathematically as

$$P = C_1/V$$

or
$$PV = C_1 \qquad (60)$$

where P is the pressure, V is the volume, and C_1 is a proportionality constant whose value depends on the (constant) temperature and the amount of gas involved. The property of gases expressed by Equation 60 is called *Boyle's law.*

● **Charles' Law** If a gas is heated, it expands so that it occupies an increased volume if the pressure is maintained constant. The French physicist Jacques Alexandre César Charles (1746–1823) found that under these circumstances a linear relation exists between the volume and the temperature, expressible by the relation

$$V = V_0(1 + \beta t) \qquad (61)$$

where V is the volume at temperature t, t is the temperature in degrees Celsius, V_0 is the volume at 0°C, and β is

119

a constant called the coefficient of expansion of the gas at constant pressure. Since the zero of temperature on the Celsius scale is arbitrarily chosen as the freezing point of water at standard atmospheric pressure, negative temperatures on this scale are perfectly meaningful, and it is interesting to note the temperature at which an extrapolation of Equation 61 indicates the volume would go to zero. This temperature is clearly $-1/\beta$, and since β is very nearly $\frac{1}{273}$ or 0.00366 per degree Celsius for all gases, Equation 61 predicts zero volume at a temperature of $-273°C$. If a new temperature scale is now defined having degrees of the same size as Celsius degrees but whose zero is at $-273°C$, we have

$$T = t + 273 \tag{62}$$

where T is the temperature on the new scale, called the absolute or Kelvin scale, and is expressed in degrees Kelvin (°K). Substitution of Equation 62 in Equation 61 with $\beta = \frac{1}{273}$ yields

$$V = \frac{V_0}{273}T = C_2 T \tag{63}$$

where C_2 is a constant which is seen to depend on V_0 and hence on the quantity of and pressure on the gas. Thus, if the temperature is given on the absolute scale, volume and temperature are not just linearly related but are proportional. The linear dependence of volume on temperature at constant pressure is called *Charles' law*.

● **Gay-Lussac's Law**   If a gas is heated but not allowed to expand (that is, maintained at constant volume), the pressure will rise, and the investigations of the French scientist Joseph Louis Gay-Lussac (1778–1850) showed that, like the volume–temperature relation at constant pressure, the pressure–temperature relation at constant volume is linear and expressible by the relation

$$P = P_0(1 + \beta t) \tag{64}$$

where P is the pressure at $t°C$, P_0 the pressure at $0°C$, and β the pressure coefficient of the gas at constant volume. Note that the same symbol is used for the pressure coefficient in Equation 64 and the coefficient of expansion in Equation 61. This is because the same value, $\frac{1}{273}$ per degree Celsius, is found for both and both have the same significance, namely that the volume and therefore also the pressure of an ideal gas should become zero at $-273°C$ or $0°K$. If Equation 62 is substituted in Equation 64 in the same manner as in Equation 61, we get

$$P = \frac{P_0}{273}T = C_3 T \tag{65}$$

where C_3 is a constant that depends on P_0 and hence on

the quantity of gas involved and the volume in which it is confined. Again, use of the absolute temperature scale makes pressure and temperature proportional. As a result of Gay-Lussac's work, the linear dependence of pressure on temperature at constant volume is usually known by his name.

● **The Ideal Gas Law**   Equations 60, 63, and 65 can be combined in the single relation

$$PV = C_4 T \tag{66}$$

where C_4 is a new constant whose value depends only on how much gas is involved. In view of Avogadro's law, which states that equal volumes of all gases at the same pressure and temperature contain the same number of molecules, stating how much gas we have by giving the actual number of molecules present allows us to write 66 as

$$PV = NkT \tag{67}$$

where $C_4 = Nk$, N is the number of molecules in question, and k is a *universal* constant applying to all gases with no dependence on temperature, pressure, volume, or quantity. It is called *Boltzmann's constant* and has the value 1.38×10^{-23} joule per molecule-degree absolute. The student should note carefully that the product PV has the dimensions of energy so that the units for Boltzmann's constant are those of energy per molecule-degree. The value of this constant thus depends on the units of energy used. Expressing the amount of gas by the number of molecules present requires inconveniently large numbers, and so the number n of moles is more usually given. Since by definition a mole contains Avogadro's number N_0 of molecules, the total number of molecules N is equal to nN_0, and Equation 67 may be written

$$PV = nRT \tag{68}$$

where $R = N_0 k$ is called the *universal gas constant*. It serves the same purpose as k but has the dimensions of energy per *mole*-degree rather than energy per *molecule*-degree. Equation 68 (or 67) is a relation between the pressure, volume, and temperature of an ideal gas and is therefore the ideal gas's equation of state. It is called *the ideal gas law*.

In the present experiment we shall check Boyle's law and the law of Gay-Lussac: We shall seek experimental verification of Equations 60 and 64. Moreover, if the linear relation predicted by Equation 64 is obtained, simple extrapolation to $P = 0$ will give the location of the absolute zero of temperature on the Celsius scale and hence the value of β. In this way Equation 65 will also be verified. Then if pressure is found to be proportional to the absolute temperature at constant volume and inversely proportional to volume at constant temperature, the proportionality of volume and temperature at constant pressure (Charles' law), and hence the ideal gas law, is implied. The experimental test of Boyle's law consists in observing a series of different volumes, measuring the corresponding pressures, and plotting P against $1/V$ to see if a straight line is obtained. The apparatus is made up of two glass tubes of uniform bore connected by a piece of flexible rubber tubing and mounted on a supporting frame to which a scale is attached (see Fig. 37). The apparatus is partially filled with mercury. One of the glass tubes is provided with a stopcock; for this part of the experiment the air upon which observations are to be made is confined in this tube between the stopcock and the mercury column. The other tube is open to the atmosphere and can be raised or lowered, thus changing the height of the mercury column

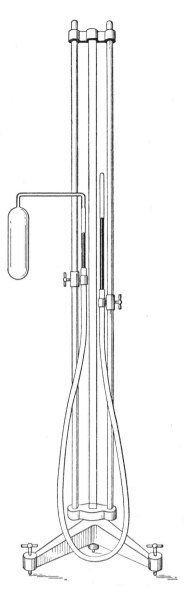

Figure 38 Air Thermometer

that confines the air. The difference in height of the mercury in the two tubes causes a pressure to be exerted on the gas trapped in the closed tube. This is called the gauge pressure because it is the pressure that would be measured by a gauge that reads zero when exposed to atmospheric pressure if it were connected to the confined air. The total or absolute pressure on the confined air is the sum of the gauge pressure and the atmospheric pressure, which is applied on top of the mercury column in the open tube. Thus the total pressure P acting on the confined air is given by

$$P = A + (h_2 - h_1) \tag{69}$$

where A is the atmospheric pressure as read on the barometer; h_2 is the level of the mercury in the open tube; and h_1 is the mercury level in the tube closed by the stopcock. Notice that Equation 69 is consistent if h_1 and h_2 are in centimeters and all pressures are expressed in

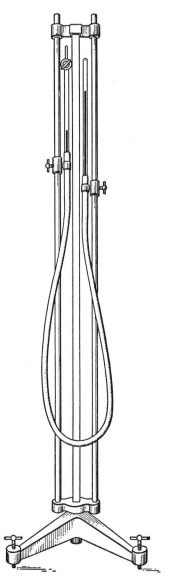

Figure 37 Boyle's Law Apparatus

centimeters of mercury. Note also that if h_1 is greater than h_2 the quantity $(h_2 - h_1)$ is negative and subtracts from atmospheric pressure to give the total pressure on the confined air.

For the test of Gay-Lussac's law, a bulb is connected to the stopcock by a short length of rubber tubing as shown in Fig. 38. The stopcock is opened so that the volume of confined air includes the volume in the glass tube above the mercury column, the volume of the rubber hose, and the volume of the bulb. The total volume is thus very much larger than that used for the test of Boyle's law, and the bulb provides for convenient heating to a known temperature. This is done by immersing the bulb in water contained in a Pyrex beaker, the temperature of the confined air being varied by heating or cooling the water bath. The total volume of confined air is kept constant by adjusting the height of the open glass tube so that the mercury level in the tube connected to the bulb is brought to a particular scale division before each set of readings is taken. Pressure readings are taken by using Equation 69 as in the test of Boyle's law and temperature readings by means of a thermometer in the water bath. Note that in the present experiment the relation in Equation 64 is being checked and β determined by measuring the gas temperature with an external thermometer and obtaining the pressure from the difference in height of the mercury columns. However, if the linear relation postulated by Gay-Lussac were correct and β known to high accuracy, the arrangement of Fig. 38 could serve as a thermometer, the difference in height of the mercury columns being a direct measure of the temperature of the gas in the bulb. In fact, the linear dependence of gas pressure on temperature at constant volume is so good and the value of β has been determined so accurately that the constant volume hydrogen thermometer — an instrument similar to the present setup but using hydrogen instead of air — is the standard thermometer in terms of which other types of thermometers are calibrated.

APPARATUS

1. Boyle's law apparatus
2. Bulb with connecting tube
3. Barometer
4. One 0–50°C thermometer
5. One 0–100°C thermometer
6. 1000-cc Pyrex beaker
7. Ring stand and clamps
8. Bunsen burner
9. Ice

PROCEDURE

1. Record the reading of the barometer and the room temperature as read on the 0–50°C thermometer. *Caution:* Suspend the thermometer in the room for several minutes to allow it to come to equilibrium and do not touch it during this period.

2. Check that the apparatus has the proper amount of mercury in it. To do this, open the stopcock and see if each tube is about half full. The mercury level will, of course, be the same in both tubes when the stopcock is open. Adjust the tube support clamps so that the top of each tube is at about the same height and the clamps are about midway up their support bars, so that the open tube may be both raised above and lowered below the closed one through as great a range as possible.

3. Close the stopcock and thereafter do not handle the closed tube or allow the enclosed air to be heated above or cooled below room temperature in any way. Record the level of the top of the closed tube (just below the stopcock) and the level of the mercury columns in the two tubes. Note that atmospheric pressure should continue to exist in the closed tube after the closing of the stopcock, so that the mercury level should remain the same in both tubes.

4. Decrease the pressure on the gas trapped in the closed tube by lowering the open tube as far as the instrument will permit. Allow it to remain in this position for a few minutes to test for leaks. If there is any leakage in the apparatus, consult the instructor. Record the level of the mercury column in the closed tube and in the open tube.

5. Vary the pressure in nine approximately equal steps from the smallest to the largest value that can be obtained so as to get ten different observations. Record the level of the mercury column in each tube for each setting. After changing the level, allow a minute or two before taking readings to permit the confined air to reach equilibrium at room temperature. Remember not to handle the closed tube or allow spurious heating or cooling of the enclosed air to take place.

6. When the Boyle's law measurements have been taken, return the open tube to the position at which the

mercury levels are equal and open the stopcock. Connect the bulb to the stopcock nipple by means of the rubber tube.

7. Mount the Pyrex beaker on the ring stand with the bunsen burner under it. Support the bulb inside the beaker by means of a clamp, not allowing the bulb to touch the sides or bottom.

8. Fill the beaker with ice and water until the bulb is entirely covered. Watch the motion of the mercury in the closed tube to make sure it does not get up to the stopcock. If the mercury appears to be rising too high, lower the open tube, being careful not to let mercury spill out of it.

9. Pick a scale division a few centimeters below the bottom of the stopcock as your constant volume mark and adjust the open tube so as to bring the mercury column in the closed tube to this level. Wait for temperature equilibrium to be reached, keeping the closed-tube mercury level at your constant volume mark until equilibrium is established. This will be indicated by the pressure stabilizing so that no further adjustment of the open tube is required. Stir the water well and measure its temperature with the 0–50°C thermometer. Record this temperature and the height of the mercury level in the open tube.

10. Remove the ice and add enough water to cover the bulb. Light the bunsen burner and heat the water slowly until its temperature is about 10°C. Remove the bunsen burner and stir the water well with the 0–50°C thermometer. Keep the mercury level in the closed tube at your constant volume mark by raising the open tube as required. When equilibrium has been reached, record the temperature of the water and the height of the mercury in the open tube.

11. Repeat Procedure 10, raising the temperature of the water in steps of about 10° until the boiling point is reached. Use the 0–100°C thermometer in place of the 0–50°C one as soon as the temperature goes above 40°C.

12. When the experiment is completed, remove the bulb from the water and lower the open tube progressively as the bulb cools so as to keep the mercury from rising up to the stopcock. Leave the apparatus in the condition it was in at the end of Procedure 6.

13. Check the barometer reading to be sure there has been no appreciable change since it was read in Procedure 1. If there is, use the earlier reading for the Boyle's law work and the later one for the test of the law of Gay-Lussac.

DATA

Barometer reading, beginning _____   Room temperature _____
                  end _____   Level of the top of the closed tube, h_3 _____

Mercury Level		Gauge Pressure $h_2 - h_1$	Volume of Air $h_3 - h_1$	Total Pressure	PV	1/V
Closed Tube h_1	Open Tube h_2					

Height of chosen constant volume mark   _____   Accepted value of the absolute zero   _____

Measured value of the absolute zero of   Percent discrepancy   _____
temperature on the Celsius scale   _____

Mercury Level in the Open Tube	Gauge Pressure	Total Pressure	Temperature of the Bulb

CALCULATIONS

1. Compute the value of $h_2 - h_1$ for each setting of the apparatus in Procedures 4 and 5.

2. Compute the value of P, the total pressure on the confined air, for each observation taken in Procedures 4 and 5. Express the results in centimeters of mercury. *Note:* For these and the following calculations, use three significant figures in the numbers representing the pressure, the volume, and the reciprocal of the volume. In calculating the product, use four significant figures.

3. Calculate the volume of the enclosed air for each of the observations of Procedures 4 and 5. *Note:* Since the tube is of uniform bore, the tube's volume is proportional to its length. The length $h_3 - h_1$ of the column of air confined in the closed tube may therefore be taken as a measure of this air's volume.

4. Calculate the product PV for each corresponding pressure and volume found in Calculations 2 and 3.

5. Calculate the value of $1/V$ for each volume found in Calculation 3.

6. Plot a curve using the values of the pressure as abscissas and the corresponding values of $1/V$ as ordinates. Use an entire sheet of graph paper for the curve. Read the instructions on plotting graphs in the Introduction.

7. For each temperature measured in Procedures 9–11, calculate the gauge pressure exerted on the air in the bulb by subtracting the height of the mercury in the closed tube (height of the constant volume mark) from the height of the mercury in the open tube. Note that this pressure will be negative in many cases. Then add the atmospheric pressure to obtain the total pressure in centimeters of mercury.

8. Plot a curve using the values of the temperature as abscissas and the values of the total pressure obtained in Calculation 7 as ordinates. For more accurate results, construct the graph of the data using as large a scale as possible; that is, utilize the whole sheet of graph paper and extend the line only to 0°C.

9. From the graph of Calculation 8 determine the position of the absolute zero of temperature on the Celsius scale; that is, calculate the temperature at which this graph predicts that the pressure would be zero. A simple proportion may be used to do this since, according to Equation 64, the curve should be a straight line. Compare your result with the accepted value by calculating the percent discrepancy.

QUESTIONS

1. Show how your results and the graph from Calculations 4–6 verify Boyle's law.

2. Explain how an error is introduced into the Boyle's law experiment when the room temperature changes.

3. How does the difference in height of the mercury in the two tubes cause a pressure to be exerted on the enclosed air?

4. Does Boyle's law hold accurately for a real gas at any temperature or pressure such as a very high or a very low one? Explain.

5. In the apparatus used to test Gay-Lussac's law, is an error introduced by the fact that the air in the rubber tubing between the bulb and the stopcock and in the glass tube between the stopcock and the mercury column is not in the water bath? Estimate how big and in which direction such an error might be.

6. What conclusion do you draw from the shape of the curve obtained in Calculation 8?

7. A spherical basketball 10 inches in diameter is pumped up to a gauge pressure of 44.1 pounds per square inch. If the normal atmospheric pressure is 14.7 pounds per square inch, what volume of air at this pressure has been pumped into the ball?

8. In an air thermometer apparatus like that used to test Gay-Lussac's law, the level of the mercury in the open tube is 20 centimeters higher than that in the closed tube when the bulb is at 0°C. At what temperature will the level in the open tube be 40 centimeters higher than that in the closed tube? Assume standard atmospheric pressure.

9. A constant volume gas thermometer indicates an absolute pressure of 50.0 centimeters of mercury at 0°C and 68.3 centimeters of mercury at 100°C. At what temperature will it read an absolute pressure of 56.1 centimeters of mercury?

10. Calculate the volume occupied by 1 mole of an ideal gas at 0°C and 1 atmosphere pressure. Will your result be true for real gases such as nitrogen or oxygen? What difference will the choice of gas make?

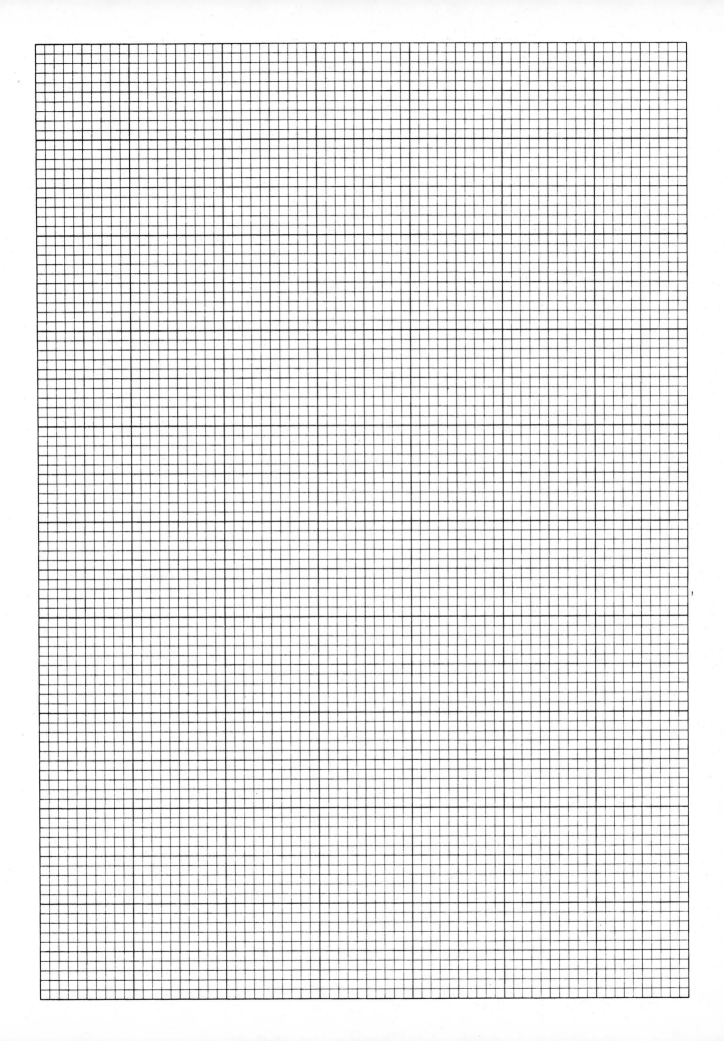

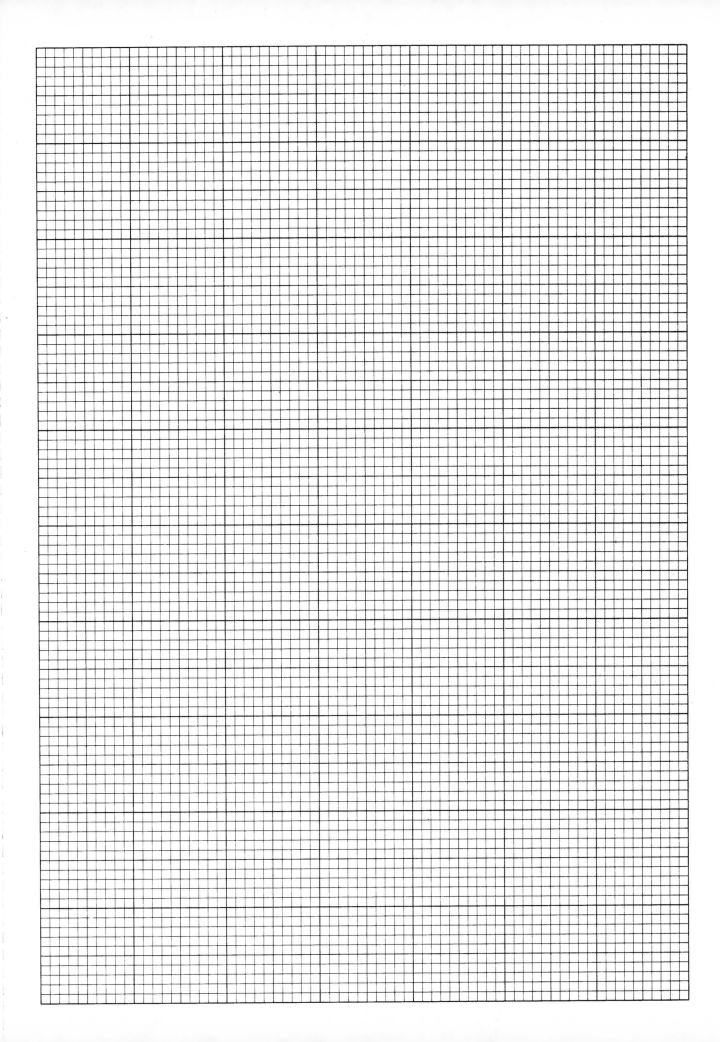

The Mechanical **21** Equivalent of Heat

Before the work of Benjamin Thompson (Count Rumford, 1753–1814) and James Prescott Joule (1818–1889, after whom the MKS unit of energy is named), mechanical energy and heat were considered to be two quite different things. It was Thompson who first noted, while boring cannon for the Elector of Bavaria, that while the large amount of work put into turning the drill seemed to disappear rather than become stored as kinetic or potential energy, the drill and cannon got hot with no apparent source of heat. He suggested that the work was the source of the heat, that mechanical work was transformed into heat by friction, and that, therefore, heat was not something separate and apart but just another form of energy. Joule put this idea on a quantitative footing by transforming a known amount of work into a measureable quantity of heat. In this way he not only showed that heat was a form of energy, so that calories and joules were really different size units of the same thing, but also measured the conversion factor between these units. This conversion factor is called the mechanical equivalent of heat. The purpose of the present experiment is to determine the mechanical equivalent of heat by doing a known amount of work against friction and measuring the amount of heat thereby produced. The apparatus used is a highly improved version of Joule's original setup.

THEORY

When two surfaces slide one over another, a frictional force arises parallel to the surfaces and must be opposed by an equal parallel applied force if motion at constant speed is to be maintained. The applied force will then do a work equal to its magnitude multiplied by the distance through which it moves in pushing one surface over the other; but this work is not stored as potential energy, nor does it appear as an increase in the kinetic energy of the moving body. The sliding surfaces do get hot, however, and Joule's and Thompson's idea was that the friction transforms the mechanical work done into heat, which thus appears as a different form of energy. Evidence in favor of this concept is obtained by showing that the amount of heat that appears is always proportional to the amount of work done against friction. Joule's experiment, like the present one, consisted of a setup by which a predetermined amount of work could be done against friction (in his case, against the viscous friction encountered by a paddle turning in a fluid) and the resulting heat measured.

Figure 39 shows the experimental arrangement to be used. A copper band is wrapped around a cylindrical copper calorimeter drum and carries a weight on its lower end. The other end is not attached to the drum but is allowed to go slack. Thus, when the drum is turned (by a hand crank in such a direction as to hoist the weight as if the drum were a windlass), the band slips on the drum surface, and the weight, while supported clear of the floor, is not raised. The value of this weight is thus equal to the frictional force between the band and the drum surface on which it slides, and the distance through which the force acts (the distance the sliding surfaces move with respect to each other) is simply the distance moved by any point on the drum surface. This is just the drum's circumference multiplied by the number of revolutions through which it is turned. The work done is thus

$$W = \pi D N M g \tag{70}$$

where W is the work in joules, D is the diameter of the drum in meters, N is the number of revolutions through which the crank is turned, M is the mass in kilograms of the weight suspended on the copper band, and g is the acceleration of gravity in meters per second^2. Notice that the units of the expression on the right in Equation 70 are newton-meters or joules, a joule being a newton-meter by definition.

The copper drum serves as a calorimeter. It is filled with water and insulated from the base of the apparatus by the plastic holder and shaft on which it is mounted. The copper band is in thermal contact with the drum and must be included with it. In operation the drum is filled with water, and a thermometer is inserted through a leakproof seal so that the water temperature may be measured. The relation between the heat delivered to the

129

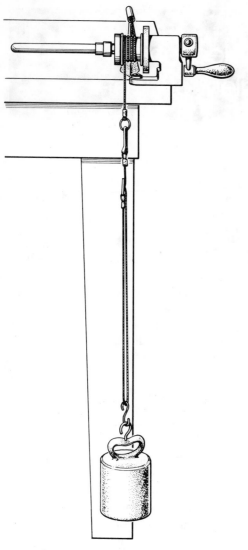

Figure 39 Apparatus for Measuring the Mechanical Equivalent of Heat

calorimeter assembly and the observed temperature rise is

$$Q = [m + (m_1 + m_2)c + C_t](t - t_0)$$

where Q is the heat delivered in calories, m is the mass of water in grams, m_1 is the mass of the calorimeter drum in grams, m_2 is the mass of the copper band in grams, c is the specific heat of copper in calories per gram-°C, C_t is the effective heat capacity of the thermometer in calories per °C, t is the temperature in °C after turning the drum, and t_0 is the initial temperature in °C. If, in fact, the work done against friction is converted into heat, then W should be proportional to Q and we may write

$$\pi DNMg = J[m + (m_1 + m_2)c + C_t](t - t_0) \quad (71)$$

where J is the proportionality constant and has the dimensions of joules per calorie. J is called the mechanical equivalent of heat and can be computed from Equation 71 if all other quantities are known. In the present experiment the proportionality of W and Q will be checked and the value of J measured.

APPARATUS

1. Mechanical-equivalent-of-heat apparatus, complete with 5-kilogram weight and special thermometer

2. Vernier caliper
3. Triple-beam balance

PROCEDURE

1. The apparatus will be set up when you come to the laboratory. Inspect it carefully, noting in particular how the copper calorimeter drum is attached to the plastic mounting disc, so that you will know how to remove it and reattach it. A special thermometer is provided which does not have a very large range but whose expanded scale is arranged to allow small temperature changes near room temperature to be easily read. This thermometer is extremely fragile. Be particularly careful when it is inserted in the calorimeter drum, as a careless knock will break it off. Also note that it is larger in diameter than most thermometers and rolls easily. Take care not to let it roll off the bench.

2. Weigh the empty copper calorimeter drum with the thermometer sealing ferule but without the thermometer. Record the mass m_1 thus measured.

3. Weigh the copper band and record its mass as m_2.

4. Measure the diameter of the calorimeter drum with the vernier caliper.

5. Fill the drum nearly full of water (about 50 or 60 grams) and again weigh it. Then push the thermometer gently through the seal, insert it into the drum, and tighten the sealing ferule. Do not overtighten. Normal tightening with the thumb and forefinger should be sufficient to prevent leaks.

6. Attach the drum to its mounting disc, being careful not to hit the protruding thermometer by accident. Wrap the copper band around the drum until almost its full length is wound up (four or five turns). Attach the 5-kilogram weight to one end so that it hangs freely and attach the other end to a spring hooked to a peg in the apparatus base. The peg may be placed in any of a number of holes in the base; its position should be chosen so that approximately the same length of copper band extends away from the drum at either end. The cord by which the weight is suspended from the band has a length adjustment in it; this should be set so that the weight clears the floor when the crank is being turned in the direction tending to raise the weight. The band should slide smoothly on the drum surface when the crank is turned at a steady rate, and the end attached to the spring should be slack so as not to detract from the force applied to the band by the suspended weight. Check that there are no leaks around the thermometer.

7. The experiment should be started with the temperature of the calorimeter drum and its contents two or three degrees below room temperature. The calorimeter temperature will then be a few degrees above room temperature at the end of the run. In this way the heat exchange between the calorimeter and its surroundings will just about balance out to zero over the running time. If the initial reading of the thermometer is more than three or four degrees below room temperature due to the calorimeter drum having been filled with cold water, turn the crank until the temperature reaches a better initial value. Notice that a lock is provided on the base that will hold the crank in a fixed starting position and that the thermometer may be turned in its seal so that the scale faces upward for easy reading when the crank is in the starting position just noted. If the thermometer is not adjusted in this way, the apparatus also has a mirror that may be swung out under the thermometer so that its scale may be read when it is facing down. Record your initial temperature value t_0 as the temperature corresponding to zero turns of the crank.

8. Start your experimental run immediately after recording t_0 and work rapidly so as to minimize the opportunity for heat exchange between the calorimeter drum and its surroundings. Turn the crank through 50 turns, then observe and record the temperature. Repeat for another 50 turns, and continue, recording the temperature after each 50-turn interval, until a total of 250 turns have been completed.

9. At the end of the experiment, detach the weight and unwind the copper band from the calorimeter drum. Remove the drum from the plastic mounting disc, loosen the sealing ferule, and carefully take out the thermometer. Replace the thermometer in its case and put it away safely. *Do not* leave it lying on the laboratory bench. Empty the calorimeter drum and leave it with the rest of the apparatus.

DATA

Mass of empty calorimeter drum and
   sealing ferule, m_1 _____

Mass of copper band, m_2 _____

Mass of calorimeter drum, sealing
   ferule, and water _____

Mass of water, m _____

Diameter of calorimeter drum, D _____

Number of Turns	0	50	100	150	200	250
Temperature						
Temperature Rise $t - t_0$	✕					

Mechanical equivalent of heat J, from
   the graph _____

Value of the mechanical equivalent of
   heat from Table I, Appendix _____

Percent error _____

CALCULATIONS

1. Subtract the mass of the empty calorimeter from its mass when full to get the mass of the water.

2. Subtract the initial temperature t_0 (corresponding to zero turns of the crank) from the temperature observed after 50 turns to get the temperature rise $t - t_0$ due to 50 turns. Repeat using the observed temperatures at 100, 150, 200, and 250 turns.

3. Plot a curve using the values of the temperature rise $t - t_0$ as abscissas and the numbers of turns as ordinates. Choose your scales so as to use an entire sheet of graph paper. Note that the origin should be a point on your curve.

4. Check that your curve is a straight line and find its slope. According to Equation 71, this slope should be $J[m + (m_1 + m_2)c + C_t]/\pi DMg$, hence J can be calculated if the other quantities are known. Look up the specific heat of copper in the Appendix and use 0.80 calories per degree Celsius for the effective heat capacity of the thermometer. *Note:* This value is correct for the thermometer supplied with the apparatus made by Leybold. If a different thermometer is used, the necessary data will be furnished by the instructor. Enter your calculated value of J on your data sheet.

5. Look up the accepted value of J and compute your percent error.

QUESTIONS

1. Explain how your work in this experiment, and in particular your graph, indicates that heat is a form of energy.

2. A clever feature of the apparatus used in this experiment is the fact that the frictional force is developed between the copper band and a *rotating* surface. Why is this a great advantage?

3. What problem would be encountered if a standard 0–50°C thermometer were used in place of the special one provided?

4. A 2000-pound car is traveling at 60 miles per hour when the engine is disengaged and the brakes are applied, bringing the car to a stop. How many calories of heat are produced in the stopping process?

5. A Prony brake is a device for measuring the power output of an engine. It consists of a brake applied to a drum attached to the engine shaft so that the engine does work against the frictional force provided by the brake. The system is arranged so that the heat produced can be measured. In some Prony brakes this is done by cooling the brake with water whose rate of flow and temperature rise is measured. If a 50°C rise is noted for the water cooling a Prony brake while the engine under test is developing 200 horsepower, how many gallons per minute must be flowing through the brake?

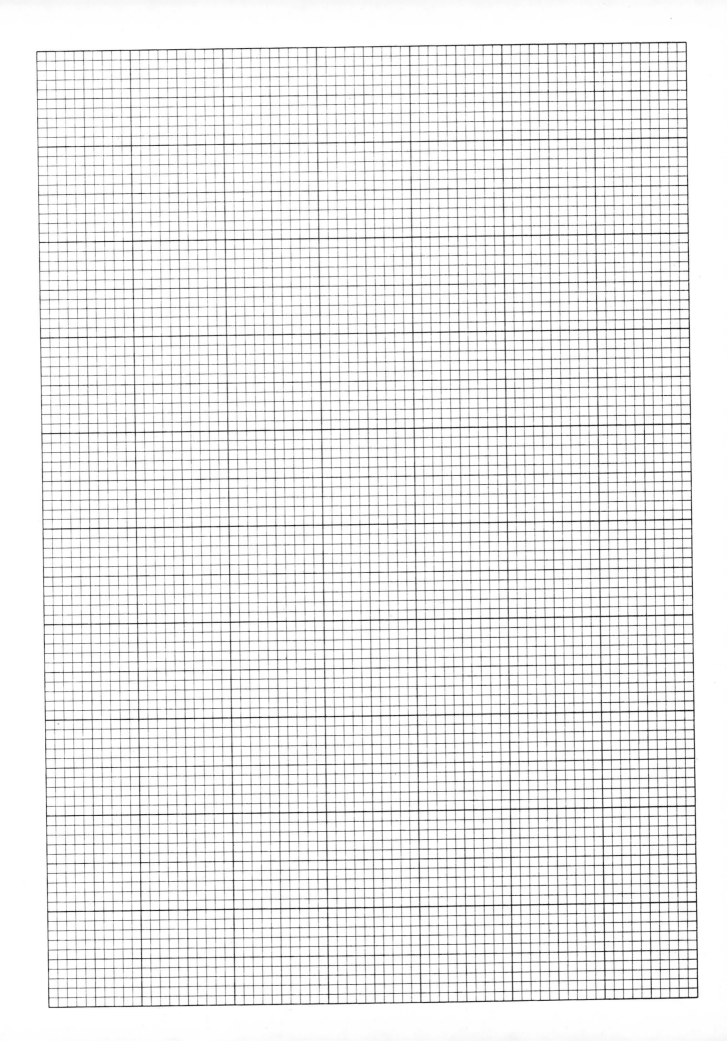

The Ratio of the 22 Specific Heats of Gases

Gases, like all substances, exhibit a specific heat, that is, a requirement that a certain amount of heat be added to a unit quantity of the gas in question to raise its temperature one degree. Although it would be possible to quote specific heats in calories/gram-°C as is usually done with solids and liquids, we shall be interested in a more fundamental unit of quantity, the mole. This is defined as Avogadro's number of molecules, so that we shall be concerned with the amount of heat required to raise the temperature of a collection of this number of molecules of a particular gas by one Celsius degree. The specific heats we will be considering may thus properly be referred to as *molar* specific heats. We will find that a gas exhibits a larger specific heat when it is allowed to expand on heating than it does when it is confined to a fixed volume. In the latter case, the gas undergoes a *constant volume process* and the specific heat is called the *specific heat at constant volume,* designated c_v. If, on the other hand, the gas is allowed to expand as it is heated in such a way that the pressure remains constant, a larger specific heat called the *specific heat at constant pressure,* c_p, is measured. The ratio c_p/c_v is generally represented by the Greek letter γ and is an important quantity in kinetic theory and thermodynamics. The purpose of this experiment is to measure γ for a monatomic and a diatomic gas by the method of Rüchhardt and to test the predictions of kinetic theory in this regard.

THEORY

When a gas is allowed to expand as it is heated, it does work in moving the walls of the containing chamber (*e.g.,* pushing out the piston in a containing cylinder), so some of the energy put in as heat comes out again as this work. Only the remainder stays in the gas to raise its internal energy and thus its temperature. The first law of thermodynamics, which is just a statement of the conservation of energy for this situation, may be written

$$\Delta Q = \Delta U + \Delta W \qquad (72)$$

where ΔQ is the amount of heat added to the gas, ΔU is the change in the gas's internal energy, and ΔW is the work done.

Now suppose the increase in internal energy is reflected by a rise ΔT in temperature. Dividing Equation 72 through by ΔT yields

$$\frac{\Delta Q}{\Delta T} = \frac{\Delta U}{\Delta T} + \frac{\Delta W}{\Delta T}$$

Now $\Delta Q/\Delta T$ is by definition the heat capacity of the particular quantity of gas that you're heating, and if you happen to choose 1 mole for this quantity, then $\Delta Q/\Delta T$ is the molar specific heat. Furthermore, if the heating is done at constant volume (the pressure, of course, rises in such a process), no work is done because the gas doesn't move anything, so that $\Delta W = 0$. Then $\Delta Q/\Delta T =$ $\Delta U/\Delta T$ and we conclude that

$$c_v = \frac{\Delta U}{\Delta T} \qquad (73)$$

However, if the gas is allowed to expand at constant pressure, ΔW is not zero and its value must be calculated. This is easy to do by writing down the ideal gas law (Equation 68) with $n = 1$ because we have 1 mole of gas. It is

$$PV = RT$$

where P is the absolute pressure on the gas, V its volume, T its absolute temperature, and R the universal gas constant. Clearly, if the pressure is held constant, a change ΔT in the temperature will cause a change ΔV in the volume that is related to ΔT by the expression

$$P\Delta V = R\Delta T$$

But $P\Delta V$ is the work done in the expansion, for suppose that the expansion is carried out by letting the gas push on a piston whose face has area A. Then the force on the piston is PA. If the piston moves through a distance Δx, the work done is by definition $PA\Delta x$. But ΔV is the volume swept out by the piston as it moves and is

135

thus equal to $A \Delta x$. Hence we have $\Delta W = PA \Delta x = P \Delta V$ and

$$\frac{\Delta W}{\Delta T} = \frac{P \Delta V}{\Delta T} = R$$

We conclude that the specific heat at constant pressure is greater than the specific heat at constant volume by the universal gas constant, or

$$c_p = c_v + R \qquad (74)$$

Note that the dimensions of R are energy/mole$-°$C, as they should be.

Classical kinetic theory has shown that a mole of any gas stores an amount of energy equal to $\frac{1}{2}RT$ for each *degree of freedom* of its molecules. A degree of freedom is defined as a way in which a molecule can store either kinetic or potential energy, and a way which is independent of other ways. A molecule that acts like a mass point has three degrees of freedom because it can move in each of three perpendicular directions and can thus store kinetic energy in each of these motions. The molecules (actually atoms) of a monatomic gas can be considered mass points, so a mole of such molecules should have an internal energy $U = \frac{3}{2}RT$. Hence for monatomic gases such as helium, neon, or argon

$$c_v = \frac{\Delta U}{\Delta T} = \frac{3}{2}R$$

$$c_p = c_v + R = \frac{5}{2}R$$

$$\gamma = \frac{c_p}{c_v} = \frac{5}{3} = 1.67$$

A diatomic gas, on the other hand, consists of molecules that may be represented by little dumbbells, the two atoms being spaced apart by the interatomic bond just as the big ends of a dumbbell are by the intervening shaft. Such a structure will have a moment of inertia when rotated about any axis perpendicular to the dumbbell's centerline and can thus store rotational kinetic energy in this motion. Since two mutually perpendicular axes can be constructed perpendicular to the centerline, there are two independent rotational modes (one for rotation about each of these so-called *principal axes*) and hence two rotational degrees of freedom. There should be a third one corresponding to rotation about the centerline itself (the third principal axis), but if the atoms are really mass points, there is no radius for any of the rotating mass in this mode and thus no moment of inertia and no possibility of storing energy. This means that we have only two rotational degrees of freedom in addition to the three translational ones that any movable object has. It follows that a diatomic molecule

has a total of five degrees of freedom, and so a mole of a diatomic gas such as hydrogen, nitrogen, or oxygen should have an internal energy $U = \frac{5}{2}RT$. Consequently, for a diatomic gas

$$c_v = \frac{\Delta U}{\Delta T} = \frac{5}{2}R$$

$$c_p = c_v + R = \frac{7}{2}R$$

and

$$\gamma = \frac{c_p}{c_v} = \frac{7}{5} = 1.40$$

Note that we have the important results: (a) all monatomic gases have the same *molar* specific heats and thus the same value of γ, namely 1.67, regardless of their atomic masses or chemical properties; and (b) the same holds for all diatomic gases, except that for them $\gamma = 1.40$. Therefore, the measurement of γ for a variety of gases is a very significant test of kinetic theory.

Although it is possible to measure c_p and c_v individually by straightforward methods for any particular gas and then to calculate their ratio, a clever procedure for obtaining γ directly was invented by F. E. Rüchhardt in 1929. His apparatus is illustrated in Fig. 40. It consists of a large flask fitted with a rubber stopper in which is inserted a length of glass tubing having a very smooth bore of precise diameter. A steel ball bearing of mass m is dropped into the tube, in which it fits closely enough to make a gas-tight seal while nevertheless sliding with negligible friction. Fig. 40 shows the ball in its equilibrium position. The downward forces are its weight mg and the force due to atmospheric pressure (P_0) on its upper surface. This force is P_0A where A is the ball's cross-sectional area. The balancing upward force is PA where P is the absolute pressure in the flask, so that at equilibrium

$$PA = P_0A + mg \qquad (75)$$

Now let the ball be displaced by a distance x from its equilibrium position as shown in Fig. 40. If it is pushed down, the volume V of the gas trapped in the flask and the tubing below the ball will be decreased by a small amount dV, the pressure in the flask will be increased by a small amount dP, and the ball will be forced back up by the increased pressure. Because of the momentum it gains, it will overshoot the equilibrium position, the pressure will drop below the equilibrium value, and the ball will come to a stop and fall back down again. In short, the ball will bounce up and down, oscillating about its equilibrium position, the pressure and volume of the gas in the flask accordingly varying above and below their equilibrium values.

Because the motion of the ball is rapid, the changes

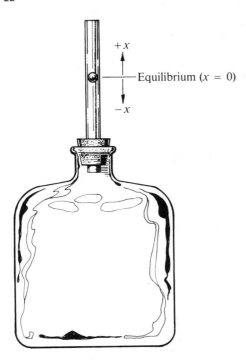

Equilibrium ($x = 0$)

+x

−x

Figure 40   Rüchhardt's Apparatus for Measuring γ

the change in volume dV is Ax. Substitution in Equation 76 gives

$$V^\gamma \frac{F}{A} + \gamma P V^{\gamma-1} A x = 0$$

which after a rearrangement of terms becomes

$$F = -\gamma \frac{A^2 P}{V} x \qquad (77)$$

We note at once that this is a linear restoring force, so that the ball must oscillate with simple harmonic motion. The proportionality constant in Equation 77 is $\gamma A^2 P/V$, and since we know that the period of simple harmonic motion is 2π times the square root of the ratio of the oscillating mass to this constant, we have

$$\tau = 2\pi \sqrt{\frac{mV}{\gamma A^2 P}}$$

where τ is the period, m is the mass of the ball, V is the volume of the flask (including the volume in the tube between the ball and the stopper), A is the cross-sectional area of the ball ($\pi/4$ times its diameter squared) and P is the equilibrium pressure in the flask (given by Equation 75). The solution for γ is

$$\gamma = \frac{4\pi^2 mV}{\tau^2 A^2 P} \qquad (78)$$

Thus by watching the bouncing ball and timing its oscillations to measure τ, γ can be obtained from this formula. The method is capable of good results, but note that the requirements that the ball make a gas-tight seal in the glass tube and at the same time slide without friction are highly contradictory, and also note that errors in the period measurement are doubled in γ as a result of τ appearing squared in Equation 78. With good technique, however, the difference between the values of γ for monatomic and diatomic gases is easily observed.

dP and dV in the gas pressure and volume take place in times too short to allow a significant heat transfer into or out of the gas through the walls of the containing vessel. The gas is thus essentially undergoing an adiabatic process, for which the relation between pressure and volume is

$$PV^\gamma = \text{constant}$$

Differentiating to obtain the relation between a small change dV in volume and the resulting small change dP in pressure yields

$$V^\gamma \, dP + \gamma \, P V^{\gamma-1} \, dV = 0 \qquad (76)$$

Now note that the net force F on the ball is $A\,dP$ and that

APPARATUS

1. Aspirator flask, about 10-liter capacity
2. Two single-hole rubber stoppers to fit flask
3. Stopcock tube for flask
4. Rubber pad
5. Precision glass tube
6. Steel ball to fit tube
7. Triple beam balance
8. Micrometer caliper
9. Wooden ramrod for cleaning tube
10. Lintless cloth patches or tissues
11. Stop watch or stop clock
12. Barometer
13. Tank of argon with regulator and connecting hose

PROCEDURE

1. Wash your hands and dry them thoroughly. Cleanliness is extremely important in this experiment, because the clearance between the steel ball and the bore of the precision glass tube is only about one ten-thousandth of an inch and no lubricant can be used. The slightest fingerprint, grease spot, or speck of dust can cause the ball to stick in the tube, requiring a thorough cleaning of both tube and ball before a successful run can be achieved. Even when your hands are clean and dry, do *not* touch the ball but handle it using the lintless cloth patches or tissues.

2. Record the volume of the aspirator flask. This will be marked on the flask or given to you by the instructor.

3. Weigh the ball on the triple-beam balance and record its mass in grams.

4. Measure the diameter of the ball using the micrometer caliper. To do this, first close the caliper jaws to obtain the zero reading. Then make three independent measurements of the diameter, rotating the ball between measurements. Handle the ball only with the cloths or tissues provided and make sure the micrometer's jaws are clean. Record all your readings.

5. Inspect the Rüchhardt apparatus, which should already be set up when you come to the laboratory. Make sure that both rubber stoppers are seated firmly in the flask. Note that the flask has a side port (not shown in Fig. 40) through which the gas pressure may be released, the flask filled with a gas other than air, or the ball removed if it falls in the flask at the end of a run. The stopper in this port is fitted with a short tube and stopcock, which you should make sure is closed. Note also that there should be a rubber pad in the flask on the bottom to prevent breakage when the ball drops on it.

6. Read the barometer and record the reading.

7. Practice your technique and test the apparatus by carefully dropping the ball into the precision glass tube. Remember that the slightest bit of dust or dirt in the tube or on the ball will prevent it from sliding properly. Pick up the ball only with the lintless cloth or tissue; do not touch it with your fingers. If the apparatus is properly clean, you should be able to observe at least 7 oscillations of the ball in the tube before the motion damps out. Thereafter the ball will either stick in the tube or fall into the flask. If it falls into the flask, remove the lower stopper and carefully extract the ball. If it sticks in the tube, remove the tube and upper stopper as a unit and recover the ball either by tipping the tube back and forth until the ball dislodges itself and rolls out (be prepared to catch it in a tissue) or by *gently* pushing it out with the wooden ramrod. CAUTION: Be extremely careful when handling the precision tube. Do not lay it on the bench (it may roll off) or make any rapid motions around it or with it. It is very easily broken, and the sharp ends of a broken glass tube are very dangerous. Also remember that because of the close tolerances to which the bore has been held, this is an expensive piece of apparatus.

8. If you cannot get 7 oscillations of the ball in the tube, these items will have to be cleaned. The ball may be wiped off with the lintless cloth or tissues. To clean the bore of the precision tube, remove it along with its stopper *carefully* from the flask and run a cloth patch or tissue through it two or three times with the wooden ramrod. Be extremely careful not to break the tube while doing this. *Note:* No lubricant has ever been found that helps with the contradictory requirements of a gas-tight seal and no friction between the ball and the tube wall. Best results are obtained when the ball and the tube are absolutely dry. This means that the cleaning patches must be dry, and no solvent can be used. If you cannot get the ball and tube clean enough with the dry patches to get at least 7 oscillations when you try the experiment, call your instructor.

9. When you are sure that the apparatus is working correctly, start your measurement of the period τ by dropping the ball into the precision tube and timing its oscillations. Do this by starting the stop watch (or stop clock) at the instant the ball gets to its lowest point and counting a cycle each time the ball returns to this position. Stop the stop watch at the end of 4 or 5 cycles and record the elapsed time and the number of cycles timed. This procedure may be conducted efficiently by two people, one of whom drops the ball in the tube while the other counts the cycles and operates the stop watch. When the ball has come to rest, remove it as described in Procedure 7, being careful not to get any dust on it or in the tube bore.

10. Repeat Procedure 9 twice to get three independent measurements of the period τ.

11. Fill the flask with argon by attaching the hose from the argon tank to the glass tube in the lower port and opening the stopcock. The instructor will show you how to operate the regulator on the tank to start the flow of argon. Flush argon through the flask for about 3 minutes or as determined by the instructor. Argon is heavier than air, so filling the flask from the bottom displaces the air up and out through the precision tube.

When the flask has been filled, shut off the regulator first and then the stopcock. Disconnect the line only after you have made sure that both the regulator and the stopcock have been turned off.

12. Repeat Procedures 9 and 10 with argon in the flask. Record all your data.

13. Read the barometer again at the end of the experiment and record the reading.

DATA

Mass of ball _____   Volume of flask _____

Diameter of ball:

    Micrometer readings, trials: 1. _____ 2. _____ 3. _____

    Average reading _____   Zero reading _____

    Diameter _____   Cross-sectional area _____

Barometer readings

    Beginning _____   End _____

Gas	Run	Time in Seconds	Number of Oscillations	Period in Seconds
Air	1			
	2			
	3			
Argon	1			
	2			
	3			

Air:

    Average period _____   Value of γ _____   Percent error _____

Argon:

    Average period _____   Value of γ _____   Percent error _____

CALCULATIONS

1. Compute an average diameter for the ball by finding the average of your 3 measurements and correcting the result for your micrometer's zero reading.

2. Compute the ball's cross-sectional area from its diameter. Remember that the area of a circle of radius r is πr^2 and that $r = d/2$ where d is the circle's diameter.

3. Convert your barometer readings to dynes/cm² by using the fact that 1333 dynes/cm² is equal to 1 millimeter of mercury. If there is any difference in the two readings, use the first reading in the calculation of γ for air and the second for argon.

4. Calculate the equilibrium pressure P in the flask from Equation 75. Use the value of A from Calculation 2 and the barometric pressure taken at the beginning of the experiment for P_0. Be careful of your units. In this experiment the CGS system is convenient because the triple-beam balance reads in grams and the micrometer caliper reads in centimeters, but you must be consistent. All pressures must be in dynes/cm², all volumes in cm³ or milliliters, and the value used for the acceleration of gravity, g, must be 980 cm/sec².

5. Compute the measured period in each of your three runs for air by dividing the time by the number of oscillations counted in each case. Then compute the average of these three results and record it as the average period for air.

6. Calculate your measured value of γ for air by using Equation 78. Here A is the cross-sectional area obtained in Calculation 2, P is the equilibrium pressure from Calculation 4, and τ is the average period found in Calculation 5.

7. Compare your result from Calculation 6 with the value of γ for air predicted by kinetic theory by finding the percent error. Note that because air is composed almost entirely of nitrogen and oxygen, it may be considered a diatomic gas.

8. Repeat Calculations 4 through 7 for argon. Use the barometric pressure taken at the end of the experiment for P_0, and take the value of γ predicted by kinetic theory as that of a monatomic gas.

QUESTIONS

1. List and discuss briefly the principal sources of error in Rüchhardt's method of measuring γ. How much error do you think each has contributed to your result? Do you think that the theoretical values for γ lie within the experimental error of your measured values?

2. A modification of Rüchhardt's method suggested by Rinkel is to release the ball in the tube from rest at a point where the pressure in the flask is equal to the atmospheric pressure outside, so that the net force on the ball is just its weight mg. The distance L through which the ball falls from the release point to the lowest point reached before it stops and starts back up is then measured instead of the period of the subsequent oscillation. (a) Derive a formula for γ in terms of L instead of τ. (b) Find the initial amplitude of oscillation, which is the distance between the *equilibrium point* and the lowest position reached.

3. (a) Find the proper relation between T and V for an adiabatic process. (b) Find the proper relation between T and P for this type of process.

The Force Between 23
Electric Currents

Although magnets appear to have poles that attract or repel other poles, in reality no magnetic poles are found to exist in the sense that electric charges do. What, then, is the source of the magnetic field if there are no poles or "magnetic charges" to fill this role? Magnetic fields result from the *motion* of electric charges, that is, from electric currents. Permanent magnets produce magnetic fields because the electrons going around in their orbits in the atoms of the material constitute tiny loops of electric current each of which is the source of a magnetic field. In materials like iron, nickel, and cobalt, out of which permanent magnets are made, it is possible to line up the atoms so that their fields are collinear and add up to the large fields found around such magnets. This process is called magnetization, and in permanent magnets the atoms will remain lined up provided the temperature is not high enough for the resulting random thermal motion to upset the arrangement (the "Curie temperature").

The most elementary source of a magnetic field is a long, straight wire carrying an electric current. The magnetic lines of force due to such a current are simply concentric circles centered on the wire, as shown in Fig. 41. In addition, because there are no magnetic-pole sources of the magnetic field, the magnetic forces are not properly described as the action of a magnetic field on a pole. Instead a force is exerted by a magnetic field on a moving electric charge, that is, on an electric current. Thus in describing magnetic fields and forces, we consider only moving electric charges and do not think in terms of poles at all. In particular, when an electric

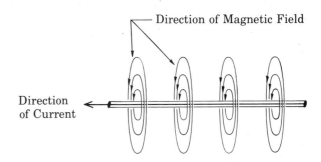

Figure 41   Magnetic Field Direction

charge moves in a magnetic field, it feels a force in a direction perpendicular both to the direction of the field lines (magnetic lines of force) and to the direction of motion. Clearly the charge will feel no force if it moves parallel to the field lines, and the force it feels will increase from zero to a maximum as the direction of motion is changed from parallel to perpendicular to these lines. In Fig. 41, if a second wire is laid parallel to the one shown, a current in it will be perpendicular to the field at its position and a force pushing it towards or away from the second wire (depending on the current direction) will result. This is the most fundamental possible arrangement for the study of magnetic forces. The force law between parallel current-carrying wires may be regarded as the fundamental law of magnetostatics, just as Coulomb's law is so regarded in the electrostatic case. The purpose of the present experiment is to study this law.

THEORY

The magnetic field of a long, straight wire is directed as shown in Fig. 41 and has the value

$$B = \frac{\mu_0 I}{2\pi r} = 2 \times 10^{-7} \frac{I}{r} \qquad (79)$$

where B is the magnetic field strength in teslas, I is the current in the wire in amperes, r is the perpendicular distance in meters from the wire to the point where B is being evaluated, and μ_0 is a constant called the permeability of free space and has the value $4\pi \times 10^{-7}$ weber per ampere-meter. Notice that for the electromagnetic work in this experiment, MKS units must be strictly adhered to. Therefore r in Equation 79 must be expressed in meters even though it will often be so small that the meter will not be the most convenient unit to use.

If a long, straight wire carrying current I' is in a magnetic field of strength B, the force on a length L of the wire is given by

$$F = BI'L \sin \theta \qquad (80)$$

where F is the force in newtons, B is the field strength in teslas, I' is the current in amperes, L is the length in meters of the wire on which force F is exerted, and θ is the angle between the directions of the current and the field. The force is perpendicular to these two directions, but its sense remains to be determined. This is most easily done with the aid of Fig. 42, which shows the directions of a current I' and a magnetic field B by means of arrows. If your right hand is then placed so that the fingers curl in the direction needed to rotate the current

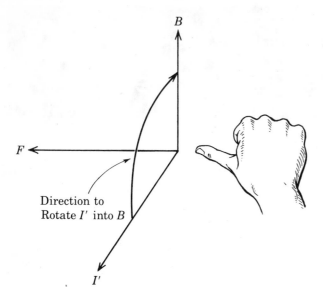

Figure 42 The Direction of the Magnetic Force

arrow into the field arrow, the extended thumb will point in the direction of F. This procedure, sometimes called "the right-hand rule," is also useful in determining the field direction around the wire of Fig. 41. In this case, the wire is grasped with the right hand with the thumb extended in the direction of current flow. The fingers will then curl naturally around the wire in the direction of the magnetic field lines.

If a second wire carrying current I' is laid parallel to the one shown in Fig. 41 at a distance d away from it, the value of B at this wire due to the current I in the first wire will, according to Equation 79, be $2 \times 10^{-7} I/d$. Moreover, Fig. 41 shows that the direction of B at the position of the second wire will be perpendicular to the wire and hence to I'. Equation 80 then says that the force on a length L of the second wire will be $BI'L$ and will be directed either toward or away from the first wire. Consideration of Fig. 42 shows that the two wires will attract each other if the currents I and I' are in the same direction and will repel if these currents are oppositely directed.

Combining Equations 79 and 80 for the case of parallel wires yields

$$F = \frac{\mu_0 I'IL}{2\pi d} = 2 \times 10^{-7} \frac{I'IL}{d} \qquad (81)$$

for the force of attraction or repulsion between the wires. Note that it makes no difference which wire is considered the source of the field and which the one on which F acts. Equation 81 is obtained in either case. In an actual experiment, one of the two parallel wires will be movable so that the force on it can be measured. The other, which is fixed in position, may then be conveniently taken as the field source.

In the present experiment a sensitive balance is used to measure the force between two parallel wires. High sen-

sitivity is needed because the factor 2×10^{-7} in Equation 81 results in very small forces unless inconveniently large currents and/or small values of the separation distance d are used. The apparatus, which is shown in Fig. 43, consists of a heavy base on which one straight wire is mounted by means of insulating posts. The second or movable wire forms one side of a rectangular frame which itself forms one arm of a beam balance. The other arm consists of a threaded rod on which counterweights may be screwed back and forth to balance the beam assembly. The beam rotates on knife edges under the side of the rectangle opposite the wire, and a mirror is mounted on this side so that small rotations of the beam assembly may be easily measured. This is done by means of an optical lever as shown in Fig. 42. A reading telescope and vertical scale are placed at a distance of about one meter from the mirror. The image of the illuminated scale as seen in the mirror is viewed through the telescope, and the position of the cross hairs upon the image of the scale is a measure of the rotation of the beam assembly and hence of the spacing d between the wires. The position of the cross hairs noted when the wires are touching is the zero reading. Because d is the distance between the wire *centers*, the zero reading corresponds to a separation $d = r_1 + r_2$, where r_1 and r_2 are the respective wire radii. Other readings give

$$d = \frac{Sa}{2D} + r_1 + r_2 \qquad (82)$$

where d is the on-center separation of the wires in meters, D is the distance in meters from the mirror to the scale, a is the width of the rectangular frame (the distance from the knife edges to the movable wire) in meters, and S is the difference between the scale reading obtained with separation d and the zero reading. The beam assembly is also provided with a damping system consisting of a conducting blade moving in the field of a permanent magnet and a beam lift which raises the beam assembly off the knife edges to prevent damage during transportation and to position the assembly correctly on the knife-edge supports. In addition, the movable wire carries a small pan in which calibrated weights in the milligram range may be placed to measure the force exerted on this wire by the magnetic field of the fixed wire.

The earth's magnetic field is about 5×10^{-5} tesla (0.5 gauss) and therefore cannot be neglected in comparison with the fields that can be produced at the movable wire by reasonable currents in the fixed one. Although the apparatus can and should be oriented with the wires parallel to the earth's field (roughly north-south) to minimize its effect, we cannot count on eliminating it entirely and steps must be taken to compensate for it. Suppose that the component of the earth's field perpendicular to the plane of the two wires is B_0. Then the total field to be used in Equation 80 is $B + B_0$, where B is the

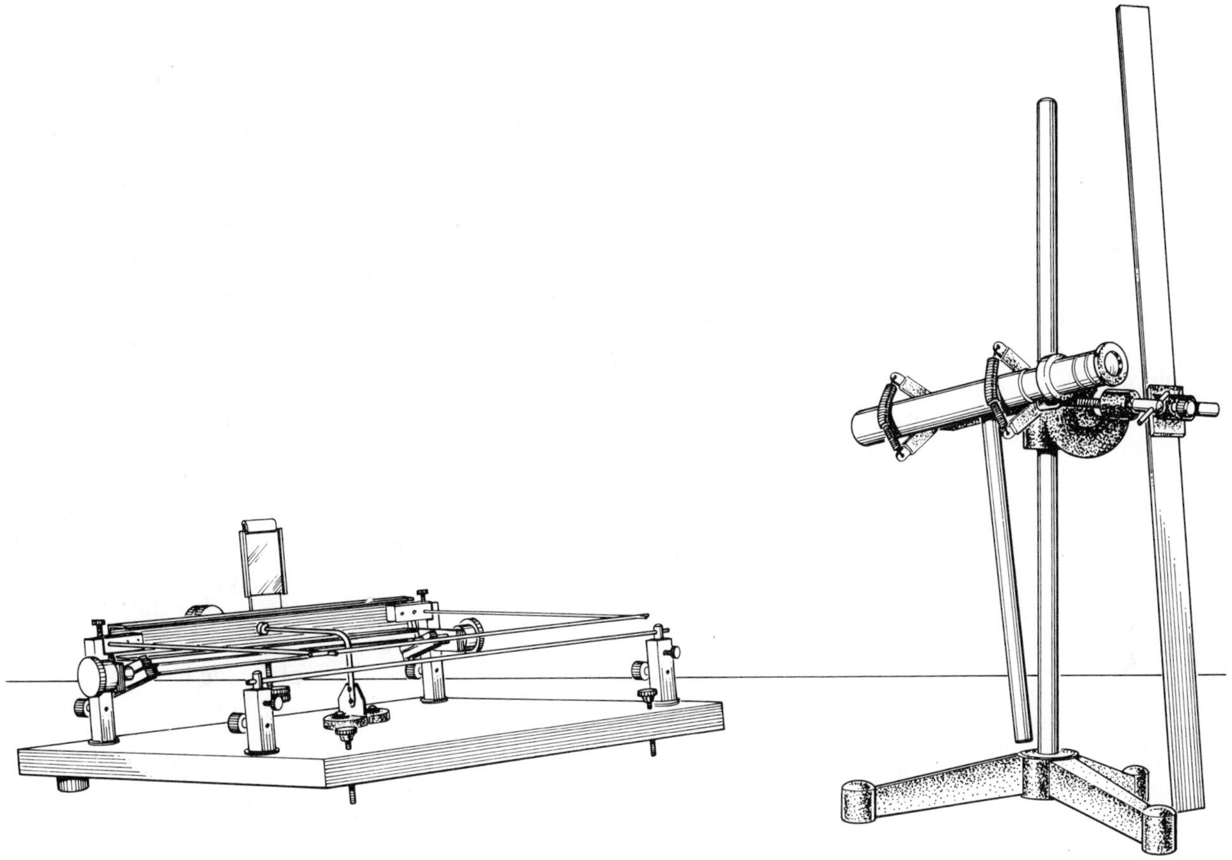

Figure 43   Current Balance

field due to current I in the fixed wire, and Equation 81 becomes

$$F = \left(\frac{\mu_0 I}{2\pi d} + B_0\right)I'L \qquad (83)$$

The apparatus is now connected with the two wires in series so that $I = I'$. Equation 83 then becomes

$$F = \left(\frac{\mu_0 I^2}{2\pi d} + B_0 I\right)L \qquad (84)$$

If the current I is then reversed, the force is given by

$$F = \left(\frac{\mu_0 I^2}{2\pi d} - B_0 I\right)L \qquad (85)$$

in which the force due to the current is unchanged because I has been reversed in both wires (the sign of I^2 doesn't change) but the force due to the earth's field has been reversed by the reversal of the current in the movable wire. Clearly the average of the two forces measured with the current in each of the two directions will be the force due to the current alone, or that given by Equation 81 with $I = I'$.

APPARATUS

1. Current balance with calibrated weights
2. Telescope and scale with support stand
3. Desk lamp or scale illuminator
4. Compass
5. Level
6. Micrometer caliper
7. Vernier caliper
8. Two-meter stick

9. Power supply capable of delivering at least 10 amperes at 12 volts D.C., or 12-volt storage battery
10. Ammeter, 0–10 amperes D.C.
11. Rheostat, 5-ohms, 10-ampere current rating
12. Resistor, 1-ohm, 100-watts
13. Two double-pole, double-throw knife switches

PROCEDURE

1. Set up the current balance on the laboratory bench with the wires extending in a north-south direction as indicated by the compass. Make sure that the beam assembly is lifted off the knife edges by the beam lift. Level the apparatus by means of the two thumb screws at the front of the base.

2. Operate the lifting mechanism to lower the beam assembly onto the knife-edge bearing surfaces. Check that the movable wire and the fixed wire below it are in the same vertical plane. If they are not, loosen the set screws holding the wire sides of the rectangle in the knife edges and push the sides in or out as required. Be sure to tighten the set screws after making this adjustment. Again lift the beam assembly off its knife edges with the beam lift, noting that an important function of the lifting mechanism is to position the wire rectangle correctly. Lower the beam onto the knife-edge supports and recheck the position of the movable wire. Check also that, when viewed from the front, the wires are straight and parallel. Ideally, if the wires are just brought together, no light should be visible between them when a white paper is held behind them. Adjustment for parallelism may be made by loosening the set screws retaining the lower (fixed) wire in its mounting posts and setting it parallel to the upper (movable) wire. However, even after this adjustment has been made, the wires will seldom be so straight as to touch at every point along their full length, but perfect straightness is not essential for good results. Consult your instructor if the wires appear seriously bent.

When the current balance is correctly adjusted, the beam assembly should pivot freely on its knife edges. In particular, the damping blade should ride in the middle of the hole cut for it in the base and should not rub on any of the damping magnet poles. If the blade comes too close to one side of the hole, lift the beam assembly off its supports and *gently* press the blade right or left as required. To replace the beam assembly, first operate the beam lift, put the assembly in position, and then lower it carefully onto the knife-edge supports. If either damping magnet is too close to the blade, loosen that magnet's retaining screw and adjust its position as necessary. With all these adjustments complete, raise the beam assembly off its supports with the beam lift.

3. Take the beam assembly out of the apparatus and measure the length and width of the wire rectangle with the vernier caliper. The length, which gives length L of the movable wire, is the distance between the centers of the supporting side wires. The width is measured from either knife edge to the center of the movable wire.

4. Measure the diameter of the movable wire and the fixed wire with the micrometer caliper. Make two measurements in different places on each wire. Then replace the beam assembly on the lifting mechanism, but do not lower it onto the knife-edge supports.

5. Wire the apparatus as shown in Fig. 44. Notice that switch SW_1 reverses the current through both wires of the current balance, whereas SW_2 reverses the current in one wire with respect to the other in order to change from repulsive to attractive force measurements. The neutral position of either switch may be used to open the circuit, but SW_1 will probably be the more convenient for this purpose. Have the instructor check

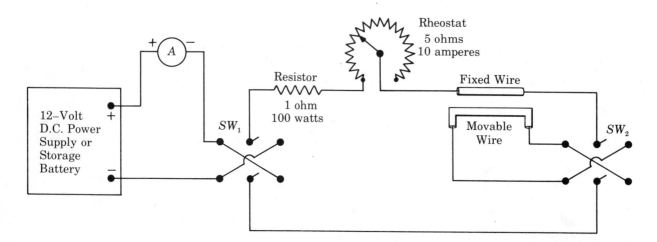

Figure 44   Circuit for the Current Balance

your wiring before you turn on the power supply, or, if a storage battery is used, before making the connections to its terminals.

6. Place the telescope and scale about 1 meter from the current balance and aimed so that the telescope is focused on the image of the scale in the mirror. Carefully measure the distance from the mirror to the scale with the meter stick.

7. With no weights in the weight pan, lower the beam assembly into operating position and adjust the counterweights behind the mirror (not the ones below it, which have been properly set and should not be touched) until the assembly balances with the movable wire a few millimeters above the fixed one. Now place just enough weight in the pan to bring the wires together. View the mirror image of the scale through the telescope and record the reading under the cross hairs to the nearest tenth of a millimeter. This will be the zero reading. An adjusting screw behind the mirror may be used to give this reading a convenient value if desired.

8. Carefully remove all weights from the weight pan and let the beam assembly come to equilibrium. Note that a good way to remove the weights without disturbing the apparatus is to blow them off. Read the scale as viewed through the telescope and record the reading.

9. Set the rheostat at maximum resistance and SW_2 so that the currents in the two wires are oppositely directed. Turn on the power supply and close SW_1. Adjust the power supply and/or the rheostat for a current of between 3 and 4 amperes. Carefully add weights to the weight pan until the beam assembly returns to approximately the same position it had in Procedure 8, as determined by looking at the image of the scale in the telescope. Then make a fine adjustment of the current so that the scale reading is exactly the same as that obtained in Procedure 8. Since the beam assembly is now in the same position it occupied with no weight and no current, the repulsive force between the parallel wires must be just equal to the added weight. Record this weight and the final current as read on the ammeter. *Note:* If you are using an adjustable power supply, its output control may provide a finer current adjustment than the rheostat. In this case the rheostat may be left at any convenient setting, usually near its minimum resistance, and all current adjustment carried out with the power supply control. If a fixed-voltage power supply is available, such a control may be provided by inserting a variable transformer between the power line and the power supply input.

10. Reverse the current through the apparatus by throwing switch SW_1. Readjust the current to again obtain the same scale reading as that found in Procedure 8. Record this new current value.

11. Repeat Procedures 9 and 10 with additional weight and a current of between 6 and 7 amperes.

12. Repeat Procedures 9 and 10 with still more weight and a current of close to 10 amperes.

13. Turn the current off by opening SW_1. Adjust the counterweights behind the mirror until the same scale reading found in Procedure 8 is obtained with the weights of Procedure 12 still in the pan. Throw switch SW_2 to give currents in the same direction in the two wires so that the force between them will be attractive.

14. Close SW_1 and remove weights from the pan until the scale reading of Procedure 8 is once more approximately obtained and make a fine adjustment of the current until this reading is obtained exactly. Under these conditions the force due to the removed weight is being provided by the magnetic attraction between the wires. Record the final value of the current and the amount of weight removed.

15. Throw switch SW_1 to reverse the current through the apparatus and readjust the current to return the scale reading to the same value as before. Record the new current value.

16. Reduce the current to between 6 and 7 amperes and add weight to bring the scale reading back to that of Procedure 8. Adjust the current to get this scale reading exactly. Record this current and the difference between the weight now in the pan and the weight in the pan in Procedure 13.

17. Reverse the current through the apparatus by throwing SW_1 and readjust it to give the scale reading of Procedure 8. Record the new current value so obtained.

18. Repeat Procedures 16 and 17 with the current set between 3 and 4 amperes.

19. Remove all weights and repeat Procedures 8–18, first readjusting the counterweight behind the mirror so that the beam assembly balances with the movable wire at a different distance from the fixed one.

DATA

Length of the movable wire, L _____

Width of the wire rectangle, a _____

Radius of the movable wire, r_1 _____
  Micrometer readings:

              Trial 1. _____

              Trial 2. _____

  Average reading _____

  Zero reading _____

Radius of the fixed wire, r_2 _____

Micrometer readings:

              Trial 1. _____

              Trial 2. _____

  Average reading _____

Distance from mirror to scale _____

Zero scale reading _____

First zero-weight, zero-current
  scale reading _____

First wire-separation distance, d _____

Relative Direction	Current			Weight on Pan	Magnetic Force
	Forward	Reverse	Squared		
Opposite (Repulsive Force)					
Same (Attractive Force)					

Second zero-weight, zero-current scale reading _____

Second wire-separation distance, d _____

Relative Direction	Current			Weight on Pan	Magnetic Force
	Forward	Reverse	Squared		
Opposite (Repulsive Forces)					

| Relative Direction | Current | | | Weight on Pan | Magnetic Force |
	Forward	Reverse	Squared		
Same (Attractive Forces)					

Value of μ_0 with first wire separation _____   Average value of μ_0 _____

Value of μ_0 with second wire separation _____   Percent error _____

CALCULATIONS

1. Compute the average of the two diameter readings for each wire. Subtract the zero reading of the micrometer to obtain the diameter in each case and divide your results by 2 to obtain the radii r_1 and r_2 of the two wires.

2. Calculate the first wire-separation distance d using Equation 82 and your data from Procedures 7 and 8.

3. Compute the square of the current readings obtained in Procedures 9–12 and 14–18.

4. From your data on the weight in the weight pan obtained in Procedures 9–12, find the repulsive magnetic force between the wires for each current. This calculation requires multiplication of the weight in the pan *in kilograms* by the acceleration of gravity in meters per second per second.

5. From your data on the weight in the weight pan obtained in Procedures 14–18, find the attractive magnetic force between the wires for each current. This calculation requires subtracting the weight in the pan in each case from the weight that was there in Procedures 12 and 13. The difference is then expressed in kilograms and multiplied by the acceleration of gravity in meters per second per second.

6. Plot a curve using the magnetic forces obtained in Calculations 4 and 5 as abscissas and the corresponding values of the square of the current as ordinates. Use an entire sheet of graph paper for this curve. Place the origin near the center of the sheet and consider attractive forces and the corresponding values of I^2 to be negative. Distinguish between points representing current through the apparatus in the forward direction and those for current in the reverse direction by using different colored pencils or marking the points in some distinctive way (for example, circles around points for forward current and squares around points for reverse current). Draw the best straight line through your points, remembering that the origin (zero force, zero current) should be a point on this line.

7. Calculate a value of μ_0 from the slope of the graph obtained in Calculation 6.

8. Repeat Calculations 2–7 for the second wire separation and the associated data called for in Procedure 19.

9. Compute an average of your two values of μ_0 and compare this result with the accepted value by finding the percent error.

QUESTIONS

1. What conclusions do you draw from your two graphs?

2. The theory section points out that the effect of the earth's magnetic field can be eliminated by averaging the two force values obtained with opposite flow directions of a particular current. Explain why this was not done in the present experiment and what was done instead. In this connection, discuss the relative positions of the points representing forward and reversed current on your graphs.

3. Explain why reversing switch SW_2 changes the sign of I^2 but reversing switch SW_1 does not.

4. Why must the weight in the weight pan be expressed in kilograms in Calculations 4 and 5?

5. Derive Equation 82 from a consideration of the apparatus geometry and the fact that when a mirror turns through a given angle, a reflected light beam is turned through twice that angle.

6. Suppose the current balance was oriented so that a horizontal component of the earth's magnetic field equal to 0.17 gauss (1.7×10^{-5} tesla) remained perpendicular to the movable wire. What weight (in milligrams) would be needed to balance out the effect of this field when a current of 10 amperes flows in the wire?

7. What is the purpose of the damping blade and magnet arrangement? How does it work?

8. A length of No. 14 bare copper wire (diameter = 0.163 cm) is laid flat on a table and connections are made to its ends with flexible leads. A second wire is mounted 1 cm above and parallel to the first one and is connected in series with it so as to produce an attractive force when a current is passed through the two wires. Find the current necessary to just raise the first wire off the table. Do you think the No. 14 wire could handle this current without overheating?

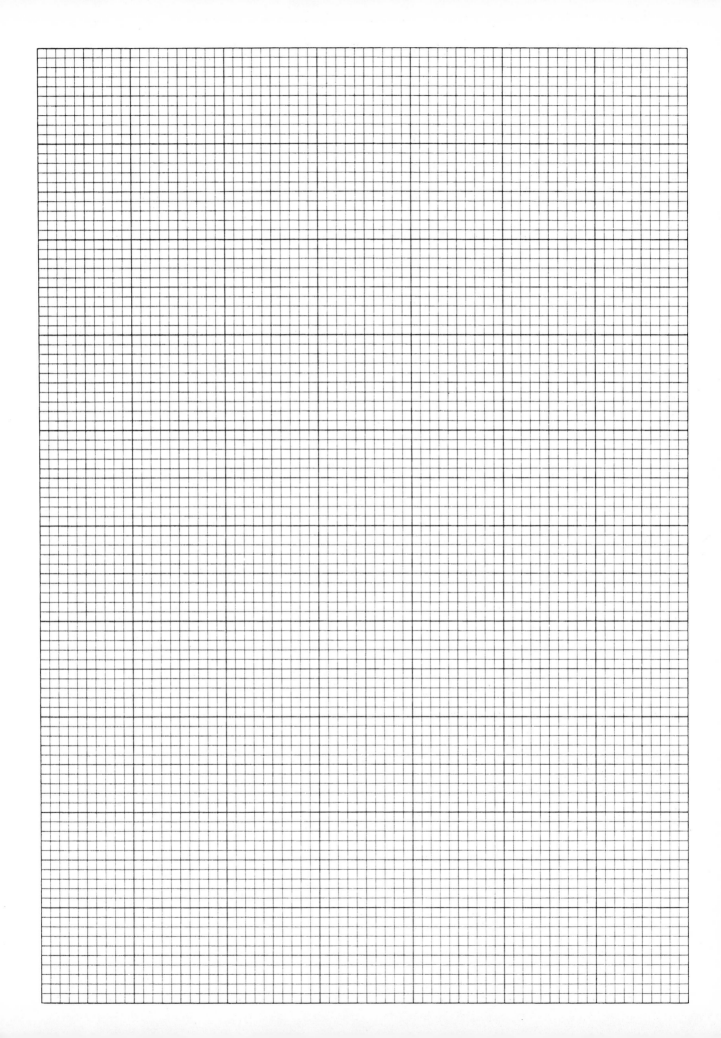

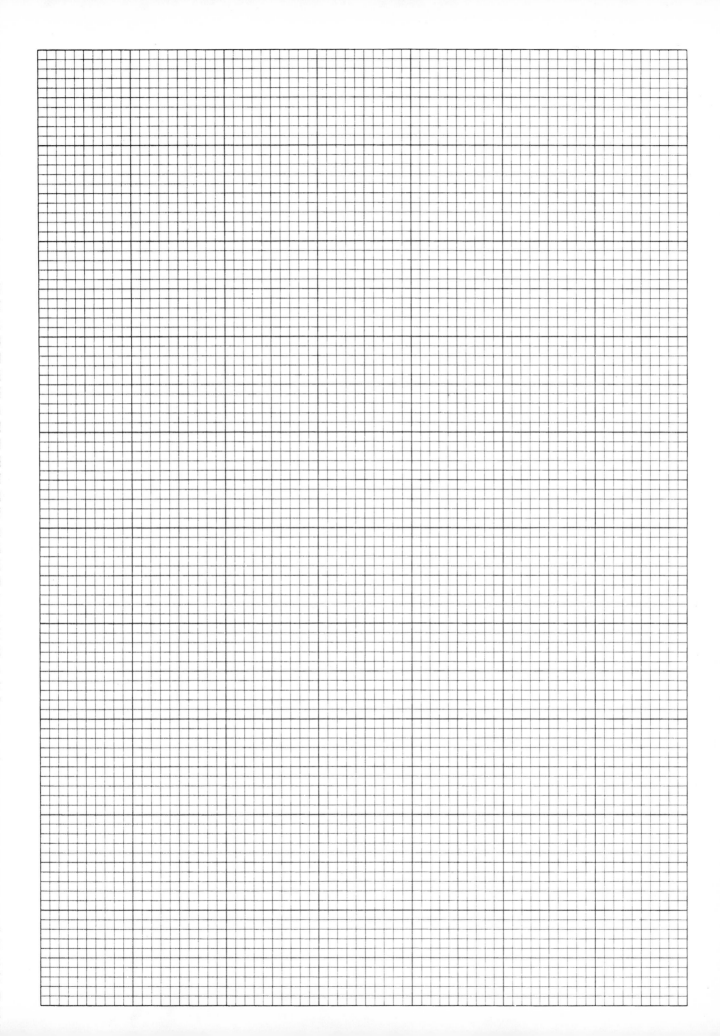

Ohm's Law **24**

The most important principle in direct-current electrical circuit theory is Ohm's law, named after its discoverer, Georg Simon Ohm (1787–1854). This principle states that the current flowing (*i.e.*, the rate at which electrical charge passes) through a conductor is proportional to the voltage, or potential difference, between that conductor's ends. It is the object of this experiment to study Ohm's law and the law relating to series and parallel direct-current circuits.

THEORY

Ohm's law states that the voltage across a conductor is proportional to the current through it, the proportionality constant being the *resistance* of the conductor. This relation may be expressed mathematically as

$$V = RI \text{ or } I = \frac{V}{R}$$

where V is the voltage across the conductor in volts and I is the current in amperes. Notice that an ampere, the unit of current, is a coulomb per second, that a volt is a joule per coulomb, and that the units of resistance R must then be joule-seconds per coulomb², or, in more fundamental units, newton-meter-seconds per coulomb². Both for convenience and in recognition of great scientists, a complex unit like this is often named after the person concerned with its development, and in the present case the unit of resistance is appropriately called the ohm. Thus R is in ohms when V is in volts and I is in amperes.

A portion of a direct-current (D.C.) circuit that contains no voltage sources but only an arrangement of various resistors with wires coming out at two points for connection to the rest of the circuit is called a *two-terminal passive network*. The voltage between the ends (terminals) of such a network is proportional to the current through it, so that the entire network may be replaced by a single resistor whose value is that of the proportionality constant. Two very usual resistor combinations of this sort are the *series* connection shown for the two resistors in Fig. 45 and the *parallel* connection shown in Fig. 46.

When resistors are connected in series the current is the same in each, because the current in any one must all go on to the next one. The total voltage across the network, however, is the sum of the voltage drops across each resistor. It is then easy to show that the network may be replaced by a single resistor whose resistance is the sum of the individual resistances. Thus

$$\begin{aligned} I &= I_1 = I_2 \\ V &= V_1 + V_2 \\ R &= R_1 + R_2 \end{aligned}$$

For the parallel connection, the total current is clearly the sum of the currents through the individual resistors. Moreover, the voltage across the network as a whole is also that across each resistor. In this case, using Ohm's law in the form $I = V/R$, we can show that the parallel network may be replaced by a single resistor the recip-

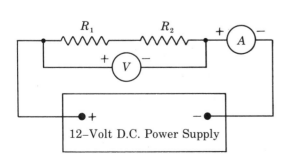

Figure 45   Resistances in Series

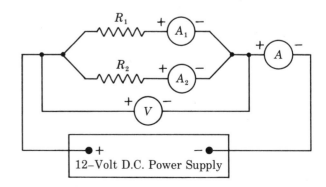

Figure 46   Resistances in Parallel

155

rocal of whose resistance is the sum of the reciprocals of the individual resistances. The pertinent relations are

$$I = I_1 + I_2$$
$$V = V_1 = V_2$$
$$\frac{1}{R} = \frac{1}{R_1} + \frac{1}{R_2}$$

In all these equations, I is the total current through the network, I_1 is the current through resistance R_1, and I_2 is the current through resistance R_2. V is the voltage across the complete network, while V_1 is the voltage across R_1, and V_2 is the voltage across R_2. R is the value of the *equivalent* resistor, which can replace the complete network without altering the relation between the network current I and its voltage drop V.

APPARATUS

1. Tubular rheostat (about 100 ohms)
2. Tubular rheostat (about 200 ohms)
3. 12-volt D.C. power supply or storage battery
4. D.C. milliammeter (0–500 milliamperes)
5. D.C. voltmeter (0–15 volts)

PROCEDURE

CAUTION: Do not plug the power supply into the supply line or turn it on until your circuit has been approved by the instructor. The milliammeter can be instantly and permanently ruined by an improper connection. Any changes in the circuit are to be made only after the power supply has been turned off. Have your circuit approved by the instructor every time you make any changes. If a storage battery is used, do not make the connections to the battery terminals until the instructor has checked your circuit. Disconnect the wires from these terminals whenever the circuit is to be changed and do not reconnect them without the instructor's approval.

1. Connect the two rheostats in series as shown in Fig. 45. Use the 100-ohm rheostat as R_1 and set it at about half resistance. Use the 200-ohm rheostat as R_2 and set it at maximum. Connect the milliammeter as shown in Fig. 45, but connect the voltmeter across R_1 alone rather than across the two rheostats in series. Have your circuit approved by the instructor.

2. Plug in and turn on the power supply, or connect the appropriate leads to the battery terminals if you are using a storage battery. Record the current through the circuit and the voltage across R_1.

3. Increase the current through the circuit in three approximately equal steps by reducing the resistance of R_2, the final setting of this rheostat being zero. Record the current through and the voltage across R_1 for each setting of R_2. Do not alter the setting of R_1 during this procedure.

4. Turn off the power supply or disconnect the battery. Reset R_2 to a resistance near its midrange value and reconnect the voltmeter across the two ends of this rheostat without disturbing the rest of the circuit. Turn on the power supply or reconnect the battery leads and record the readings of the voltmeter and the milliammeter.

5. With the power off, reconnect the voltmeter across the two rheostats in series as shown in Fig. 45. Apply power to your circuit and record both meter readings.

6. Connect the two rheostats in parallel as shown in Fig. 46, with the milliammeter at the point marked A and the voltmeter connected as shown. Do not change the settings of the rheostats.

7. Before applying power, have the circuit approved by the instructor. Record the total current through the circuit and the voltage drop across the parallel resistance combination.

8. Switch off the power supply or disconnect the storage battery. Connect the milliammeter in the circuit at point A_1 without disturbing the rest of the circuit. Apply power and record the current through R_1 and the voltmeter reading.

9. Repeat Procedure 8 for R_2, connecting the milliammeter at A_2 instead of A_1.

DATA

Current, milliamperes				
Voltage, volts				

Resistance value from graph   _____

	Current (Amperes)			Voltage (Volts)			Resistance (Ohms)		
	I	I_1	I_2	V	V_1	V_2	R	R_1	R_2
Series Connection									
Parallel Connection									

CALCULATIONS

1. Plot a curve using the values of current obtained in Procedures 2 and 3 as abscissas and the corresponding values of voltage as ordinates. Remember that the origin is a fifth point on this curve.

2. From the slope of your graph, find the resistance at which R_1 was set in Procedure 1. Remember that your currents are in milliamperes.

3. Compute the value of R_2 and also the value of the total resistance in the series combination by the application of Ohm's law, that is, from the measured values of the current and the voltage drops in the data of Procedures 4 and 5. The value of R_1 has already been found in Calculation 2.

4. Calculate the values of the two resistances of the parallel connection and also of the total resistance by the application of Ohm's law, that is, from the measured values of the currents and the voltage drop in the data of Procedures 7–9.

5. Compute the value of the total resistance of the parallel combination by the application of the equation for parallel circuits, using the values of the resistances obtained in Calculation 3.

QUESTIONS

1. Explain how your graph demonstrates Ohm's law.

2. Show how your results in Calculations 3–5 check the formulas for the current, voltage, and resistance in series and parallel circuits.

3. In Figs. 45 and 46, voltmeter V is shown connected so as not to include ammeter A in the network. What error in finding the network resistance from the readings of V and A results from this connection? What error would be made if V was connected to the other side of A? How can these errors be minimized?

4. The resistance included between two points in a circuit is 10 ohms. How much resistance must be placed in parallel to make the total resistance 4 ohms?

5. A lamp whose resistance is 60 ohms and one whose resistance is 40 ohms are connected in parallel to a 120-volt line. (a) What is the resistance of the combination? (b) What is the current in each lamp? (c) What would be the current if they were connected in series? (d) What then would be the voltage drop across each lamp?

6. A lamp is connected to a 120-volt line. When the switch is first turned on and the filament is cold, the current through the lamp is 8 amperes, but this drops quickly to 0.5 ampere when the filament is hot. What is the percent increase relative to the cold resistance due to the change in temperature?

7. Prove that the resistance of n identical resistors in series is n times the resistance of one resistor and that the resistance of n identical resistors in parallel is $1/n$ times the resistance of one resistor.

8. Show that in a series combination of resistors the total resistance can never be less than that of the largest resistor whereas in a parallel combination the total resistance can never be more than that of the smallest.

9. The conductance G of a resistor is defined as the reciprocal $1/R$ of its resistance. Express Ohm's law and the formulas for series and parallel resistors in terms of conductances instead of resistances. Find out what the unit of conductance is commonly called and write it below.

10. You are given three identical resistors, each of 100 ohms resistance. Show how by using them singly or in series, parallel, or series–parallel combinations you can obtain resistances of $33\frac{1}{3}$, 50, $66\frac{2}{3}$, 100, 150, 200, or 300 ohms.

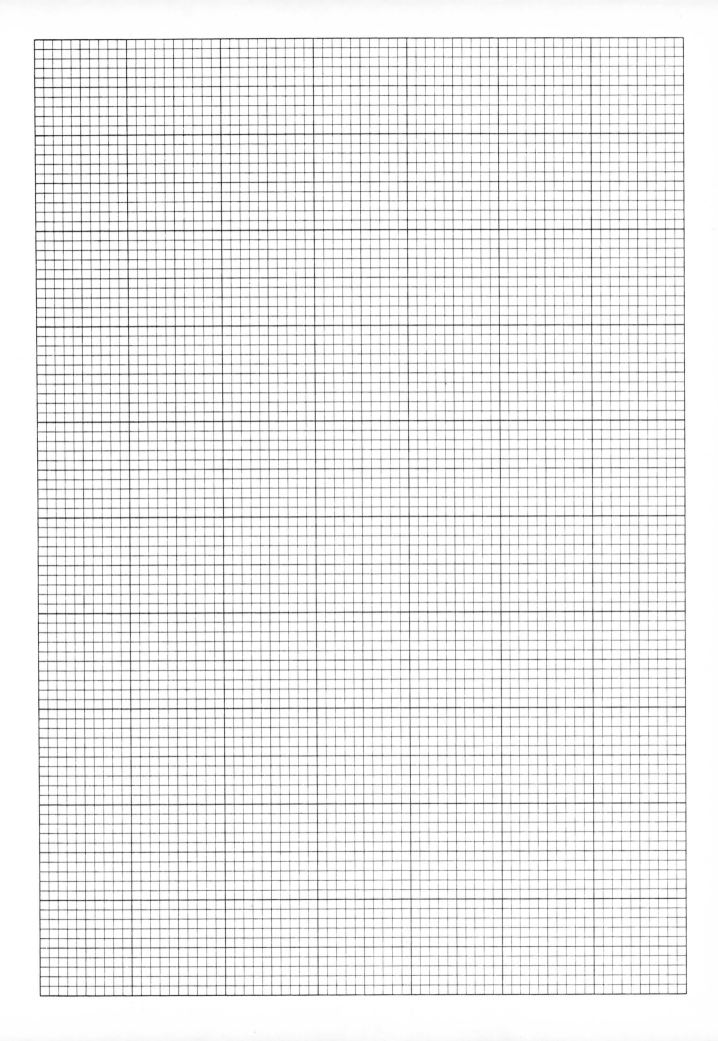

The Wheatstone Bridge and 25
Specific Resistance

The most accurate method of measuring resistances of widely different values is by means of the Wheatstone bridge. The purpose of this experiment is to learn to use the slide-wire form of the Wheatstone bridge and to measure the resistance of several metallic conductors with it. Experiment shows that if the temperature of a substance is kept constant, then its resistance depends on its dimensions. Thus the resistance of a wire varies directly with its length and inversely with its cross-sectional area. For a piece of wire or bar of unit length and unit cross-sectional area, the resistance is called the specific resistance or the resistivity of the material of which the wire is made. A further purpose of this experiment is therefore to determine the specific resistance of the metals making up the sample wires used.

THEORY

A Wheatstone bridge is a circuit consisting of four resistors arranged as shown in Fig. 47. It is used for finding the value of an unknown resistance by comparing it with a known one. Three known resistances are connected with the unknown resistance, a galvanometer, a dry cell, and a key, as shown in Fig. 47.

For a condition of balance, no current flows through the galvanometer. Hence the current through R_1 is the same as the current through R_2, and the current through R_3 is the same as that through R_4. Moreover, because there is no current through G, there must be no voltage across it, hence the potential drop across R_1 is equal to that across R_3. This requires that

$$i_1 R_1 = i_2 R_3$$

Similarly, the potential drop across R_2 must equal that across R_4, so that

$$i_1 R_2 = i_2 R_4$$

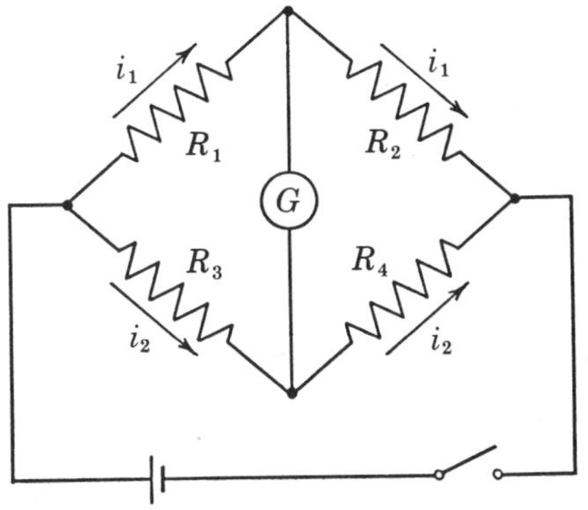

Dividing the first equation by the second yields

$$\frac{R_1}{R_2} = \frac{R_3}{R_4} \qquad (86)$$

Therefore, if three of the resistances are known, the fourth may be calculated from Equation 86.

In the slide-wire form of the bridge, as shown in Fig. 48, the resistances R_3 and R_4 are replaced by the uniform wire AB with a sliding contact key at C. Since the wire is uniform, the resistances of the two portions are proportional to the lengths, hence the ratio R_3/R_4 is equal to the ratio AC/CB. If R_1 is represented by a known resistance R, and R_2 by the unknown resistance X, Equation 86 becomes:

$$\frac{R}{X} = \frac{AC}{CB}$$

From this the value of X may be calculated.

It is found that the resistance of a conductor depends on the material of which it is made, its length, its cross-

Figure 47   The Wheatstone Bridge Circuit Balanced

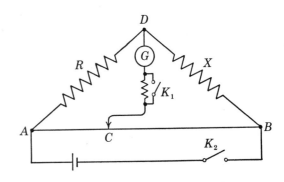

Figure 48   The Slide-Wire Wheatstone Bridge

sectional area, and the temperature. For constant temperature the resistance is given by

$$R = \rho \frac{L}{A} \qquad (87)$$

where R is the resistance in ohms, L is the length in centimeters, and A is the cross-sectional area in square centimeters. The proportionality factor ρ is known as the specific resistance or resistivity of the conductor. It is constant for a given temperature and depends only on the material used. Thus the specific resistance of a substance is the resistance per centimeter of length when the cross section is 1 square centimeter. Its units are ohm-centimeters.

In order to find ρ it is necessary only to determine the resistance of a measured length of wire of known diameter. The specific resistance is then calculated from Equation 87.

APPARATUS

1. Slide-wire form of Wheatstone bridge
2. Dry cell
3. Galvanometer
4. Standard decade resistance box (1000 ohms)
5. Single-pole, single-throw switch
6. Key
7. 4700-ohm, 1-watt carbon resistor
8. Board with seven lengths of wire of unknown resistance. The board is about 1 meter long and the wires are mounted between binding posts. The first five lengths are a single piece of copper wire. The sixth is a piece of manganin wire and the seventh a piece of nichrome.
9. Micrometer caliper
10. Meter stick with caliper jaws

PROCEDURE

1. Hook up the slide-wire form of the bridge as shown in Fig. 48. Let X be the resistance of the five lengths of copper wire in series, and let R be the standard resistance box. Before closing any switches, have the circuit approved by the instructor.

2. Measure the resistance of the copper wire with your bridge. To obtain a balance, first set R to 10 ohms and move sliding contact C to the middle of the bridge wire. This contact should be moved only when its edge is not touching the wire. Leave switch K_1 open, hold down key K_2, and then tap the contact key C. If the galvanometer shows a deflection, change the value of R and repeat. Continue this until the least possible galvanometer deflection is observed. Final adjustment is made by shifting the sliding contact until no deflection occurs when key C is tapped. Key K_2 should be held down only when an observation is being made.

3. For the most sensitive adjustment of the bridge, the 4700-ohm protective resistor is short-circuited. When the galvanometer shows no deflection on tapping the contact key C, close switch K_1. Adjust the sliding contact very carefully, moving it no more than a fraction of a millimeter at a time, until the galvanometer again shows no deflection when key C is tapped. Record the setting of the standard resistance R and of the sliding key.

4. Repeat Procedures 2 and 3 with the sixth length of wire (the manganin wire) connected at X in place of the copper wire.

5. Repeat Procedures 2 and 3 with the seventh length of wire (the nichrome wire) substituted at X.

6. Measure the length of each wire by means of the meter stick and caliper jaws, adjusting the jaws so that they measure the distance between the binding posts where the wire is mounted. Remember that the total length of copper wire is the sum of the five measured lengths.

7. Measure the diameter of each type of wire at four different points with the micrometer caliper. For the copper wire, the four points should be spread out along the total length. Record all four readings of the diameter of each wire, estimating to one tenth of the smallest scale division. The readings should be to 0.0001 centimeter.

8. In a similar manner, take and record four independent readings of the zero setting of the micrometer to 0.0001 centimeter. Be sure to note the algebraic sign of the zero correction.

DATA

Wire	R Ohms	AC Cm	CB Cm	X Ohms	Resistivity Ohm–Cm	Resistivity From Table VII, Appendix
Copper						
Manganin						
Nichrome						

Wire	Micrometer Readings					Diameter Cm	Area Cm²	Length Cm
	1	2	3	4	Average			
Copper								
Manganin								
Nichrome								
Zero Reading								

CALCULATIONS

1. Compute the average of the readings for the diameter of each wire, as well as for the zero reading of the micrometer. Calculate and record the diameter of each wire, properly corrected for the zero reading.

2. Compute the cross section of each wire in square centimeters.

3. Calculate the values of the resistances measured.

4. Compute the specific resistance for each metal used.

5. Look up the values of the specific resistances in Appendix Table VII and record them. Why might they differ somewhat from the ones you have just measured?

QUESTIONS

1. Why is the bridge method not adapted to measuring the hot resistance of an electric lamp?

2. Name three sources of error in using a slide-wire bridge.

3. Why is it more accurate to set the sliding contact near the center of the bridge instead of near one end when the balance is obtained?

4. How is the operation of the Wheatstone bridge affected by changes in the cell voltage? Would there be an advantage in using a higher voltage? What limits the voltage that can be used?

5. Assuming that there is an error of 1 mm in locating the position of the contact on the slide wire, calculate the percentage error in determining the value of X when the contact was (a) at 50 cm, (b) at 5 cm. Suppose that $R = 10.0$ ohms in both cases, and $AB = 100.0$ cm.

6. Draw a diagram of the bridge as actually used in one of your measurements.

7. How does the resistance of a wire vary with the resistivity?

8. What would be the effect of a change of temperature on the resistivity?

9. What is the resistance of a copper bus bar 10 meters long and 0.3 × 0.3 cm in cross section?

10. Calculate the resistance of 40 miles of No. 22 B&S gauge copper wire. The diameter of the wire is 0.0644 cm. The value of the specific resistance is given in Table VII, Appendix.

11. The temperature coefficient of resistance for copper is 0.00393 per degree Celsius. Assuming room temperature to be 22°C, compute the specific resistance of copper at 0°C from your measurements.

12. Assume that there was an error of 0.001 cm in measuring the diameter of the copper wire used in the experiment. Calculate the error that this would introduce into your value of the specific resistance.

Meter Sensitivity: **26**
Voltmeter and Ammeter

It is extremely important to understand the operation of electrical measuring instruments because they are constantly used in making electrical measurements. A galvanometer is an instrument for measuring very small electric currents. It forms the basis of all non-digital electrical measuring instruments, hence an understanding of its working principle and the type of measurements that can be made with it is essential for quantitative electrical work. In the present experiment the resistance and sensitivity of a galvanometer will be measured, and the instrument will be converted into a voltmeter that gives full-scale deflection for 3 volts. In addition, it will be converted into an ammeter giving full-scale deflection for 1 ampere.

THEORY

A galvanometer is an instrument for measuring small currents. It essentially consists of a coil of fine wire mounted so that it can rotate about a diameter in the field of a permanent magnet. When a current flows through the coil, magnetic forces are produced that exert a torque on it. It therefore rotates until equilibrium is established between the torque due to the field and that exerted by a restoring spring. The angle of rotation and hence the deflection of the attached indicating needle is directly proportional to the current flowing through the coil, provided the magnetic field is arranged so that the torque it produces depends only on coil current and does not vary as the coil rotates through its full-scale range.

The sensitivity of a galvanometer is usually expressed as the current that must be passed through the instrument to produce a deflection of one scale division. The so-called full-scale current sensitivity, which is the current necessary for full-scale deflection, is also often used. If the full-scale sensitivity is known, the sensitivity in amperes per division can be found by dividing the full-scale sensitivity by the number of divisions. This result is called the galvanometer constant K. Clearly

$$i = Kd$$

where i is the current passing through the galvanometer in amperes and d is the galvanometer deflection in scale divisions. Notice that the more sensitive a galvanometer is, the less current is required to cause a given deflection and the smaller the galvanometer constant K.

Another important characteristic of a galvanometer is

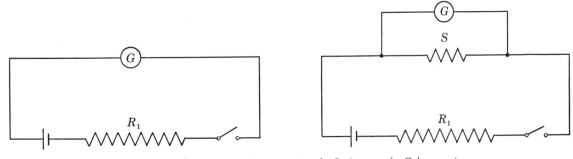

Figures 49 and 50   Circuits for Measuring the Resistance of a Galvanometer

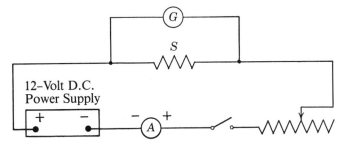

Figure 51   Circuit for Using a Galvanometer as an Ammeter

167

the resistance r of its coil. Both K and r will be measured in the present experiment.

A galvanometer may be converted into an ammeter by placing a suitable shunt in parallel with the coil to permit the flow of a larger current through the complete ammeter than the galvanometer coil will carry alone. This also makes the meter read in amperes with a predetermined full-scale value. Let S be the resistance of such a shunt (see Fig. 50) and r the resistance of the galvanometer. Now let I_g be the current through the galvanometer, I_s the current through the shunt, and I the total current through the complete ammeter. Then, since r and S are in parallel, the current through the galvanometer is given by

$$I_g = I\frac{S}{r + S}$$

If the ammeter is to display a deflection of n divisions when I amperes pass through it, then I_g must equal Kn, and we get

$$I\frac{S}{r + S} = Kn \tag{88}$$

If I is the full-scale current reading desired and n the total number of divisions on the meter scale, Equation 88 can be used to calculate the necessary shunt resistance S. In most cases, where a sensitive galvanometer (small K) is being used to measure a large current I, S will be much smaller than r and can thus be neglected in the denominator of this equation. S is then given to a very good approximation by

$$S \approx \frac{rKn}{I} \tag{89}$$

A galvanometer may also be converted into a voltmeter by placing a large resistor (called a multiplier) in series with the coil. If R is the multiplier resistance, then

$$I_g = \frac{V}{R + r} \tag{90}$$

where V is the voltage impressed across the complete voltmeter (galvanometer and multiplier) terminals. Note that the voltmeter draws current I_g from the source whose voltage is being measured and that because the galvanometer coil and the multiplier resistor are in series, I_g flows through both. Now if V is the desired full-scale voltage reading, the corresponding I_g must equal Kn, and Equation 90 becomes

$$Kn = \frac{V}{R + r} \tag{91}$$

from which the necessary multiplier resistance can be calculated. In most cases R will be much greater than r, and so Equation 91 may be approximated by

$$R \approx \frac{V}{Kn} \tag{92}$$

This simple formula is just Ohm's law, saying that the resistance R is given by the voltage V across it divided by the corresponding current Kn through it, the small additional series resistance r having been neglected.

Much of the above theory requires a knowledge of the galvanometer constant K and coil resistance r, and in this experiment a special method will be used to measure these parameters. The problem is to measure them without passing an excessive current through the galvanometer coil. This is done by connecting a known e.m.f. V in series with a large resistance R_1 and the coil as shown in Fig. 49. The value of R_1 is adjusted so that an exact full-scale deflection is obtained. Then by Ohm's law

$$I_g = Kn = \frac{V}{R_1 + r} \approx \frac{V}{R_1} \tag{93}$$

so that the galvanometer constant is given by

$$K \approx \frac{V}{nR_1} \tag{94}$$

A shunt resistance S is now connected in parallel with r (see Fig. 50) and set so that the new deflection is exactly one-half scale. Under these conditions

$$I_g = \frac{1}{2}Kn = \frac{SV}{R_1(r + S)} \tag{95}$$

The insertion of the shunt does not alter the fact that $Kn \approx V/R_1$, and substitution of this relation from Equation 93 in Equation 95 leads to the conclusion that $S = r$. This result is not surprising if we note that the combination of R_1 and the e.m.f. source constitutes what is called a "current source", a source of a fixed current V/R_1. The current is fixed only to the approximation that resistances such as r or the parallel combination of r and S inserted in series with R_1 are so much smaller than R_1 that they make a negligible change in the total circuit resistance. This approximation is usually very good, however.

APPARATUS

1. Portable galvanometer
2. D.C. voltmeter (0–3 volts)
3. D.C. ammeter (0–1 ampere)
4. Dry cell
5. Decade resistance box (10,000 ohms)
6. Decade resistance box (1000 ohms)

7. Single-pole, single-throw switch
8. 12-volt D.C. power supply or storage battery

9. 100-ohm, 50-watt rheostat
10. Roll of No. 28 copper wire

PROCEDURE AND CALCULATIONS

1. Measure the voltage of the dry cell with the voltmeter and record it.

2. Connect the dry cell in series with the 10,000-ohm resistance box, the galvanometer, and the switch as shown in Fig. 49. The resistance box, which serves as R_1, should be set at 3000 ohms.

3. Before closing the switch, have the circuit approved by the instructor. Now close the switch and note the galvanometer deflection. Carefully adjust R_1 until the galvanometer reads exactly full scale. Do not allow the galvanometer deflection to exceed the full-scale limit. Record the value obtained for R_1.

4. Use Equation 94 to find the galvanometer constant K.

5. Open the switch. Connect the 1000-ohm resistance box in parallel with the galvanometer as shown in Fig. 50, making S equal to 300 ohms and leaving R_1 connected and set as before. Close the switch and adjust S until the galvanometer reads exactly half scale. Record this value of S as your galvanometer resistance r.

6. Calculate the series resistance R (the multiplier) that will convert the galvanometer into a voltmeter giving full-scale deflection for 3 volts.

7. Connect the galvanometer in series with the 10,000-ohm decade resistance box and set the box to your calculated value of R. This combination is now a voltmeter. Use it to measure the voltage of the dry cell, making sure that your series multiplier resistance R is included in the circuit. Record this voltage.

8. Calculate the shunt resistance S that will convert the galvanometer into an ammeter requiring 1 ampere for full-scale deflection. Compute the length of No. 28 copper wire needed for this resistance. Cut a length 1 inch greater than this from the supply roll and clean $\frac{1}{2}$ inch at each end for connection to the galvanometer terminals. Connect your shunt in parallel with the galvanometer by squeezing the bared $\frac{1}{2}$-inch lengths under the terminal screws. Make sure that the full $\frac{1}{2}$-inch is used at each terminal so that the effective length of the shunt is that computed above. Also check that the connections are tight so that no extraneous resistance is introduced. This combination is now an ammeter.

9. Connect your ammeter combination in series with the regular ammeter, the rheostat set with all of its resistance in the circuit, the switch, and the power supply or storage battery, as shown in Fig. 51. When making connections to your meter, be careful not to disturb the shunt connections made in Procedure 8. Before closing the switch, have your circuit approved by the instructor.

10. Turn on the power supply if you are using one and close the switch. Adjust the rheostat until the standard ammeter reads 0.2 ampere, and record the reading of your ammeter. Now adjust the rheostat to make the standard ammeter read 0.4, 0.6, 0.8, and 1.0 ampere in turn, recording your ammeter's reading each time. Should your ammeter read off scale (due to your having made S a little large) when the standard ammeter indicates 1 ampere, try to estimate the reading, or, if necessary, note that the galvanometer needle is against the stop at the high end.

DATA

● **Galvanometer Characteristics**

Dry cell voltage V _____   Number of scale divisions n _____

Value of R_1 for full-scale deflection _____   Galvanometer constant K _____

Galvanometer resistance r _____

● **Voltmeter**

Resistance R for full-scale deflection of 3 volts _____   Reading of this voltmeter for the voltage of the dry cell _____

● **Ammeter**

Shunt resistance S for full-scale deflection of 1 ampere _____   Length of No. 28 copper wire required _____

Current in Standard Ammeter	0.2 Amp	0.4 Amp	0.6 Amp	0.8 Amp	1.0 Amp
Reading of Galvanometer Used as Ammeter					

QUESTIONS

1. Does your data justify neglecting r in comparison with R_1 in Equation 93?

2. (a) Derive Equation 95. (b) Show that when $Kn = V/R_1$ is substituted in this equation, the conclusion $S = r$ follows.

3. The sensitivity of a voltmeter is often expressed in "ohms per volt," which means the resistance of the complete meter (galvanometer and multiplier in series) divided by the full-scale voltage reading. Derive a relation between sensitivity in ohms per volt and the galvanometer constant K, and find the ohms-per-volt sensitivity of the meter used in this experiment.

4. Calculate the series resistance R that will convert your galvanometer into a voltmeter reading 150 volts for full-scale deflection.

5. Calculate the shunt resistance S that will convert your galvanometer into an ammeter reading 0.3 ampere for full-scale deflection.

6. How should an ammeter be connected in a circuit?

7. How should a voltmeter be connected in a circuit?

8. A 1.5-volt dry cell furnishes current to two resistances of 2 ohms and 16,000 ohms in series. Across the 2-ohm resistance is connected a galvanometer whose resistance is 200 ohms. The deflection produced is 10 scale divisions. Calculate the current through the galvanometer and the current sensitivity.

9. A certain 3-volt voltmeter requires a current of 10 milliamperes to produce a full-scale deflection. How may it be converted into a voltmeter with a range of 150 volts?

10. Find the sensitivity of the voltmeter of Question 9 in ohms per volt.

Joule's Law **27**

Whenever an electric current flows through a conductor, a certain amount of electrical energy is transformed into heat energy, just as mechanical energy may be transformed into heat energy by friction. The object of this experiment is to measure the mechanical equivalent of heat by the electrical method. The electrical energy used may be calculated by measuring the current, the voltage, and the time; while the number of calories liberated in the process is measured by means of a calorimeter.

THEORY

Energy exists in many forms and is measured in many different units. The purpose of this experiment is to find the ratio between the mechanical unit and the heat unit of energy by using an electric current as a source of heat and equating the energy supplied by the current, expressed in joules, to the same energy expressed in calories.

Whenever a current flows through a conductor, a certain amount of heat is developed. The work done by the electric current is given by

$$W = VIt$$

where W is the work expressed in joules; V is the potential difference between the terminals of the conductor in volts; I is the current in amperes; and t is the time that the current flows in seconds. But, by Ohm's law

$$V = IR$$

where R is the resistance of the conductor, so that the preceding relation becomes

$$W = I^2Rt$$

This relation is known as Joule's law of heating. Expressed in words, the law states that when an electric current flows through a pure resistance, the quantity of heat produced in the conductor is proportional to the time of current flow, to the resistance, and to the square of the current. Note that W is the electrical energy dissipated in the conductor expressed in joules. This energy appears as heat, which can be measured in calories by calorimetric methods.

From the principle of conservation of energy, whenever work is done in developing heat, the quantity of heat produced is always proportional to the work done. Thus

$$W = JH$$

where W is the work expressed in joules; H is the heat energy produced, in calories; and J is the mechanical equivalent of heat, which is the number of joules required to produce one calorie.

If the heat energy is used to raise the temperature of the water contained in a calorimeter, then

$$H = (m + m_1c_1 + m_2c_2)(t_2 - t_1)$$

where H is the heat energy produced in calories, m is the mass of water in grams, m_1 is the mass and c_1 is the specific heat of the calorimeter, m_2 is the mass and c_2 is the specific heat of the immersion heater, t_1 is the initial temperature of the water, and t_2 is the final temperature.

If the heat energy developed in a conductor is to be expressed in calories, then the mathematical statement of Joule's law becomes

$$H = \frac{W}{J} = \frac{I^2Rt}{J} = \frac{I^2Rt}{4.186}$$

where H is the heat developed in the conductor.

APPARATUS

1. Immersion heater for 12-volt operation
2. Calorimeter
3. Thermometer (0–50°C)
4. Stop watch or stop clock
5. D.C. voltmeter (0–15 volts)
6. D.C. ammeter (0–5 amperes)
7. Equal-arm balance and weights
8. 5-ohm rheostat
9. 12-volt D.C. power supply or storage battery

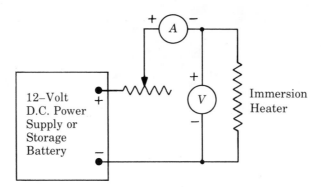

Figure 52   Circuit for Electrical Measurement of the Mechanical Equivalent of Heat

PROCEDURE

1. Weigh the immersion heater. Then connect it along with the ammeter, voltmeter, and control rheostat as shown in Fig. 52. Have the circuit approved by the instructor. If you are using a power supply, you may connect it in the circuit, but make sure its power switch is off and/or it is not plugged into the line. If you are using a storage battery, leave the battery leads unconnected.

2. Weigh the empty calorimeter. Fill it about two thirds full of cold water (as cold as it will run from the tap). Replace the calorimeter in its container and immerse the heater.

3. Before connecting power to your circuit, make sure the heater is completely immersed in the water and that the rheostat is set for maximum resistance. Plug in the power supply and turn it on, or make the appropriate connections to the storage battery. Adjust the current to the value suggested by your instructor by means of the control rheostat. *Note:* If your power supply has a voltage control, you may find it provides a smoother and more precise way of setting the current. In this case set the rheostat to about 1 ohm and leave it there. When the current has been properly adjusted, turn off the power switch or disconnect the storage battery.

4. Stir the water and measure its temperature, estimating to the nearest tenth of a degree.

5. Turn on the power supply or connect the storage battery. At the same instant, start the stop watch. Record the voltmeter and ammeter readings. Keep the current constant using either the rheostat or the power supply control (if provided). Record the voltage every minute, and take the average of these readings as the effective voltage across the heater. Stir the water from time to time with either the heater or the thermometer until its temperature is about 10°C above room temperature.

6. Shut off the power source and stop the stop watch at the same instant. Record the time. Continue stirring the water until the maximum temperature is reached. Record this temperature to the nearest tenth of a degree.

7. Weigh the calorimeter with the water in it, but without the immersion heater or thermometer.

DATA

Temperature of the room	_____	Average voltage	_____
Weight of the immersion heater	_____	Time	_____
Specific heat of the immersion heater	_____	Weight of the calorimeter with water	_____
		Weight of the water	_____
Weight of the empty calorimeter	_____	Electrical energy used	_____
Specific heat of the calorimeter	_____	Heat energy developed	_____
Current	_____	Calculated value of the mechanical equivalent of heat	_____
Initial temperature of the water	_____	Value of the mechanical equiv- alent of heat from Table I, Appendix	_____
Final temperature of the water	_____		
Temperature rise	_____	Percent error	_____

CALCULATIONS

1. Calculate the electrical energy used in joules and the heat energy developed in calories.

2. Compute the mechanical equivalent of heat. Compare your value with the accepted value by calculating the percent error.

QUESTIONS

1. Discuss the principal sources of error in this experiment and justify the difference between your value of the mechanical equivalent of heat and the correct value.

2. Using your data, calculate how long it would take to heat the water from 20°C to 100°C by sending a current of 2 amperes through the same coil immersed in the water.

3. In Question 2 how long would it take if a current of 4 amperes were used?

4. Give three examples of desirable heating effects of an electric current and three examples of undesirable ones.

5. State the factors on which the heating effect of an electric current depends and how it varies with each.

6. What must be the resistance of a coil of wire in order that a current of 2 amperes flowing through the coil may produce 1000 calories of heat per minute?

7. What is the purpose of starting with the temperature of the water lower than room temperature and ending about the same amount above room temperature?

8. How long after it is turned on will a 100-watt electric heater take to bring a quart of water to a boil from room temperature (20°C)? Assume normal sea-level atmospheric pressure.

Electrolysis 28

An important effect of an electric current is the chemical decomposition of a conducting solution through which such a current is passed. This decomposition results in the depositing or ''plating out'' of solid or gaseous elements at the electrodes. The object of the present experiment is to study the chemical action of an electric current and to determine the electrochemical equivalent of hydrogen and oxygen.

THEORY

Certain liquids, called electrolytes, have the property of conducting an electric current. The most common of these are water solutions of acids, bases, or salts. When an electric current passes through an electrolyte, chemical decomposition takes place accompanied by the liberation of a negative valent element at the positive electrode (anode) and a positive valent element at the negative electrode (cathode). This process is called electrolysis. The laws governing electrolysis were enunciated by Michael Faraday (1791–1867) and may be stated as follows:

1. The mass of a substance liberated at an electrode of an electrolytic cell is proportional to the quantity of electricity passed through the cell. This quantity in coulombs is the product of the current in amperes and the time in seconds during which the current flows.

2. When the same quantity of electricity is passed through different electrolytic cells, the masses of the substances liberated are proportional to their chemical equivalents. The chemical equivalent of a substance is defined as its atomic weight divided by its valence.

It is easy to understand the theory behind these laws. For an atom to be liberated at either electrode it must receive a number of electronic charges (positive or negative depending on whether the action is taking place at the anode or the cathode) equal to its valence. Thus the charge necessary to liberate one atom is le coulombs, where e is the electronic charge in coulombs and l the valence. Because there are Avogadro's number N_0 of atoms in a mole, the amount of electricity needed to liberate one mole of atoms of the substance is $N_0 le$. But a mole is a gram atomic weight, that is, an amount of the substance whose mass in grams is numerically equal to its atomic mass number A. Therefore, $N_0 le$ coulombs will ''plate out'' A grams of the substance and $N_0 le/A$ coulombs will plate out 1 gram. Because It is the number of coulombs passed through the electrolyte, the number of grams of material liberated will be

$$M = \frac{A}{N_0 le} It \qquad (96)$$

where M is the mass in grams, I is the current in amperes, and t is the time in seconds during which the current flowed; the other symbols have already been defined. The quantity A/l should be recognized as the chemical equivalent, and Equation 96 bears out the statement that M is proportional to $(A/l)It$. The proportionality constant $1/N_0 e$ is a universal constant, being composed only of the fundamental constants N_0 and e and thus depending neither on the material being deposited nor on the amount of charge passed through the electrolyte. Its reciprocal $N_0 e$ is the amount of charge required to liberate one chemical equivalent of the substance being plated out. This quantity is called the faraday and has the value 96,500 coulombs. Thus Equation 96 could be written

$$M = \frac{C}{F} It$$

where C is the chemical equivalent in grams and F is the faraday. More often it is written as

$$M = ZIt \qquad (97)$$

where Z represents the entire proportionality constant between the mass liberated and the amount of electricity passed through the cell. Z is called the *electrochemical equivalent* of the substance in question and is expressed in grams per coulomb. It is equal to the substance's *chemical equivalent* divided by the faraday, as can be confirmed by a study of the values given in Appendix Table VIII.

In the decomposition of water by electrolysis, an electric current is passed through a cell containing a dilute solution of sulphuric acid. Hydrogen gas is evolved in one arm of the apparatus and oxygen gas in the other. The volumes of the gases formed when a known current flows for a known period of time are measured and the pressure on each volume of gas is determined. Room temperature is also measured and converted to the absolute, or Kelvin, scale. The number of moles and thus the mass of each gas can then be calculated from the ideal gas law, Equation 68.

Finally, the electrochemical equivalent is obtained by solving Equation 97 for Z.

APPARATUS

1. Electrolytic cell
2. D.C. ammeter (0–1 ampere)
3. 5-ohm rheostat
4. Stop watch or stop clock

5. Meter stick with caliper jaws
6. 12-volt D.C. power supply or storage battery
7. Barometer
8. Single-pole, single-throw switch
9. 0–50°C thermometer

PROCEDURE

1. Connect the electrolytic cell in series with the switch, ammeter, rheostat, and power supply or storage battery as shown in Fig. 53. The electrodes are alike, so that either one may be used as the anode. Have the circuit approved by the instructor before making final connections to the storage battery or plugging in the power supply.

2. Begin with all the resistance of the rheostat in the circuit. Plug in and turn on the power supply or connect the storage battery. Close the switch and adjust the rheostat to produce a current of exactly 0.5 ampere. Open the switch and ask the instructor to let out the gases that have formed in the cell. The apparatus is now ready for operation.

3. Turn on the current and start the stop watch at the same instant. Keep the current at exactly 0.5 ampere by means of the control rheostat. If your power supply has a voltage control, it may provide a better control of the current. In this case leave the rheostat set at about 1 ohm.

4. Keep the current flowing until 50 cm³ of hydrogen have collected. Open the switch and stop the watch at the same instant. Record the exact time to the second.

5. Record the exact volumes of hydrogen and oxygen formed to 0.1 cm³.

6. Measure h_H and h_O (see Fig. 53) with the meter stick and caliper jaws. Record these readings.

7. Record the barometric pressure.

8. Use the 0–50°C thermometer to measure the room temperature. Leave it suspended in the laboratory for several minutes before reading it so that it will have come to equilibrium, and do not touch it during this period.

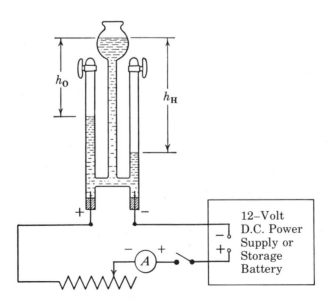

Figure 53   Electrolysis of Water

DATA

	Current	Time	h_H	h_O	h_H	h_O	Barometric Pressure	Room Temp.	Vapor Pressure of Water
Units					cm Hg	cm Hg			
Value									

	Volume	Pressure	Number of Moles	Atomic Mass	Mass	Electro-Chemical Equivalent	Value of Z from Table VIII, Appendix	% Error
Units								
Hydrogen								
Oxygen								

CALCULATIONS

1. Compute the values of h_H and h_O in cm of mercury. The density of mercury is 13.6 grams per cubic centimeter.

2. Compute the pressure on the hydrogen; the value of h_H in cm of mercury has to be added to the barometric pressure and the vapor pressure of water must be subtracted from the total. Similarly, compute the pressure on the oxygen.

3. Calculate the number of moles of hydrogen gas evolved by substituting your known values of pressure, volume, and temperature in Equation 68. Be careful of the units. The product of pressure and volume is an energy, which is why the dimensions of the universal gas constant are energy per mole-degree absolute. The best way to handle the present situation is to convert volume to liters (1000 cm^3 = 1 liter) and pressure to atmospheres (76.0 cm Hg = 1 atmosphere) so that PV will be in liter-atmospheres. Look up R in liter-atmospheres per mole-degree absolute and don't forget to convert the room temperature to the absolute scale for use in Equation 68. In a similar manner, calculate the number of moles of oxygen gas evolved.

4. Calculate the mass of the hydrogen and that of the oxygen. Remember that these gases are diatomic; the molecular mass number is twice the atomic mass number in each case.

5. Calculate the electrochemical equivalent for both hydrogen and oxygen. Compare your values with the known values in Table VIII, Appendix, by computing the percent error.

QUESTIONS

1. Compute the time it would require a current of one ampere to liberate 22.4 liters of hydrogen under standard conditions of temperature and pressure.

2. How many grams of water would be decomposed by 1 coulomb?

3. Is the sulfuric acid used up in the process of the electrolysis of water? Explain.

4. An ammeter reading 0.5 ampere is connected in series with a coulomb-meter that deposits 1.50 grams of silver in 40 minutes. What is the percentage error of the reading of the ammeter?

5. A current flowing through copper sulfate and silver nitrate coulomb-meters connected in series deposits 0.200 gram of copper in 10 minutes. How much silver is deposited?

6. Assume a spoon has a total area of 20 square centimeters. It is to be coated with silver 0.01 mm thick. The current is 0.05 ampere. Compute the mass of silver needed and the time required to deposit this amount of silver.

7. Starting with their respective atomic mass numbers, calculate the densities of hydrogen and oxygen under standard conditions (0°C and 1 atmosphere pressure). Remember that these gases are diatomic. Check your results with the values given in Appendix Table III.

8. What error would have been made in your results if you had neglected to take into account the vapor pressure of water in the two arms of the cell?

9. Another way of finding the masses of hydrogen and oxygen liberated in the present experiment is to use the gas densities given in Appendix Table III. Because these densities are for gases at standard conditions, however, your volumes must be appropriately converted by means of the ideal gas law, which says that PV/T is a constant for any fixed quantity of gas. Find your hydrogen and oxygen masses by this method and see if the results agree with those obtained in Calculations 3 and 4.

10. Can an electrolysis experiment provide you with a way of determining Avogadro's number? Explain.

The Potentiometer **29**

One of the most important instruments used in the electrical laboratory is the potentiometer, which is an instrument for measuring voltages, or potential differences. The potentiometer possesses a great advantage over the voltmeter because, under operating conditions, it draws no current from the source whose voltage is being measured and thus does not disturb the value of that voltage. The potentiometer is capable of very high precision and is easy to handle. One of its most important uses is the calibration of electrical measuring instruments such as voltmeters and ammeters. The purpose of this experiment is to study the slide-wire form of the potentiometer and to measure several voltages with it.

THEORY

The electromotive force (e.m.f.) of a cell is its terminal voltage when no current is flowing through it. Practical voltage sources are never "ideal"; that is, they cannot maintain a fixed potential difference between their terminals regardless of the current drawn from them. The terminal voltage always falls somewhat below the "no-load" or "open circuit" voltage whenever a load, *i.e.*, some circuit drawing current, is connected. The open-circuit voltage of a voltage source can therefore not be measured accurately with an ordinary voltmeter because the voltmeter itself draws some current. This problem may be minimized by using a voltmeter of high sensitivity or even an electronic voltmeter (EVM). In this latter instrument the actual meter movement is driven by a separate power source, and only the very high input resistance (of the order of 10 million ohms in ordinary EVM's) of the associated electronic amplifier is connected across the voltage to be measured. All such devices require calibration, however. The potentiometer measures the voltage of a given source by a null method, so that when a balance is reached and the reading is being taken, no current is drawn from the source. Standard cells are available that produce an accurately known open-circuit voltage. The potentiometer allows precise measurements of other voltages to be made by comparing them with the known standard voltage.

In the potentiometer a continuously variable potential difference is produced by sending a steady current from an appropriate power supply through a uniform bare slide wire. The voltage source to be measured is connected through a galvanometer and a protective resistance between one end of the wire and a sliding contact in such a way that its polarity opposes that of the wire's potential difference. For balance the contact is moved along the wire until no current flows through the galvanometer. The voltage across this portion of the wire is now equal to that of the source being measured. This process is repeated with a standard cell. Because the wire is uniform, the length of the wire spanned is proportional to the potential drop. Thus the open-circuit voltage (sometimes called the *electromotive force* or e.m.f.) of any voltage source can be compared with that of a standard cell.

The circuit used in this experiment is shown in Fig. 54. E_s is the standard cell, E_x is a cell whose e.m.f is to

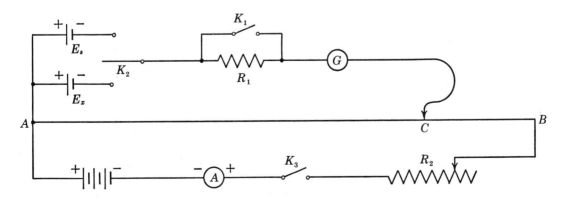

Figure 54 The Potentiometer Circuit

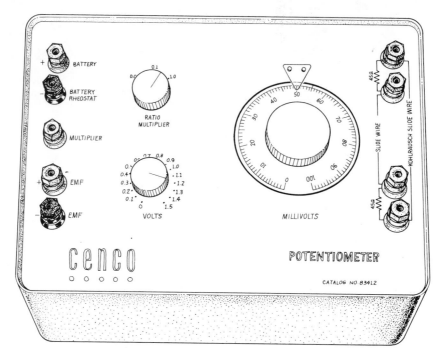

Figure 55 Typical Student Potentiometer (Cenco)

be measured, K_2 is a double-throw switch that can connect either cell in the circuit, G is the galvanometer, R_1 is the protective resistance, and R_2 is a rheostat used to adjust the current in the slide wire. Because the electromotive force of the standard cell is equal to the potential drop in the length of the wire spanned for a condition of balance and the same is true for the unknown cell, the e.m.f.'s are proportional to the lengths of wire spanned. Thus

$$\frac{E_x}{E_s} = \frac{L_x}{L_s}$$

The unknown e.m.f. is then given by

$$E_x = E_s \frac{L_x}{L_s} \qquad (98)$$

where E_x is the unknown e.m.f., E_s is the e.m.f. of the standard cell, L_x is the length of the wire (AC) used for the unknown cell, and L_s is the length of the wire used for the standard cell for a condition of balance.

The scale measuring length along slide wire AB can be made to read voltage directly by proper adjustment of the rheostat R_2. Thus if the total length AB of the slide wire is 1 meter and R_2 is set so that the e.m.f. across this length is 1 volt, then an unknown voltage can be read directly from the measured position of contact C at balance. To establish exactly 1 volt across AB, the standard cell is selected with switch K_2, contact C is set at a scale reading equal to the known standard cell voltage and balance is obtained by adjustment of R_2.

If the potential across AB is set at 1 volt, the potentiometer is limited to measuring voltages equal to or less than 1 volt. Since in this experiment voltages up to 1.5 or 1.6 volts are to be measured, the voltage across AB will be set at 2 volts and unknown voltages read from the scale by taking the scale reading at contact C and doubling it.

Commercial potentiometers are available with slide wires calibrated directly in volts and with various auxiliary resistors designed to give the unit several voltage ranges included in a single instrument assembly. A simple example of such an instrument is shown in Fig. 55.

APPARATUS

1. Slide-wire potentiometer
2. Galvanometer
3. 4700-ohm, 1-watt carbon resistor
4. 100-ohm, 50-watt rheostat
5. 12-volt D.C. power supply or storage battery
6. Standard cell
7. Good dry cell, or dry cell which tests 1.5 volts
8. Old dry cell, or a dry cell which tests about 1.0 volts
9. D.C. ammeter (0–1 ampere)
10. D.C. voltmeter (0–3 volts)
11. Two knife switches, single-pole, single-throw
12. Single-pole, double-throw switch

PROCEDURE

1. Connect the apparatus as shown in Fig. 54 using the good dry cell for E_x. R_1 is the 4700-ohm resistor and R_2 the rheostat. Keep all switches open while connections are being made. Have your circuit approved by the instructor.

2. *Leave switch K_1 open.* Set R_2 for maximum resistance and then close switch K_3. Adjust the current as read on the ammeter to about 0.6 ampere and keep it constant at this value for the next four steps by means of the control rheostat. Read the exact value chosen.

3. Close switch K_2 in the direction to put the standard cell E_s in the circuit and find the setting of the contact point C for a condition of balance. For this initial adjustment, the galvanometer is connected in series with a high protective resistance. Contact C is pressed and the galvanometer deflection observed. This contact is then moved along the wire until there is no galvanometer deflection when the key is pressed. CAUTION: Do not slide contact C along the wire while pressing it into contact. This will cause the knife edge to scrape the wire and thereby damage it.

4. Final adjustments are made with the protective resistance R_1 short-circuited by switch K_1. When the galvanometer shows no deflection on pressing C, close K_1. Again locate C so that the galvanometer shows no deflection. Only a *very small* displacement of the contact point will be needed to do this. Closing K_1 increases the sensitivity of the galvanometer many times and allows point C to be located more accurately. Record the final setting of the contact point and the known value of the standard cell voltage.

5. *Open switch K_1.* Move K_2 to put the good dry cell E_x in the circuit instead of the standard cell. Obtain a balance as described in Procedures 3 and 4 to measure the e.m.f. of the good dry cell. Record the final setting of contact point C.

6. Again *open switch K_1.* Replace the good dry cell with the old dry cell and repeat Procedure 5 to measure the old dry cell's e.m.f. Record the final contact point setting.

7. *Open switch K_1* and return switch K_2 to its original position putting the standard cell in the circuit. Set contact point C to a scale reading exactly equal to one half the known e.m.f. of the standard cell and adjust the rheostat for zero galvanometer deflection. Close K_1 for final adjustments of the rheostat but *be very careful* not to drive the galvanometer off scale when the protective resistance is short-circuited. Record the ammeter reading and make sure it does not change for the remainder of the experiment.

8. *Open switch K_1.* Move K_2 to put the old dry cell back in the circuit. Obtain a balance as described in Procedures 3 and 4. Read contact point C's position on the scale and record it.

9. Again *open switch K_1.* Replace the old dry cell with the good dry cell and repeat Procedure 8. Record the contact point setting so obtained.

10. Measure the terminal voltage of the good dry cell and of the old dry cell with the voltmeter and record the values. Do not measure the voltage of the standard cell with the voltmeter because the current taken by the meter will cause the cell to change so that it will no longer be standard.

DATA

Slide-wire current, Procedure 2 _____   Slide-wire current, Procedure 7 _____

Cell Used	E.M.F. from Equation 98		E.M.F. Direct from Scale		Voltmeter Reading
	AC, cm	E.M.F., Volts	AC, cm	E.M.F., Volts	
Standard Cell					
Old Dry Cell					
Good Dry Cell					

CALCULATIONS

1. Compute the e.m.f. of the good dry cell from the data of Procedures 4 and 5.

2. Compute the e.m.f. of the old dry cell from the data of Procedure 6.

3. Obtain the e.m.f.'s of the good and the old dry cells again by multiplying the distances AC measured in Procedures 8 and 9 by two.

QUESTIONS

1. Compare the e.m.f. of the good dry cell with that of the old dry cell. Compare their voltages as measured with the voltmeter. Explain these results.

2. If you use a meter wire of uniform cross section having a resistance of 5 ohms with a storage battery whose e.m.f. is 6 volts and whose internal resistance is 0.2 ohm, what resistance must be placed in series with the wire in order that the potential drop per millimeter shall be exactly 1 millivolt?

3. The voltage of the power supply or storage battery was not measured or used in the calculations. Why is the value of this voltage not needed?

4. The voltage of the unknown cell as indicated by the voltmeter should be less than that found by the potentiometer. Explain why.

5. A cell whose e.m.f. is 1.52 volts is connected to a voltmeter which records 1.48 volts. If the resistance of the voltmeter is 200 ohms, what is the internal resistance of the cell?

6. Find the resistance of your slide wire from data taken in this experiment.

7. Suggest modifications to the circuit of Fig. 54 that would increase the voltage range of the potentiometer.

8. Eight small flashlight cells in series give an e.m.f. of 12 volts, just as the storage battery in your car does. Could you start your car with the flashlight cells? Explain.

Measurement of Capacitance 30 by the Bridge Method

Electrical charge can be stored in a device called a capacitor in a manner completely analogous to the storing of water in a tank. Many uses for capacitors in providing short but intense pulses of electric current for various purposes are encountered in practice. Practical capacitors exist in many forms, but each basically consists of two conducting sheets or plates separated by an insulator. If a potential difference is applied between the plates, these will acquire equal charges of opposite sign, the value of each charge being proportional to the applied voltage. The constant of proportionality, which is characteristic of the particular capacitor, is called its capacitance. The object of this experiment is to study the bridge method of measuring capacitance and to check the formulas for the capacitance of capacitors connected in series and in parallel.

THEORY

A capacitor basically consists of two metal plates separated by an insulating material called a dielectric. Such an arrangement stores electric charge by reason of the fact that, if a voltage source is connected between the plates, positive charge flows onto one plate and is taken from the other, leaving it negative. This process continues until the field in the dielectric is strong enough to stop the further flow of charging current. At this point, a certain amount of charge (positive in one case and negative in the other) is stored on each plate, and a potential difference equal to the source voltage exists between them. The ratio between this charge and this potential difference is a constant for a given capacitor and is called its capacitance. Thus

$$C = \frac{Q}{V} \qquad (99)$$

where C is the capacitance in farads, Q is the charge in coulombs, and V is the potential difference in volts. Note that the unit of capacitance is actually a coulomb² per newton-meter, but it is called a farad both to honor Michael Faraday and for the sake of convenience. Because of the size of the coulomb, the farad is too large a unit for practical use, hence the microfarad (a millionth of a farad) is the unit commonly encountered.

When a capacitor is connected in an A.C. circuit, the alternate charging and discharging make it appear to pass an A.C. current, and a linear relationship between the A.C. voltage applied and the apparent current passed that is very similar to Ohm's law for resistance is found to hold. The proportionality factor, X_c, between this voltage and current is called the *capacitative reactance* of the capacitor and is measured in ohms as is resistance. However, it is related to the capacitance by

$$X_c = \frac{1}{2\pi fC}$$

where X_c is the reactance of the capacitor in ohms, C is the capacitance in farads as before, and f is the frequency in cycles per second (or hertz) of the applied A.C. Thus the reactance, unlike resistance, depends on frequency and becomes infinite as the frequency goes to zero, an expression of the fact that for D.C. ($f = 0$), a capacitor is in fact an open circuit. However, for A.C. of a particular frequency, a capacitor behaves in many ways like a resistor and can be measured by a bridge circuit similar to the Wheatstone bridge of Experiment 25 except that capacitors replace the resistors on one side of the bridge, an A.C. source is used instead of a battery, and an appropriate detector of A.C. (in this case a telephone receiver) replaces the galvanometer as shown in Fig. 56. Comparing this figure with Fig. 47, we note that if C_1 and C_2 are substituted for R_1 and R_2, then reactances $X_{c1} = 1/(2\pi fC_1)$ and $X_{c2} = 1/(2\pi fC_2)$ appear in place of the respective resistances in Equation 86, which thus becomes

$$\frac{2\pi fC_2}{2\pi fC_1} = \frac{C_2}{C_1} = \frac{R_3}{R_4}$$

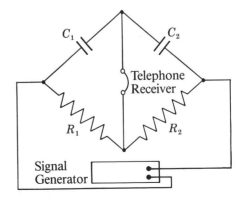

Figure 56 Bridge Method of Comparing Capacitances

189

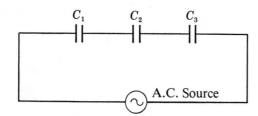

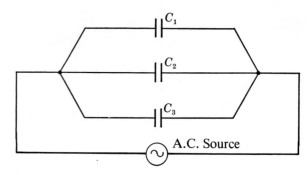

Figure 57 Capacitors in Series

Figure 58 Capacitors in Parallel

This relation, as described in Experiment 25, holds when the balance condition, *i.e.*, a null indication of the A.C. detector that replaces the galvanometer, has been set up. With R_3 and R_4 known, C_2 can be found in terms of C_1, a standard capacitor whose value is known to reasonably high precision. Notice the inverse relationship between the capacitances and the resistances, due to the inverse proportionality of capacitance and its associated reactance.

This inverse proportionality can be used to obtain the formulas for the capacitance of a series or parallel arrangement of capacitors. Suppose three capacitors are connected in series as shown in Fig. 57. In this A.C. circuit they exhibit reactances X_{c1}, X_{c2}, and X_{c3}, and the total reactance of the combination is the sum of these individual reactances. Thus

$$X_c = X_{c1} + X_{c2} + X_{c3}$$

Substitution of the expression for reactance and multiplying through by $2\pi f$ leads to the conclusion that the capacitance C exhibited by the three capacitors in series is given by

$$\frac{1}{C} = \frac{1}{C_1} + \frac{1}{C_2} + \frac{1}{C_3} \tag{100}$$

Similarly, with capacitors in parallel as in Fig. 58, we should find that

$$\frac{1}{X_c} = \frac{1}{X_{c1}} + \frac{1}{X_{c2}} + \frac{1}{X_{c3}}$$

and that there is an equivalent capacitance

$$C = C_1 + C_2 + C_3 \tag{101}$$

Note that these equations are just like those for series and parallel resistors but are reversed with respect to which

equation goes with which combination, again because of the inverse relationship between capacitance and reactance.

The capacitance bridge of Fig. 56 can also be used to measure the so-called *dielectric constant* of an insulating material. The capacitance of a parallel-plate capacitor is given by the expression

$$C = \frac{\epsilon_0 K A}{d} \tag{102}$$

where C is the capacity in farads (multiply by 1 million or 10^6 for microfarads), A is the area of either plate in square meters, d is the plate separation in meters, ϵ_0 is the permittivity of free space (see Appendix Table I), and K is the dielectric constant of the material between the plates. Note that C varies directly with K. If in our bridge we use two parallel-plate capacitors of identical dimensions, one of which (C_1) has air (K very nearly equal to 1) or, even better, a vacuum ($K \equiv 1$) between the plates while the other (C_2) has the space between the plates filled with the insulator under test, K for that material will be equal to C_2/C_1 and will be given directly by the ratio R_3/R_4 at balance.

In this experiment a standard capacitor will be used as C_1 and an unknown capacitance will be inserted at C_2. Resistors R_3 and R_4 will be decade resistance boxes, which will be adjusted to obtain the balance condition. The unknown capacitance can then be calculated from the known value of C_1 and the values of R_3 and R_4 read from the boxes. Two unknown capacitors will be measured in this manner, after which series and parallel combinations of these capacitors will be measured to check the validity of Equations 100 and 101.

APPARATUS

1. Standard capacitor (C_1) in the range 0.1 to 1.0 mfd
2. Two unknown capacitors (C_2) in the above range
3. Decade resistance box (1000 ohms)
4. Decade resistance box (10,000 ohms)
5. Audio signal generator
6. Telephone receiver or high-impedance headphones

PROCEDURE

1. Begin by connecting the telephone receiver directly to the signal generator output terminals. Turn the signal generator on, set the output level for comfortable listening, and tune the generator frequency to the optimum value. This will be the setting for which the sound heard in the receiver is loudest. Telephone receivers have a limited frequency response range and are made to be most sensitive in the range occupied by the human voice. What you have just done is to find the frequency for which your receiver is most sensitive and will thus perform best as a null detector. Remember that you are going to adjust R_3 and R_4 to give no audible sound in the receiver, and the more sensitive the receiver is the more precise that adjustment can be. For ordinary telephone receivers and communication (not high-fidelity) headsets, the optimum frequency is around 800 to 1000 hertz.

2. Wire your capacitance bridge as shown in Fig. 56. Use the 10,000-ohm decade box for R_3 and set it to 200 ohms. Use the 1000-ohm decade box for R_4, the standard capacitor for C_1, and one of the unknown capacitors for C_2. Set the signal generator output at maximum, check that it is still tuned to the optimum frequency found in Procedure 1, and listen for the tone in the telephone receiver. Leave R_3 set at 200 ohms and adjust R_4 for minimum audible signal. Record the value so obtained along with the value of R_3 and that of the standard capacitor (C_1).

3. Repeat Procedure 2 with the other unknown capacitor substituted at C_2. Record R_3, R_4, and C_1 for this case.

4. Repeat Procedure 2 with the two unknown capacitors in series at C_2. Record R_3, R_4, and C_1 as before.

5. Repeat Procedure 2 with the two unknown capacitors in parallel at C_2. Again record R_3, R_4, and C_1.

DATA

Value of R_3 _____   Value of standard capacitor (C_1) _____

Capacitors Measured	R_4	Measured Capacitance (C_2)	Calculated Capacitance	Percent Difference
Unknown No. 1				
Unknown No. 2				
Series Combination				
Parallel Combination				

CALCULATIONS

1. Compute the capacitance of each of the unknown capacitors from the data of Procedures 2 and 3, respectively. *Note:* If ordinary commercial capacitors are used for the unknowns, their values will be marked on them and can be compared with your measured values. Do not expect close agreement, however, because these capacitors have a tolerance of $\pm 10\%$ or more from the marked values.

2. From the data of Procedure 4, find the capacitance of the series combination of the two unknown capacitors.

3. Use Equation 100 and your measured values of the unknown capacitors from Calculation 1 to obtain a calculated value for the capacitance of the series combination. Compare this value with that obtained in Calculation 2 by finding the percent difference.

4. From the data of Procedure 5, find the capacitance of the parallel combination of the unknown capacitors.

5. Use Equation 101 and your measured capacitance values to obtain a calculated value for the capacitance of the parallel combination. Compare this value with that obtained in Calculation 4 by finding the percent difference.

QUESTIONS

1. Do the results you obtained experimentally for the series and parallel connections of capacitors check the laws for such connections within experimental error?

2. What would be the effect on the capacitance of a parallel-plate capacitor if a sheet of glass were introduced between the plates? Assume the dielectric to be air ($K \approx 1$) before the glass was inserted.

3. Assuming that the charge remains the same, what would be the effect on the potential difference between the plates of a parallel-plate capacitor if a sheet of glass were introduced between the plates? Take $K = 1$ before insertion of the glass as in Question 2.

4. Which factor in Equation 102 is varied in the tuning capacitors used in radio sets? Explain how this is done.

5. Three capacitors with capacitances of 2, 3, and 5 microfarads respectively are joined in parallel. The combination is connected in series with a 5-microfarad capacitor. What is the capacitance of the final arrangement?

6. A parallel-plate capacitor is made up of two circular plates separated by a sheet of mica. The plates have a diameter of 20 centimeters, and the mica is 0.1 millimeter thick and has a dielectric constant of 3. Calculate the capacitance so obtained in microfarads.

7. A 4-microfarad and a 6-microfarad capacitor are connected in parallel and charged from a source of 120 volts D.C. They are then disconnected from the source and from each other and are reconnected to each other in parallel but with the leads from the 4-microfarad capacitor interchanged. Find the final voltage across the parallel combination.

8. Derive Equation 101, the formula for the capacitance of a combination of capacitors in parallel, using Equation 99 and the fact that in any parallel combination of circuit elements the voltage across the combination is the same as that across each element.

9. If a number of uncharged capacitors are connected in series and a charging current is sent into the combination, each capacitor must receive the same charge. Use this fact and the basic definition of capacitance given by Equation 99 to derive Equation 100, the formula for the capacitance of capacitors in series.

10. Why did you use the maximum output of the signal generator to power your bridge? Discuss the factors affecting the choice of the voltage used to supply a bridge.

Measurement of Capacitance 31
by the Ballistic Galvanometer

There are two general methods of measuring capacitance. One involves the comparison of the capacitance of the unknown capacitor with that of a standard capacitor, and is similar in principle to the comparison of resistances by the Wheatstone bridge method. The other involves the use of the relation defining capacitance, that is, the ratio between the quantity of charge on a capacitor and the potential difference between the plates (Equation 99). The quantity of charge is determined by the use of a ballistic galvanometer. The object of this experiment is to measure the ballistic constant of a galvanometer, to determine the capacitance of a capacitor from the measurement of the charge with a ballistic galvanometer, and to check the laws of series and parallel connections of capacitors.

THEORY

A capacitor consists of two metal plates separated by an insulating material, which is called a dielectric. If the plates are connected to two points between which there is a difference of potential, the two plates will acquire equal and opposite charges, the magnitude of each charge being directly proportional to the applied voltage. This relation may be expressed in the form

$$Q = CV \qquad (99)$$

where Q is the quantity of electricity in coulombs, V is the potential difference in volts, and C is a constant of proportionality known as the capacitance of the capacitor. Thus the capacitance of a capacitor is the ratio between the quantity of charge on one plate and the potential difference between the plates. In the above expression, C is the capacitance in farads.

The farad is the unit of capacitance. A capacitor has a capacitance of 1 farad if it acquires a charge of 1 coulomb when 1 volt is impressed across it. The farad is too large a unit for practical purposes; hence the microfarad, or millionth of a farad, is used as a practical unit.

When capacitors are connected in series (see Fig. 59), the reciprocal of the capacitance exhibited by the combination is equal to the sum of the reciprocals of the individual capacitances. Thus

$$\frac{1}{C} = \frac{1}{C_1} + \frac{1}{C_2} + \frac{1}{C_3} \qquad (100)$$

where C is now the capacitance of the series combination.

When capacitors are connected in parallel (see Fig. 60), the total capacitance is equal to the sum of the individual ones. Thus

$$C = C_1 + C_2 + C_3 \qquad (101)$$

where C is now the capacitance of the parallel combination.

The capacitance of a parallel-plate capacitor is given by the expression

$$C = \frac{\epsilon_0 K A}{d} \qquad (102)$$

where C is in farads, A is the area of one plate in square meters, d is the distance between the plates in meters, ϵ_0 is the permittivity of free space (see Appendix Table I),

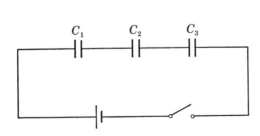

Figure 59 Capacitors in Series

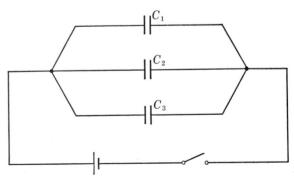

Figure 60 Capacitors in Parallel

195

and K is the dielectric constant of the material between the plates. To get the capacitance in microfarads, the above result has to be multiplied by 1 million (10^6).

When a capacitor is connected to a source of voltage, a charge is transferred from one plate to the other and the potential difference between the plates rises from the initial value zero to the final value of the impressed voltage. In transferring this charge, work has to be done. The total work done in this charging process is given by the expression

$$W = \tfrac{1}{2}QV \qquad (103)$$

where W is the work done in joules, Q is the charge on the capacitor in coulombs, and V is the potential difference between the plates in volts. This work is stored as electrical energy in the capacitor. In view of Equation 99, this energy can also be expressed as

$$W = \frac{1}{2}CV^2 = \frac{1}{2}\frac{Q^2}{C} \qquad (104)$$

where W, V, and Q are the same as before and C is the capacitance of the capacitor in farads.

In this experiment the capacitance of a capacitor is determined by measuring the charge on it and the potential difference between its plates. The capacitor is charged to a known voltage and then discharged through a ballistic galvanometer to measure the quantity of charge on the plates.

A ballistic galvanometer is used for the measurement of quantity of electricity. This type of galvanometer is so constructed that its suspended coil has a relatively large moment of inertia and a long period of vibration. As a result, when a capacitor is discharged through it, the pulse of current produced rises through a maximum and decreases to zero in a time that is short compared to the period of oscillation of the suspended coil. Thus all the charge passes through the coil before it has been able to move appreciably, and the coil receives an impulsive torque. Instead of producing a steady deflection, this type of torque produces a throw; that is, the coil swings through an angle and then comes back to rest. The reading of the galvanometer, therefore, is the maximum deflection obtained and not a steady deflection as in the ordinary type of galvanometer.

It can be shown that this maximum deflection of the galvanometer is proportional to the quantity of electricity that has passed through the coil. Thus

$$Q = kd$$

where Q is the quantity of electricity, d is the maximum deflection, and k is known as the ballistic constant of the galvanometer. It will be observed that the ballistic constant is numerically equal to the quantity of electricity necessary to produce a unit scale deflection. If Q is expressed in microcoulombs and d in millimeters, then k will be expressed in microcoulombs per millimeter.

APPARATUS

1. Ballistic wall galvanometer
2. Two unknown capacitors in the range of 0.2 to 2.0 mfd
3. A board on which are mounted eight half-microfarad capacitors with individual switches, connected in parallel to two binding posts
4. Single-pole, double-throw switch
5. Two 1.5-volt dry cells
6. D.C. voltmeter (0–3 volts)

PROCEDURE

1. Connect the apparatus in the circuit shown in Fig. 61, using the capacitor board and one dry cell. Do not connect the cell until the instructor has checked your circuit. The galvanometer has already been adjusted and should be ready for use. If it is not, consult the instructor. Do not try to adjust it yourself.

2. Observe that when switch SW is closed in the direction A, the capacitor is being charged by the dry cell, and when it is closed in the direction B, the capacitor discharges through the galvanometer. Connect the first capacitor on the board in the circuit by closing the first switch on the board. Then close switch SW in the direction A, thus charging the capacitor. Wait a few seconds for it to become fully charged and record the voltage read on the voltmeter. Now close the switch in the direction B, thus discharging the capacitor through the galvanometer. Observe the maximum deflection of the galvanometer, estimating the reading to a half millimeter. The first time this observation is made, it may be necessary to take two or three readings of the galvanometer deflection to practice catching the very end of the first throw of the galvanometer coil.

3. Connect the second capacitor on the board in parallel with the first by closing the second switch on the board. Repeat Procedure 2, observing the maximum deflection of the galvanometer when two capacitors are discharged. Keep increasing the number of capacitors in parallel until all of the eight are being used. Observe the maximum deflection of the galvanometer for each group when the capacitors are discharged, and record the readings.

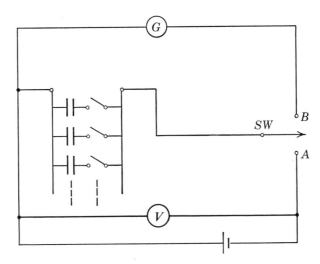

Figure 61 Circuit for the Measurement of Capacitance

4. Replace the capacitor board by one of the unknown capacitors. Charge it using one dry cell and record the voltage read on the voltmeter. Then discharge it through the galvanometer. Observe the maximum deflection of the galvanometer, estimating the reading to a half millimeter.

5. Repeat Procedure 4, using the same capacitor but with two dry cells in series instead of just the one shown in Fig. 61. If the dry cells are new and thus give slightly more than 1.5 volts each, the voltmeter will read slightly off-scale. If this happens, either estimate the reading by noting the approximate number of small scale divisions by which the pointer has passed the high end of the scale, or disconnect the voltmeter from the circuit and measure the voltage of each dry cell separately. Add these two measurements to get the total charging voltage. In either case record the charging voltage in the space provided.

6. Repeat Procedures 4 and 5, using the other unknown capacitor.

7. Connect the two unknown capacitors in parallel and charge them with only one dry cell. Record the charging voltage. Then discharge them through the galvanometer and observe the deflection.

8. Connect the two unknown capacitors in series and charge them, using two dry cells in series. Measure and record the charging voltage as discussed in Procedure 5. Then discharge the series combination through the galvanometer and observe the deflection.

DATA

	Capacitance Used	Charging Voltage	Galvanometer Deflection	Total Charge
1.	0.5 Microfarad			
2.	1.0 Microfarad			
3.	1.5 Microfarads			
4.	2.0 Microfarads			
5.	2.5 Microfarads			
6.	3.0 Microfarads			
7.	3.5 Microfarads			
8.	4.0 Microfarads			

Capacitance Used	Charging Voltage	Galvanometer Deflection	Total Charge	Value of Capacitance	Capacitance Calculated from Formula
Capacitor No. 1					
Capacitor No. 1					
Capacitor No. 2					
Capacitor No. 2					
Capacitors 1–2 in Parallel					
Capacitors 1–2 in Series					

Ballistic constant of galvanometer _____

CALCULATIONS

1. Calculate the value of the total charge on the capacitors used in Procedures 2 and 3 by using Equation 99. Note that if V is the charging voltage in volts and Q is the charge in microcoulombs, the capacitance C will come out in microfarads.

2. Using the data of Procedures 2 and 3 and the results of Calculation 1, plot the galvanometer deflections along the x axis and the corresponding values of the charge along the y axis. Draw a straight line through the plotted points in such a way as to fit them as closely as possible. Note that the origin should be a point on your line. This graph will be the calibration curve for the galvanometer. Calculate its slope. This will be the value of the ballistic constant of the galvanometer in microcoulombs per scale division.

3. Using the values of the galvanometer deflections obtained in Procedures 4–8, read the corresponding values of the total charge in microcoulombs from your graph.

4. From the data of Procedures 4 and 5, plot the voltage along the *x* axis and the corresponding values of the charge along the *y* axis. Draw a straight line through the origin, and as nearly as possible through the two plotted points. Determine the slope of this line. The value of the slope is the capacitance of the capacitor in microfarads.

5. Repeat Calculation 4 using the data of Procedure 6 and plotting the points on the same graph paper, using the same axes.

6. From the data of Procedure 7, determine the capacitance of the two unknown capacitors in parallel by dividing the total charge on them by the charging voltage.

7. Calculate the capacitance of the two unknown capacitors in parallel by using the formula for capacitors in parallel and the values of the individual capacitances obtained in Calculations 4 and 5. Compare this value with the one obtained in Calculation 6.

8. From the data of Procedure 8 determine the capacitance of the two unknown capacitors in series by dividing the measured charge by the charging voltage.

9. Calculate the capacitance of the two unknown capacitors in series by using the formula for capacitors in series and the values of the individual capacitances obtained in Calculations 4 and 5. Compare this value with the one obtained in Calculation 8.

QUESTIONS

1. Do the results you obtained experimentally for the series and parallel connections of capacitors check the laws for such connections within experimental error?

2. A capacitor is charged by a 45-volt battery and is then discharged through a galvanometer whose ballistic constant is 1.5 microcoulombs per scale division. The galvanometer deflection produced is 20 divisions. What is the capacitance of the capacitor?

3. Two capacitors with capacitances of 4 and 5 microfarads respectively are connected in parallel. The combination is connected in series with a 3-microfarad capacitor. What is the total capacitance of the final arrangement?

4. Which factor in Equation 102 is varied in the tuning capacitors used in radio sets? Explain how this is done.

5. A parallel-plate capacitor is made up of two circular plates separated by a thin sheet of mica. The plates have a diameter of 18 centimeters, and the mica is 0.2 millimeter thick and has a dielectric constant of 5.7. Calculate the capacitance of this capacitor in microfarads.

6. Compute the energy in joules in one of your standard capacitors on the capacitor board when it is charged with one dry cell. What happens to this energy when the capacitor is discharged through the galvanometer?

7. A parallel-plate capacitor has square plates 24 centimeters on a side separated by 1 millimeter. The dielectric can be air, or a 1-millimeter thick sheet of glass can be slipped between the plates without friction. The capacitor is connected to a 1000-volt D.C. source with the glass sheet between the plates. How much work is needed to withdraw the sheet (a) if the capacitor is first disconnected from the voltage source, and (b) if the capacitor is left connected to the voltage source during the withdrawal process?

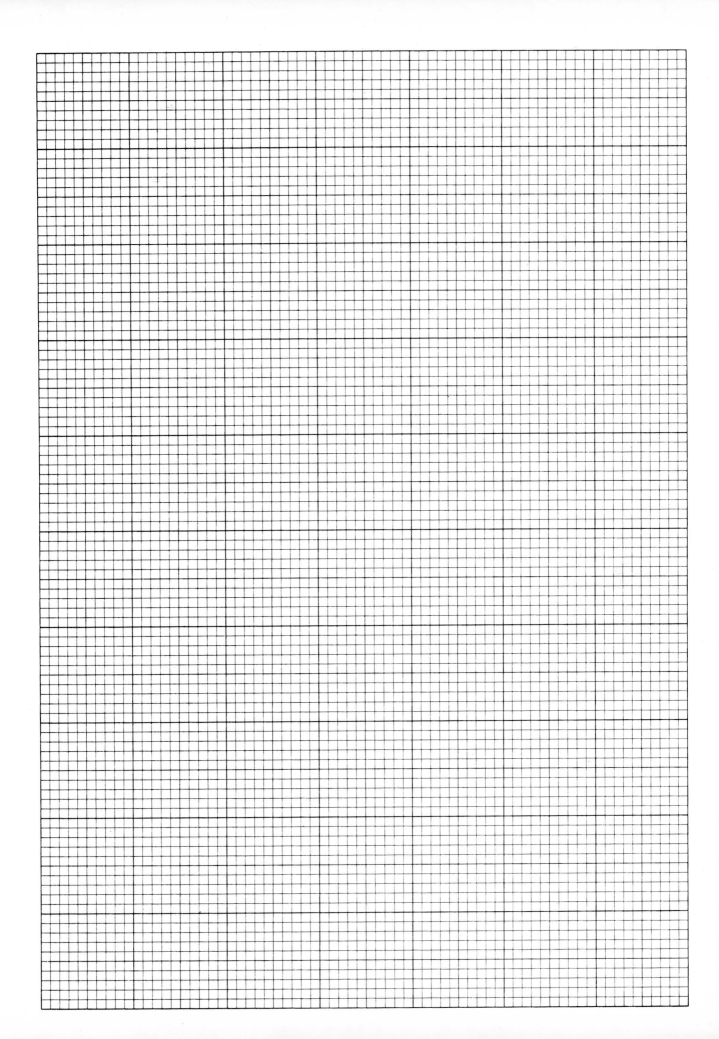

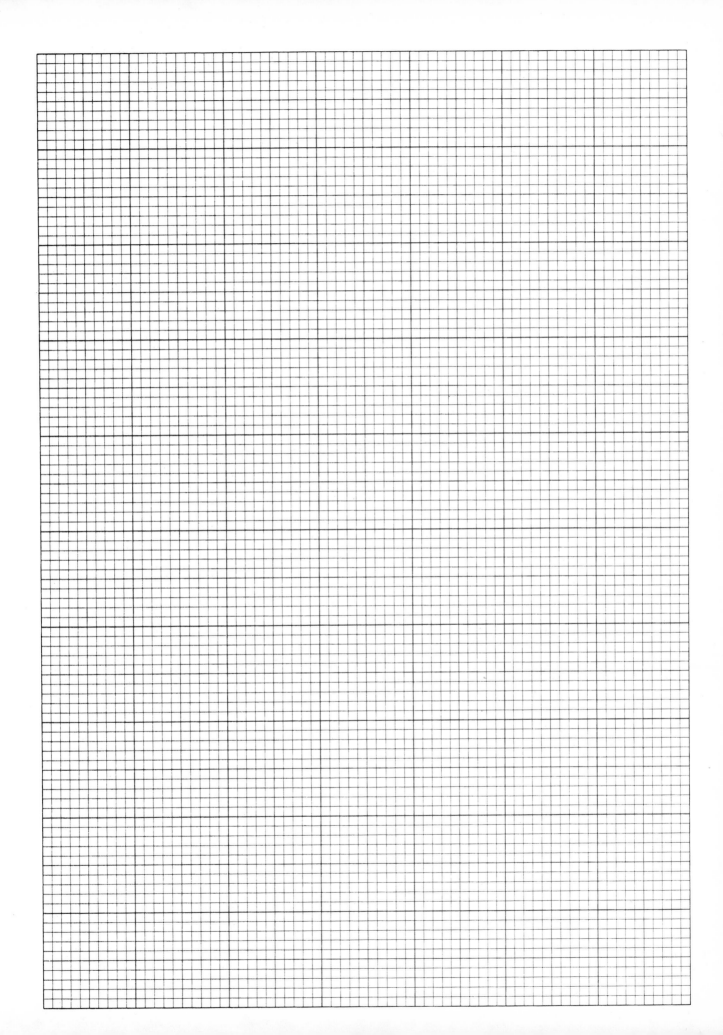

The *RC* Circuit **32**

When a resistor is connected in parallel with a charged capacitor, a current flows through the resistor, draining charge out of the capacitor. The capacitor is thus discharged over a period of time. The relation between the voltage remaining across the capacitor and the time is a very useful one in electronics and electrical engineering, and the purpose of this experiment is to study it. In doing so, we shall find that it depends on the resistance R of the resistor and the capacitance C of the capacitor only through their product RC and that another method of measuring capacitance is presented.

THEORY

Fig. 62 shows a capacitor C and a resistor R connected in parallel. Let the capacitor be initially charged to a voltage V_0 by some external voltage source which does not appear in Fig. 62 and has been removed at time $t = 0$. The capacitor then discharges through resistor R, the voltage V across the capacitor at any particular time causing a current $i = V/R$ to flow at that instant according to Ohm's law. Current i is the rate at which charge is leaving the capacitor (recall that an ampere is a coulomb per second), so that voltage V decreases steadily. Note, however, that the smaller V gets, the smaller i is and hence the slower the rate of discharge. Situations like this in which the rate at which something changes is proportional to that thing's value at every instant are common in physics and are always described by the exponential function, which is characterized by the property of having its rate of change proportional to its value. The exponential function is simply the base of natural logarithms e raised to a power proportional to the time. For example, in the absence of inhibiting effects, the rate of growth of the population of any given species is proportional to the number already present. The population as a function of time is then described by an exponential in which the exponent is t times a constant expressing the probability that any particular specimen will reproduce. In the present case, the voltage across the capacitor *decreases* as time goes on, which means that its rate of change is negative. Consequently, the exponent in the exponential must be negative. It can be shown that the voltage across the capacitor is given as a function of time by

$$V = V_0 e^{-t/RC} \qquad (105)$$

where V is the voltage in volts at time t, V_0 is the initial voltage (at $t = 0$), R is the resistance of the resistor in ohms, C is the capacitance of the capacitor in farads, and t is the time in seconds measured from the instant at which V was equal to V_0. It is interesting to note that R and C occur only as the product RC. Moreover, the exponent as a whole cannot have a dimension, and so RC must have the units of time. Indeed multiplication of ohms by farads gives seconds. The product RC is therefore called the circuit's *time constant*. Note that when $t = RC$, $V = V_0/e = 0.368V_0$ or somewhat less than half of the original voltage. Equation 105 shows that theoretically V never gets to zero (the discharge rate getting slower and slower the smaller V becomes), but in a practical case V becomes negligibly small after a period equal to a few time constants has elapsed.

In the present experiment the validity of Equation 105 will be investigated by plotting the discharge of a capacitor by a parallel resistor as a function of time. Since practical capacitors have capacitances measured in microfarads, the discharging resistor must have a value of several megohms (millions of ohms) if a time constant of a few seconds is to be realized. Although electronic equipment often uses RC circuits with time constants in the microsecond range, we are not now prepared to measure such short intervals and must therefore have time constants sufficiently large so that we can time the voltage decrease with a stop watch. To measure V, how-

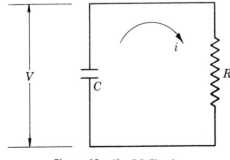

Figure 62   The *RC* Circuit

203

ever, a voltmeter must be connected across the capacitor, and an ordinary voltmeter will itself have a resistance less than the several megohms desired. To overcome this difficulty, an electronic voltmeter or EVM (sometimes called a VTVM for vacuum-tube voltmeter) will be used. An electronic voltmeter contains an amplifier that drives the actual meter with power obtained either from batteries or from the power line. Ideally this means that the meter actuating current does not have to come from the circuit under test so that the EVM input terminals can have an infinite resistance between them. Practical EVM's usually have an input resistance of about 10 megohms, but this is already so large that most circuits are not affected by it. Thus the EVM, like the potentiometer, has the advantage of drawing negligible current from the circuit under test, but it is much more convenient for general voltage-measuring purposes.

In this experiment an EVM will be used both to measure the voltage across a capacitor and to serve as the parallel resistor. A plot of V against t will be made on semilogarithmic paper and should turn out to be a straight line if Equation 105 is correct. The time constant RC can be obtained from the slope of this line, as discussed in the section of the Introduction on plotting exponentials. The values of R and C cannot, however, be found individually. Another plot will therefore be made using a different value of R obtained by connecting a large known resistance in parallel with the EVM, and a new value of RC will be calculated from its slope. Since the two values of R are related by the formula for resistances in parallel, both the value of C and the EVM input resistance can be found from the data. The nominal value of C is marked on the capacitor and serves as a check on your results.

APPARATUS

1. Power supply capable of delivering 300 volts D.C.
2. Electronic voltmeter with 0–300 volt or higher scale
3. 4-microfarad capacitor (may be the capacitor board of Experiment 31 with all switches closed)
4. 10-megohm precision resistor (1% tolerance)
5. Single-pole, single-throw switch
6. Stop watch or stop clock

PROCEDURE

1. Wire the power supply, capacitor, electronic voltmeter, and switch in the circuit shown in Fig. 63. Omit the 10-megohm resistor R_s in this step. Be sure the EVM is set to an appropriate range (0–300 volts or higher). Leave switch K_1 open and do not plug in or turn on the power supply until your circuit has been approved by the instructor. CAUTION: The output of the power supply is high enough to give a painful shock. Be careful around the apparatus and do not make changes in your circuit with the power on. Notice that opening switch K_1 is not enough. The power supply itself must be turned off with its own switch or unplugged from the power line before any wiring changes are attempted.

2. After the instructor has checked your circuit, turn on the power supply, close switch K_1, and adjust the power supply output to 300 volts as read on the EVM.

3. Open K_1 and at the same instant start the stop watch. Stop the stop watch at the instant the EVM passes through 250 volts. Record the time thus measured.

4. Close switch K_1 and readjust the power supply if necessary to bring the voltage across the capacitor to 300 volts. Then again open K_1 and start the stop watch simultaneously. Stop the stop watch when the EVM passes through 200 volts and record the indicated time.

5. Repeat Procedure 4, stopping the stop watch at 150, 100, and 50 volts.

6. Connect the 10-megohm resistor at R_s in Fig. 63. Repeat Procedures 2–5 with this setup.

7. Be sure to turn the power supply off and unplug it when you have finished the experiment.

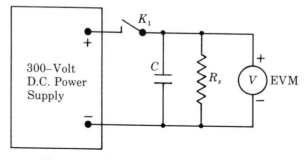

Figure 63   Circuit for Measuring Time Constants

DATA

Value of R_s _____

Voltage Across the Capacitor	Ratio V/V_0	Time (Seconds), EVM Alone	Time (Seconds), EVM in Parallel with R_s
300	1.00	0.00	0.00
250			
200			
150			
100			
50			

Value of R_mC from graph _____   Calculated value of capacitor C _____

Value of R_1C from graph _____   Nominal value of C _____

Value of meter resistance R_m _____

CALCULATIONS

1. Divide each of your values of the capacitor voltage V by the initial voltage V_0 to obtain the ratio V/V_0 in each case. Note that this will allow you to plot Equation 105 in the form

$$\frac{V}{V_0} = e^{-t/RC}$$

2. On semilogarithmic paper, plot the values of time obtained in Procedures 3–5 against the corresponding values of the voltage ratio V/V_0. Plot this ratio on the logarithmic scale and the time on the linear one. Read the instructions for plotting exponential functions in the Introduction. Draw the best straight line through your plotted points.

3. From the slope of your graph, calculate R_mC where R_m is the resistance of the EVM.

4. On the same sheet of semilogarithmic graph paper, plot the values of time obtained in Procedure 6 against the corresponding values of V/V_0. Use the logarithmic scale for the voltage ratio and the linear scale for time as before. Draw the best straight line through this set of plotted points.

5. From the slope of the graph drawn in Calculation 4, compute R_1C, where R_1 is the resistance of R_m and R_s in parallel.

6. You now have values for R_mC and R_1C, and you know that $(1/R_1) = (1/R_m) + (1/R_s)$ where the value of R_s is given. On this basis calculate the capacitance C and the meter resistance R_m.

7. Compare your calculated value of C with the nominal value marked on the capacitor. *Note:* Commercial capacitors have a tolerance of $\pm 10\%$ or more from the nominal value. You should therefore not expect remarkably good agreement between this and your measured capacitance. The instructor may, however, give you a more precise value of C for you to compare your results with.

QUESTIONS

1. Explain how your graphs check the validity of Equation 105.

2. Why was it better to plot the ratio V/V_0 rather than simply the voltage V against t in Calculations 2 and 4?

3. Show that the product RC has the dimensions of seconds when R is in ohms and C is in farads.

4. Let $R_mC = A$ and $R_1C = B$, where A and B are the measured values of these time constants obtained from the slopes of your graphs in Calculations 3 and 5. Recalling that R_1 is the resistance of R_m and R_s in parallel, derive expressions for R_m and C in terms of the known quantities A, B, and R_s.

5. A charged 2-microfarad capacitor is connected in parallel with a 500,000-ohm resistor. How long after the connection is made will the capacitor voltage fall to (a) 50 percent of its initial value; (b) 30 percent of its initial value; (c) 10 percent of its initial value; (d) 5 percent of its initial value?

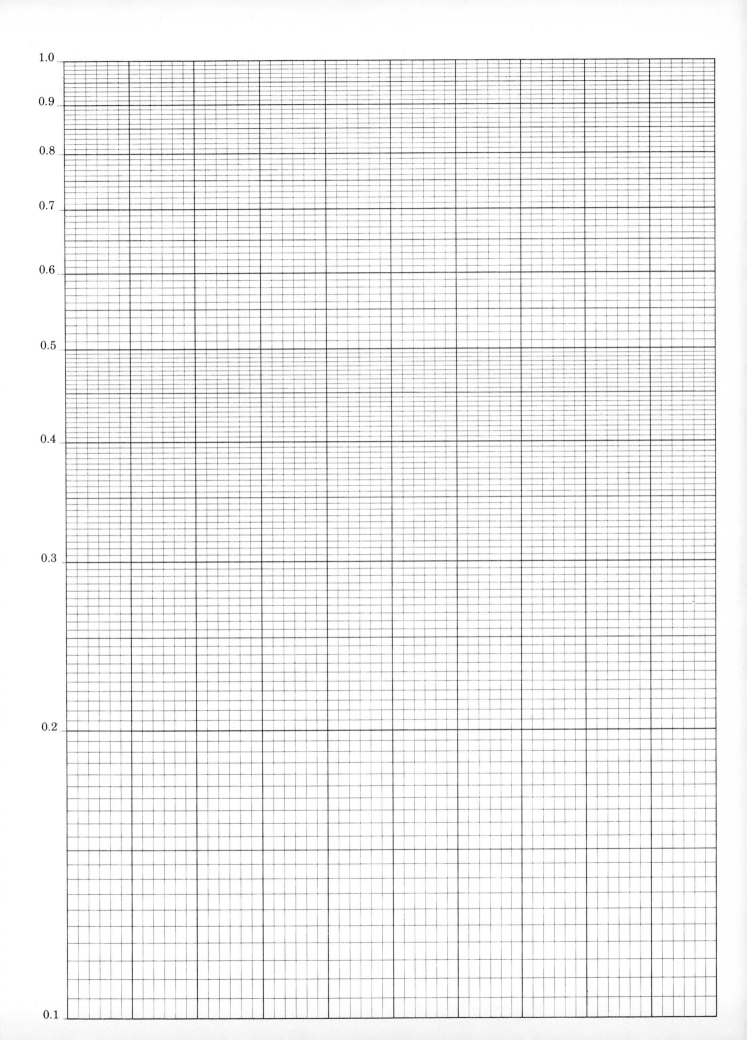

Electromagnetic Induction: **33**
The Transformer and the Generator

The production of induced currents is one of the most important branches of the study of electricity, for it is the basis of operation of both the transformer and the generator and thus forms one of the foundations of the electrical industry. The electric generator has revolutionized both modern industry and home life by furnishing us with a source of cheap electricity. Generators convert mechanical energy directly into electrical energy by electromagnetic induction. Thus, the enormous power of steam engines, gas engines, and water wheels can be transformed into electricity that can be conveniently used in motors to turn all kinds of machinery. The principle of electromagnetic induction is also fundamental to the operation of the transformer, which, by providing a simple and efficient means of stepping voltage up and down, has made possible long-distance transmission of electric power.

The present experiment deals with some aspects of electromagnetic induction. Much of the work is qualitative, so that a complete record of the observations must be kept. The object of the experiment is to study Faraday's law of electromagnetic induction, Lenz's law, and the general features of a magnetic circuit. In particular, simple examples of a transformer and an alternating current generator will be studied. In each case the electromotive force induced in a coil of wire by a varying magnetic flux will be investigated.

THEORY

An electromotive force is induced in a loop of wire whenever there is a change in the magnetic flux passing through the loop. Magnetic flux is the product of the magnetic field intensity B and the area bounded by the loop. If the direction of B is not perpendicular to the plane of the loop, then the perpendicular component must be taken. Mathematically the flux is given by

$$\Phi = B_\perp A \qquad (106)$$

where Φ is the flux in webers, $B_\perp$ is the perpendicular component of the magnetic field in teslas, and A is the area bounded by the loop. Faraday's law of electromagnetic induction then states that the e.m.f. induced in a loop by a changing magnetic flux is given by

$$E_l = -\frac{\Delta\Phi}{\Delta t}$$

where E_l is the average value of the induced e.m.f. in volts; $\Delta\Phi$ is the change in flux in webers; and Δt is the time interval in seconds during which the change in flux takes place. If instead of a one-turn loop we have a coil of N turns, voltage E_l is induced in each turn, and since the coil is in effect N turns in series, the total e.m.f. at the coil terminals is

$$E = NE_l = -N\frac{\Delta\Phi}{\Delta t} \qquad (107)$$

From the definition of flux we note that a change in Φ may be produced in any of three different ways: the area of the loop may be changed, thus changing the factor A in Equation 106; the strength of the magnetic field B may be altered; or the coil may be rotated so that the field's perpendicular component changes. All of these methods of changing Φ are used in practical applications of Faraday's law.

The minus sign in Equation 107 gives the polarity of the coil voltage relative to the sign of the change in flux. It is best understood in terms of Lenz's law, which states that the polarity of E will be such as to oppose the change in Φ that produced it. Thus, suppose the terminals of a certain coil were connected together so that an induced e.m.f. would cause a current to flow around the coil. Now suppose that a magnetic field directed through the coil from an external source were increased. Lenz's law requires that the current induced in the coil flow in such a direction as to result in a magnetic field opposing the applied one, that is, bucking the increase in B that induced it. Similarly, should the externally applied field decrease, the induced current will flow around the coil in the other direction so as to produce a field in the same direction as the applied field and thus oppose the decrease. Note that the relation between the magnetic field produced by a current flowing around a coil and the direction of this current is given by the right-hand rule. When the fingers of the right hand curl around the coil in the direction of the current flow, the extended thumb

209

points in the direction of the resulting magnetic field.

An important application of electromagnetic induction is the transformer. A transformer consists of two interwound coils called respectively the primary and secondary coils. These coils are often wound on a core of magnetic material, and this core is extended to form one or more complete loops of the material passing through and around the coils. Such a loop forms a magnetic circuit that concentrates the magnetic field due to current in either coil in the loop of material passing through the coils.

If a current is now sent through the primary coil, a magnetic field passing through the plane of both coils and hence a magnetic flux through these coils will be produced. Any change in the current will result in a change in this flux. The primary coil will therefore have induced between its terminals an e.m.f.

$$E_1 = -N_1 \frac{\Delta \Phi}{\Delta t} \qquad (108)$$

and an e.m.f.

$$E_2 = -N_2 \frac{\Delta \Phi}{\Delta t} \qquad (109)$$

will be induced between the terminals of the secondary coil, where E_1 and E_2 are in volts, N_1 is the number of turns in the primary coil, and N_2 the number of turns in the secondary coil. Note that even though the primary coil is wound of heavy copper wire so that it has negligible resistance, a voltage nevertheless develops across it when a current is sent through it, *but only when the current is changing*. Hence with direct current, once the current is on, $\Delta \Phi$ and therefore both E_1 and E_2 will be zero. This explains why transformers work only on alter-

nating current, which is by definition a current that is continually changing. The voltage E_1 appearing across the primary coil as a result of the changing primary current is sometimes called a *back e.m.f.* because of the opposing polarity given it by Lenz's law as expressed by the minus sign in Equation 108.

The most important feature of the transformer is revealed when we divide Equation 108 by Equation 109 to obtain

$$\frac{E_1}{E_2} = \frac{N_1}{N_2} \qquad (110)$$

This result says that the ratio of the primary to the secondary voltage is equal to the turns ratio. Thus in an A.C. circuit a transformer may be used to change the voltage to any desired value by simply winding the coils with the appropriate numbers of turns. This simple arrangement is used throughout the electrical industry from household appliances to huge transmission-line substations. Much of the electrical equipment in common use today would not work without it, and the fact that the transformer works on A.C. only is a principal reason for the widespread use of A.C. rather than D.C. power.

Another example of an important application of Faraday's law of induction is the electric generator. A generator is a machine which, when driven mechanically by an engine or turbine, converts the mechanical power thus supplied into electrical power.

The essential parts of a generator are the magnetic field, the armature, and some device, such as slip rings or a commutator, for connecting the armature to the external circuit.

The magnetic field is produced by permanent magnets or by electromagnets. The armature consists of a rotating coil wound on a soft iron core to obtain an intense and

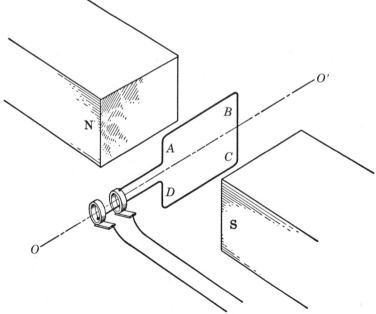

Figure 64   Principle of a Simple Generator

uniform field. As the armature rotates, the flux through the coil changes, and an electromotive force is induced across its terminals.

A simplified diagram of a generator is shown in Fig. 64, in which a rectangular coil of wire *ABCD* is mounted on a longitudinal axis *OO'* between the magnetic poles N and S. The two ends of the coil are connected to two insulated rings mounted on the axis and rotating with the coil. Two blocks of carbon, called brushes, press against these rings as they rotate and provide electrical contact with the external circuit.

A cross section of the coil is represented in Fig. 65. The e.m.f. induced in the coil will be zero at the position *AD*, or when the angle θ is zero, because at that instant the conductors are moving in a direction parallel to the magnetic field so that no change in flux occurs. The e.m.f. will be at a maximum when θ is 90°, because at that instant the conductors are moving in a direction perpendicular to the magnetic field and the rate at which the flux is changing is at its maximum. The situation is sometimes expressed in terms of the conductors "cutting" the magnetic field lines. When the conductors move parallel to *B*, no lines are cut and no e.m.f. is induced, whereas when they move perpendicular to *B* they cut straight across the field lines and the induced e.m.f. is maximum. Since these lines extend from the north pole to the south pole in Fig. 65, the component of the field *B* perpendicular to the plane of the coil is $B \cos \theta$ and the flux through the coil is $BA \cos \theta$, where *A* is the area that the coil bounds. The e.m.f. induced in one loop is just the rate of change of this flux as θ changes with rotation of the coil, and if the coil consists of *N* loops, the total e.m.f. appearing at its terminals will be *N* times this quantity. The rate of change of $\cos \theta$ with θ is $-\sin \theta$ times the rate of change of θ, and if the coil in Fig. 65 is rotated clockwise at a constant angular velocity of ω radians per second, $\theta = \omega t$ and the rate of change of θ is ω. Hence the induced e.m.f. at the coil terminals is given by

$$E = NBA\omega \sin \omega t \qquad (111)$$

The maximum value of E is seen to be $NBA\omega$ and is usually called the amplitude E_m of E. With this substitution, Equation 111 becomes

$$E = E_m \sin \omega t \qquad (112)$$

where E is the instantaneous value of the induced e.m.f. in volts, E_m is its maximum value, ω is the angular velocity, and t is the time in seconds. If instead of rotating at a constant angular velocity ω, the coil is rotated through a small angle $\Delta\theta$ from an orientation θ, the change of flux is $-BA\,\Delta\theta \sin \theta$; and if the rotation takes place in time Δt, the rate of change is $-BA(\Delta\theta/\Delta t) \sin \theta = -BA\omega \sin \theta$, where ω is the angular velocity of rotation through the small angle $\Delta\theta$. Thus, for rotation through a small angle, the induced e.m.f. is given by

$$E = NBA\omega \sin \theta = E_m \sin \theta \qquad (113)$$

where θ is the angle through which the plane of the coil is displaced (from a position at right angles to the field) when the small rotation takes place.

Therefore, if the coil in this simple form of a generator is rotated at a constant speed, the induced e.m.f. is at any instant proportional to $\sin \theta$. During the first half-revolution the e.m.f. will be in one direction, or positive, while during the second half-revolution it will be in the opposite direction, or negative. This simple form of generator is an alternating current (or A.C.) generator.

By using a model generator, it can be easily shown that the curve of the induced e.m.f. has the form of a sine wave. The model generator used in this experiment is illustrated in Fig. 66. It consists of a flat, rectangular coil mounted so that it can rotate in a uniform magnetic field. The field is produced by two sets of permanent horseshoe magnets. The coil can be rotated in equal steps by means of a spring-and-ratchet mechanism. The spring provides a constant torque, while a ratchet wheel and a stop allow the coil to rotate through a 10° interval each time the ratchet is released. The terminals of the coil are con-

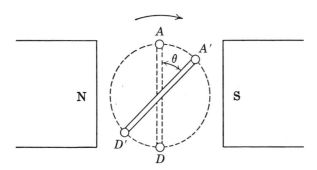

Figure 65   Armature Coil Rotating in a Uniform Magnetic Field

Figure 66   The Model Generator

nected by means of slip rings and brushes to a pair of binding posts.

If the coil is rotated through a series of small equal angles, the induced e.m.f.'s can be measured by connecting it to a ballistic galvanometer. As discussed in Experiment 31, this instrument measures the quantity of charge passed through it. We can easily show that in the present application the maximum galvanometer deflection will be proportional to the e.m.f. induced during each step $\Delta\theta$. Thus if this e.m.f. is given by Equation 113 and the total circuit resistance (coil and galvanometer) is R, then the current flowing during the step, $i.e.$, as the generator coil is rotating through $\Delta\theta$, will be

$$i = \frac{E}{R} = NBA\,\frac{\omega}{R}\sin\theta = NBA\,\frac{\Delta e\,\sin\theta}{\Delta tR}$$

The quantity of charge passing through the galvanometer is then $i\,\Delta t = NBA\,(\Delta\theta/R)\sin\theta$ and is thus proportional to the E of Equation 113. It follows that the galvanometer deflection will be proportional to the e.m.f. induced in the coil in each of the intervals. Thus the model generator shows the relative values of the e.m.f.'s that would be produced at any position of the coil if it was rotated with uniform speed.

APPARATUS

1. Primary coil of about 100 turns of copper wire
2. Secondary coil of about 1000 turns of copper wire
3. U-shaped piece of iron
4. Straight piece of iron
5. Portable galvanometer
6. Ballistic wall galvanometer
7. D.C. ammeter (0–1 ampere)
8. Multimeter with A.C. voltage ranges to 150 volts

9. Transformer delivering 6.3 volts at 3 amperes
10. Model generator
11. 12-volt power supply or storage battery
12. Tubular rheostat (about 100 ohms)
13. Decade resistance box (1000 ohms)
14. A double-pole, double-throw reversing switch
15. Permanent bar magnet

PROCEDURE

1. Connect the secondary coil to the portable galvanometer. Thrust the north pole of the magnet into the coil. Record the magnitude and direction of the deflection of the pointer, for a fast speed of motion and for a slow speed. Observe whether the deflection is to the left or to the right; a deflection of the pointer to the right means that the current enters the galvanometer at its + terminal. Determine the direction of the current induced in the coil by noting the direction of the current through the galvanometer and the direction in which the coil is wound.

2. Repeat Procedure 1, withdrawing the north pole from the coil, after having inserted it as in Procedure 1. Observe the direction of the current induced in the coil when the north pole is being withdrawn.

3. Repeat Procedures 1 and 2, using the south pole of the magnet.

4. Connect the primary coil in series with the rheostat, the ammeter, the double-pole, double-throw reversing switch, and the power supply or storage battery, as shown in Fig. 67. Have the circuit approved by

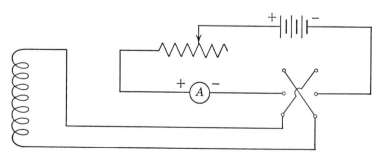

Figure 67   Primary Circuit Used in Studying Electromagnetic Induction

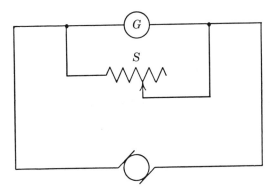

Figure 68   Circuit for the Model Generator

the instructor. Mount the two coils in a complete square of iron, made up of the U-piece and the straight piece. Using a current of 0.2 ampere through the primary coil, observe and record the magnitude and direction of the deflection when the primary circuit is opened and when it is closed. Repeat the observations using primary currents of 0.4, 0.6, 0.8, and 1.0 ampere.

5. Using a current of 1 ampere, observe the effect of reversing the current in the primary coil. Notice the direction of the induced current when the primary circuit is opened and when it is closed.

6. Keep the current through the primary coil at some constant value, say 0.5 ampere. Observe the galvanometer deflection upon opening and closing the primary circuit, using the complete square of iron.

7. Repeat Procedure 6, using the straight piece only.

8. Repeat Procedure 6, using no iron at all.

9. Repeat Procedure 8, placing the two coils twice as far apart as in 8.

10. Test your two coils as a transformer by connecting the primary coil to the 6.3-volt A.C. output of the commercial transformer and the secondary coil to the multimeter. Be sure the multimeter is switched to its highest A.C. voltage range before connecting it in the circuit. CAUTION: Multimeters are easily destroyed by being connected in circuits while switched to an improper range. Always check the position of the range switch before connecting the meter leads. Leave the line cord from the 6.3-volt transformer unplugged until your setup has been checked by the instructor.

11. Begin by mounting the two coils in the complete square of iron made up of the U-piece and the straight piece. Plug in the 6.3-volt transformer and adjust the multimeter range switch to the lowest A.C. voltage range for which the pointer remains on scale. Record the reading. Disconnect the multimeter and reconnect it to read the primary voltage. This should be close to 6.3 volts, but the commercial transformer rating is nominal and obviously depends on the line voltage that happens to be present in the laboratory at the time you do the experiment. The primary voltage must therefore be checked as described. Record your result.

12. Repeat Procedure 11 using the straight piece only.

13. Repeat Procedure 11 using no iron at all.

14. Connect the armature terminals of the model generator to the ballistic galvanometer, and then connect the resistance box in parallel with the galvanometer, as shown in Fig. 68. Set the armature coil in the position for maximum induced e.m.f., that is, with the plane of the coil parallel to the direction of the magnetic field. Adjust the shunt resistance so that the maximum deflection of the galvanometer is about 20 divisions when the coil is rotated through a 10° interval.

15. Begin your observations with the armature coil set at right angles to the magnetic field. Rotate the coil through a 10° interval and observe the maximum deflection of the galvanometer. Take observations of the galvanometer deflection for each 10° movement of the coil over one complete revolution. Reset the spring after every second rotation of the coil to make sure that the spring tension remains the same, and therefore that the speed of rotation of the coil remains the same. Record the galvanometer deflection observed for each movement of the coil. Notice that the deflection of the galvanometer corresponds to the position of the coil at the middle of its movement.

DATA

	Direction of Current	Galvanometer Deflection	
		Slow Motion	Fast Motion
North Pole into the Coil			
North Pole out			
South Pole into the Coil			
South Pole out			

Primary Current	Direction of Secondary Current		Galvanometer Deflection		Magnetic Circuit
	Opening	Closing	Opening	Closing	
0.2 Amp					Complete Square of Iron
0.4 Amp					Complete Square of Iron
0.6 Amp					Complete Square of Iron
0.8 Amp					Complete Square of Iron
1.0 Amp					Complete Square of Iron
0.5 Amp					Complete Square of Iron
0.5 Amp					Straight Piece Only
0.5 Amp					No Iron
0.5 Amp					No Iron, Double Distance Between Coils

Core Material	A.C. Voltage		Ratio Primary/Secondary
	Primary Coil	Secondary Coil	
Complete Square of Iron			
Straight Piece Only			
No Iron			

Number of turns on primary coil   _____   Ratio primary turns/secondary turns   _____

Number of turns on secondary coil   _____

Setting of the Coil	Galvanometer Deflection	Sin θ	Product of the Maximum Deflection and Sin θ
0°			
10°			
20°			
30°			
40°			
50°			
60°			
70°			
80°			
90°			
100°			
110°			
120°			
130°			
140°			
150°			
160°			
170°			
180°			
190°			
200°			
210°			
220°			
230°			
240°			
250°			
260°			
270°			
280°			
290°			
300°			
310°			
320°			
330°			
340°			
350°			
360°			

CALCULATIONS

1. Plot the data of Procedure 4, using primary currents as abscissas and galvanometer deflections as ordinates.

2. Calculate the ratios of the primary and corresponding secondary voltages measured in Procedures 11–13. Compare these voltage ratios with the turns ratio of your two coils, which you should calculate if the number of turns on each coil is known.

3. Plot a curve, putting the angular displacements of the model generator's armature coil along the x axis and the corresponding ballistic galvanometer deflections obtained in Procedure 15 along the y axis. Notice that some of the deflections are positive and some are negative, hence the x axis should be chosen in the middle of the paper. The scales on the two axes should be so chosen as to utilize, as nearly as possible, the whole sheet of graph paper. Draw a smooth curve through the plotted points, using a solid line for this part of the graph.

4. Look up the value of sin θ in the trigonometric tables in the Appendix for each angular displacement of the armature coil, and record the values. Multiply each of these by the maximum ballistic galvanometer deflection for the corresponding half cycle; that is, multiply the positive values of sin θ by the maximum positive deflection and the negative values by the maximum negative deflection.

5. Plot a second curve on the same sheet of graph paper, using the same axes. Put the angular displacements of the armature coil along the x axis and the product of the maximum deflection and sin θ along the y axis. Draw a smooth curve through the plotted points, using a dotted line for this part of the graph. This one is a sine curve. Notice how it compares with the curve of Calculation 3.

QUESTIONS

1. How does the experiment verify Lenz's law?

2. Show how the laws of electromagnetic induction were verified.

3. Discuss the essentials of a good magnetic circuit from the observations of Procedures 6–9 and 11–13.

4. Name three practical applications of electromagnetic induction.

5. By using Lenz's law, how can you predict the direction of flow of current in a secondary coil when the current in the primary is turned on? Assume a direction for the primary current and assume that the two coils are placed parallel to each other.

6. (a) Can an induced e.m.f. be present in a circuit without an induced current flowing? (b) Can an induced current flow without an induced e.m.f.?

7. A circular coil of 200 turns of fine wire has a diameter of 4 centimeters. It is held in a magnetic field of 0.5 tesla with its plane perpendicular to the field direction. If it is then pulled out of the field in a tenth of a second, what is the induced e.m.f. in volts?

8. Explain what happens when a transformer is plugged into a D.C. line by mistake.

9. Suppose that the resistance of the coil of the model generator were doubled, while the size, shape, and number of turns remained the same. (a) How would this affect the e.m.f. produced in the coil? (b) How would it affect the galvanometer deflection? Explain.

10. Compare the power required to drive a generator on open circuit with no current in the armature with the power required to drive a loaded generator. Explain the difference.

11. Referring to Fig. 64, assume that the coil *ABCD* is rotating clockwise around the axis OO', looking at the coil in the direction OO'. Predict the direction of flow of the induced current at the instant when the coil has rotated through an angle of 90° from the position shown in the figure.

12. The current induced in the coils of an armature is always an alternating current. Sometimes it is desired to change the alternating current developed in the armature to current that always flows in the same direction in the outside circuit, that is, to make a direct current generator. Explain how this is accomplished by using a commutator.

13. A coil is composed of 50 turns of wire and has an area of 100 square centimeters. It is rotated as the armature coil of a generator in a uniform magnetic field of 0.2 tesla at a rate of 180 revolutions per minute. Find the amplitude in volts and the frequency in cycles per second of the induced e.m.f.

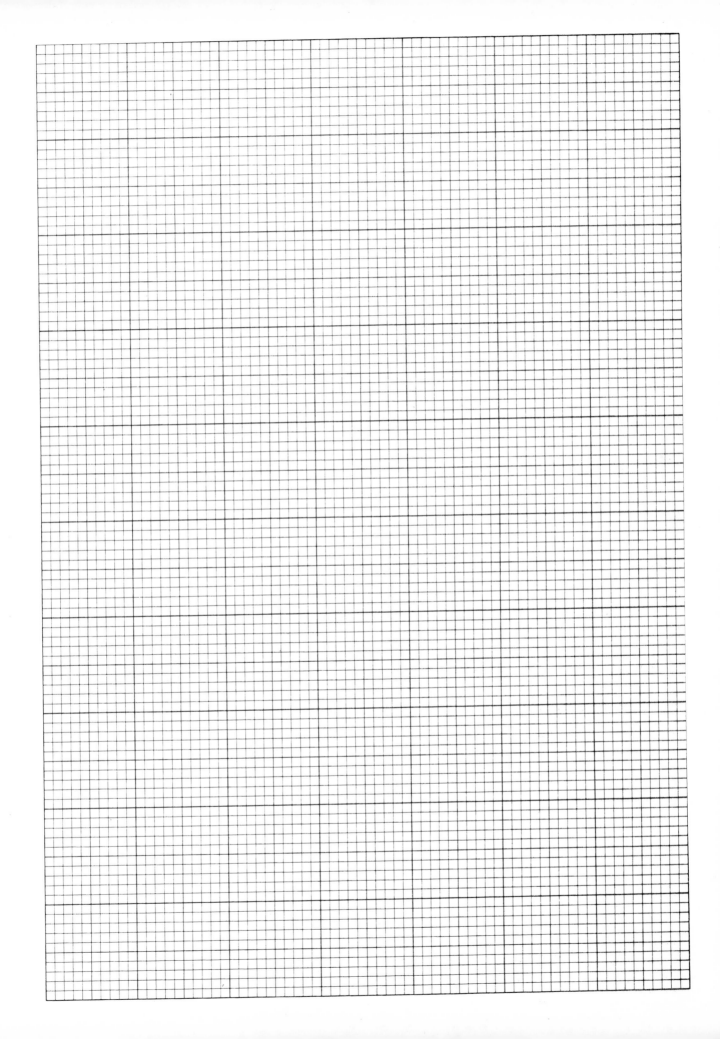

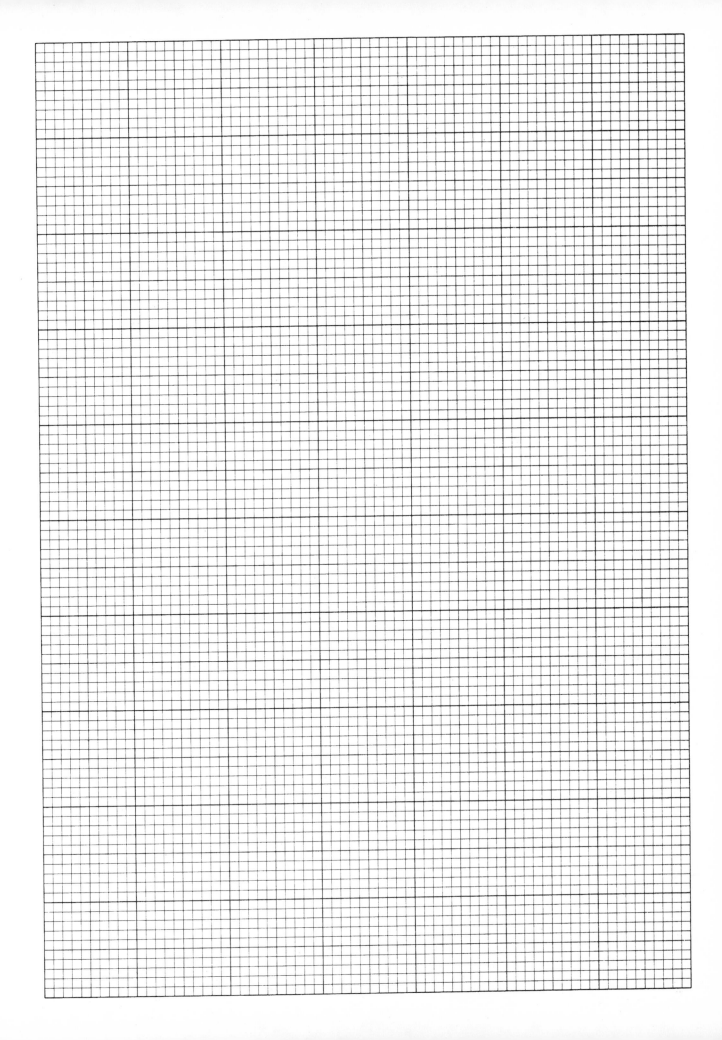

Alternating Current **34**

Because of the usefulness of the transformer, which does not work on direct current, alternating current has dominated the electrical industry for many years. Strictly speaking, the term "alternating current" should mean any current that varies with time so that it flows sometimes in one direction and sometimes in the other. However, as usually used, the term refers specifically to just one particularly simple form of such variation, namely a sinusoidal dependence on time. In this experiment only currents and voltages that are sinusoidal functions of time will be dealt with and the performance of circuit elements under these conditions studied.

THEORY

An A.C. current that is a sinusoidal function of time may be written

$$i = I_m \sin (\omega t + \phi_i) \tag{114}$$

where i is the instantaneous value of the current in amperes, I_m is the maximum value or amplitude of this current, ω is the so-called angular frequency in radians per second, t is the time in seconds, and ϕ_i is the phase angle in radians. The purpose of the phase angle is to set the value of i at $t = 0$. Thus, at this initial instant, $i = I_m \sin \phi_i$, which may have any value from 0 to $\pm I_m$ depending on the value chosen for ϕ_i. Figure 69 shows i plotted as a function of t for the particular case when $i = 0$ at $t = 0$, so that $\phi_i = 0$ and $i = I_m \sin \omega t$. Notice that the current returns to zero after flowing in the positive direction when $t = \pi/\omega$ and again after its reverse (negative) flow when $t = 2\pi/\omega$. After $t = 2\pi/\omega$, the current repeats the performance it has just gone through in the interval 0 to $2\pi/\omega$, and this complete oscillation, which is repeated over and over, is called a *cycle*. The time $2\pi/\omega$ seconds required for one cycle is called the *period* of the alternating current, and its reciprocal $\omega/2\pi$ is the number of cycles per second and is known simply as the frequency f. Equation 114 is therefore often written

$$i = I_m \sin (2\pi f t + \phi_i) \tag{115}$$

where f is in cycles per second (cps) or, according to currently accepted terminology, hertz.

The question now arises as to what happens when a sinusoidal current passes through a circuit element such as a resistor. For alternating current, two other circuit elements, the inductor and the capacitor, are of interest.

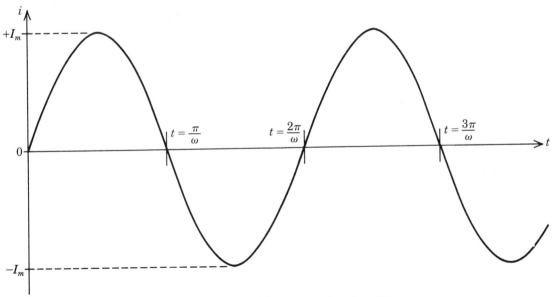

Figure 69   A Sinusoidal Current as a Function of Time

221

Notice that for direct current the ideal inductor, a coil of resistanceless wire, is simply a short circuit, whereas an ideal capacitor, two conducting plates separated by a perfect insulator, is an open circuit. On alternating current, however, the current through the inductor changes, hence the magnetic field changes and an induced e.m.f. appears across the coil terminals. In the case of the capacitor, charge flows onto its plates with one polarity when the current is in one direction and the voltage across the capacitor builds up, but then the current reverses, discharging the capacitor and recharging it in the opposite direction. We shall see that in each case there results a sinusoidal voltage across the circuit element that is proportional to the current, just as in the case of the resistance, so that a sort of Ohm's law holds for inductors and capacitors as well. The proportionality constant, however, is called the reactance instead of the resistance in the case of inductors and capacitors. The reason for this will become apparent below.

● Resistance in A.C. Circuits

If the current i flows through a resistance R, Ohm's law holds at every instant so that the voltage across the resistance is

$$v_R = iR = I_m R \sin 2\pi ft = V_{mR} \sin 2\pi ft \quad (116)$$

where we have taken zero phase angle for the current and have written $V_{mR} = I_m R$ since V_{mR} is clearly the amplitude of the sinusoidal voltage that develops across the resistance. Notice that the power dissipated in the resistance is $iv_R = I_m^2 R \sin^2 2\pi ft$ and is a function of time but is never negative. This is because the current varies but energy is transformed into heat in the resistor no matter which way the current goes through it. Power is the rate at which energy is delivered, but in this case we cannot simply multiply $I_m^2 R \sin^2 2\pi ft$ by a time interval to get the total energy because this power is not constant. An average value of the time-varying power must be calculated for this purpose, and the average of $\sin^2 2\pi ft$ over one cycle (and hence over any integral number of cycles) is $\frac{1}{2}$. Thus

$$P_{ave} = \tfrac{1}{2} I_m^2 R$$

This average power can be multiplied by a time to give an energy delivered just as the power $I^2 R$ can in the D.C. case. For this reason a quantity I_{rms} called the root mean square current (because it is the square root of the average of the square of i) is defined as $I_m/\sqrt{2}$. The root mean square (r.m.s.) current can then be used to calculate average power and energy delivered in the same way that direct current can. This concept is so useful that it is the r.m.s. value of A.C. currents and voltages that is usually quoted. Thus when the ordinary household A.C. power line is said to be 120 volts, it is the r.m.s. value, not the amplitude, that is meant.

● Inductance in A.C. Circuits

When an inductor (a coil of wire) carries a current, a magnetic field is produced. If the current and hence the field varies, there is a rate of change in the magnetic flux through the coil and an induced e.m.f. appears across its terminals even though the resistance of the coil may be zero. If the current through an inductor is given by Equation 115 with $\phi_i = 0$, a flux proportional to this current will result, so that we can write

$$\Phi = lI_m \sin 2\pi ft \quad (117)$$

where l is the proportionality constant and has the dimensions of webers per ampere. Now Faraday's induction law states that the induced voltage around a single loop is the negative of the rate of change of magnetic flux through that loop. The flux given by Equation 117 is certainly changing with time, and the rate of change of the sine of an angle is known to be the cosine of the angle multiplied by the angle's own rate of change. Thus the voltage induced around one turn of the coil is $-2\pi flI_m \cos 2\pi ft$ and across the coil terminals is $-2\pi fNlI_m \cos 2\pi ft$, where N is the number of turns. A voltage equal and opposite to this must therefore be applied to the coil terminals to keep the current flowing, and this will be

$$v_L = 2\pi fLI_m \cos 2\pi ft = V_{mL} \sin (2\pi ft + 90°) \quad (118)$$

where $L = Nl$ is called the *inductance* of the coil and is measured in henrys, in honor of the American physicist Joseph Henry (1797–1878), who was one of the early workers in electromagnetism. The inductance is a quantity that characterizes the coil and depends on the permeability of the core material (if any), the number of turns, and geometrical factors. Equation 118 suggests that L is the proportionality constant between the rate of change of current through an inductor and the resulting voltage induced across the inductor's terminals, and it is often so defined.

The important thing about the result stated by Equation 118 is that the voltage across the inductor is sinusoidal just as the current through it is. The amplitude V_{mL} of the voltage is equal to $2\pi fLI_m$ and is thus proportional to the amplitude I_m of the current. The proportionality constant $X_L = 2\pi fL$ is called the *inductive reactance* of the coil and appears to play the role of resistance in relating I_m and V_{mL}. Indeed it has the dimensions of ohms. There are several differences, however, between this case and that of the resistor. Here, while the current goes as $\sin 2\pi ft$, the voltage goes as $\cos 2\pi ft$, that is, as the current with a phase shift of 90° or $\pi/2$ radians. Thus, while the current through and the voltage across a resistor are *in phase*, the current and voltage for an inductor are *90° out of phase with the voltage leading the current*. Plots of i and v_L are shown in Fig. 70. Notice that when i is at its maxima and minima so that for an instant it and the

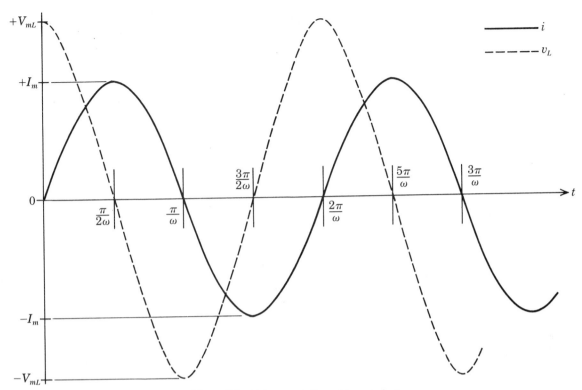

Figure 70 Voltage and Current in an Inductor

resulting magnetic flux are not changing, v_L is zero, whereas v_L has its maxima and minima at the instants when i is zero, at the points where the sine curve changes most rapidly. Note also that, unlike R, X_L is proportional to the frequency and becomes zero when $f = 0$. This is to be expected since a coil of resistanceless wire is a short circuit for direct current. Finally, the power delivered to the inductor cannot be obtained from the expression $i^2 X_L$ because v_L is not equal to iX_L. Instead the power delivered must be calculated from the original definition of electric power, namely the current times the voltage. But if i is multiplied by v_L we get $2\pi f L I_m^2 \sin 2\pi ft \cos 2\pi ft$, which is sometimes negative and sometimes positive. Moreover, the expression $\sin 2\pi ft \cos 2\pi ft$ averages to zero, showing that on the average no energy is delivered to the inductor. This is hardly surprising, since energy delivered to the inductor has to go somewhere, and the only place it can go is into the magnetic field. This is in fact what happens when the factor $\sin 2\pi ft \cos 2\pi ft$ is positive. When it is negative the magnetic field is collapsing and the stored energy is being returned to the source. Thus unlike the resistor, the inductor does not dissipate but rather temporarily stores energy.

● **Capacitance in A.C. Circuits** When an electric current flows into a capacitor, the plates become charged and the terminal voltage rises according to the relation

$$v_C = \frac{Q}{C} \tag{99}$$

where Q is the charge on the capacitor in coulombs, C is the capacity of the capacitor in farads, and v_C is the capacitor voltage. If the charging current is D.C., the capacitor voltage will simply build up to the source voltage, after which current flow will cease, for the capacitor is in fact an open circuit. However, if alternating current is used, the capacitor will charge up until the current reverses, after which it will discharge and then charge again with the opposite polarity. Thus, current flows alternately in and out of the capacitor through the connecting wires, so that alternating current appears to pass through it, although no charges can actually travel through the dielectric from one plate to the other.

The situation is illustrated in Fig. 71. Looking at the first half-cycle of the current, we note that i is positive throughout that half-period, and so the capacitor will be charging. It must reach its maximum voltage at the end of the half-cycle because thereafter i reverses and the capacitor discharges. This process continues during the second half-cycle, the capacitor attaining maximum charge in the reverse direction and hence maximum negative voltage at the end of a full cycle of the A.C. current. A detailed study shows that the capacitor voltage, like the voltage across the inductor, is sinusoidal but is shifted by $-90°$ ($-\pi/2$ radians) with respect to the current. In other words, the phase shift is again 90°, but the current leads the voltage. Thus v_C will be proportional to $\sin(2\pi ft - 90°)$ or $-\cos 2\pi ft$.

The amplitude V_{mC} of the capacitor voltage must surely be proportional to I_m, for this represents the

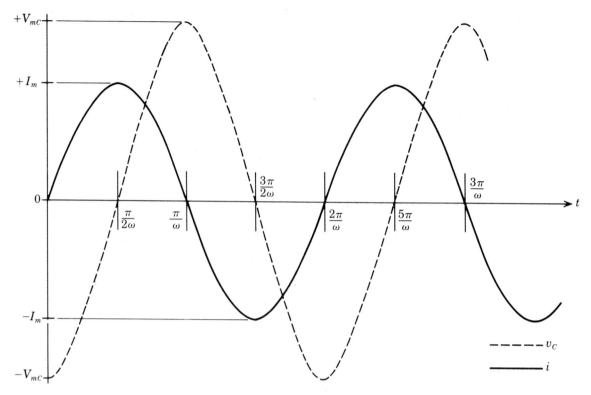

Figure 71   Voltage and Current in a Capacitor

maximum charging rate. Moreover, Equation 99 shows that V_{mC} should be inversely proportional to C, since the greater the capacity the lower the voltage developed by a given amount of charge. We also expect the voltage amplitude to be proportional to the period of the A.C. current, for half of this period is just the time allowed for charging the capacitor in any one direction. In terms of the frequency, the reciprocal of the period, we should find $V_{mC} \sim 1/f$. Detailed calculation yields

$$v_C = -\frac{I_m}{2\pi fC} \cos 2\pi ft = V_{mC} \sin (2\pi ft - 90°) \quad (119)$$

The quantity $1/(2\pi fC)$ which multiplies I_m to give V_{mc} is called the *capacitative reactance* of the capacitor. Like the inductive reactance, it is the ratio of the voltage amplitude to the current amplitude and therefore has the dimensions of ohms. It is also dependent on the frequency but goes to infinity rather than zero as the frequency decreases, indicating that the capacitor is an open circuit rather than a short circuit for direct current. Finally, as in the case of the inductor, the 90° phase difference between the voltage and the current results in energy being alternately delivered to and returned from the capacitor. In the capacitor, it is the electric field between the plates rather than the magnetic field of the coil that stores the energy; but as in the inductor, no energy is dissipated.

● **The Series *RLC* Circuit**   Suppose a resistor, an in-

ductor, and a capacitor are connected in series and a current $i = I_m \sin 2\pi ft$ sent through them as shown in Fig. 72. The total voltage v across the combination must then be the sum of the voltages across the three series elements, and from Equations 116, 118, and 119 we have

$$v = v_R + v_L + v_C = I_m[R \sin 2\pi ft \quad (120)$$
$$+ (X_L - X_C) \cos 2\pi ft]$$

$i = I_m \sin 2\pi ft$

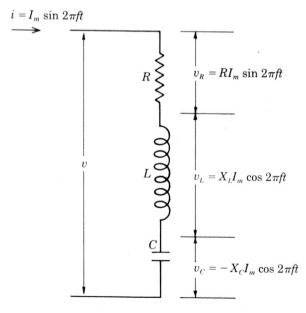

Figure 72   The Series *RLC* Circuit

This expression is not in a convenient form, for it does not display either the amplitude or the phase of v, nor does it really show that v is sinusoidal. It can, however, be rewritten in the form

$$v = I_m Z \sin (2\pi ft + \phi_v) \qquad (121)$$

That Equations 120 and 121 are equivalent is easily shown by expanding the latter with the usual formula for the sine of the sum of two angles. Doing this gives

$$v = I_m(Z \cos \phi_v \sin 2\pi ft + Z \sin \phi_v \cos 2\pi ft) \qquad (122)$$

Equations 122 and 120 are identical if

$$R = Z \cos \phi_v \qquad (123)$$

and $$(X_L - X_C) \equiv X = Z \sin \phi_v \qquad (124)$$

where X is simply defined as the reactance of the circuit and ϕ_v is the phase angle of the voltage with respect to the current.

The quantity Z is called the *impedance* of the circuit. It multiplies the current amplitude I_m to give the amplitude $I_m Z$ of the voltage across the complete circuit. Like the resistance and reactance, it is measured in ohms and is given by

$$Z = \sqrt{R^2 + X^2} \qquad (125)$$

This solution for Z is obtained by squaring and adding Equations 123 and 124. These equations can also be solved for ϕ_v by dividing the second by the first. This gives

$$\tan \phi_v = \frac{X}{R} \qquad (126)$$

These last two results suggest that Z be represented by a vector drawn on a set of coordinate axes designated X and R as shown in Fig. 73. On the left X is positive because $X_L > X_C$, and we say that the net reactance is inductive. On the right is a case of net capacitative reactance (X negative). A most interesting possibility is that of $X = 0$ by reason of X_L and X_C having the same magnitude. In this case, which is called *series resonance,* the circuit acts as if only R were present. The reason is that $v_L = X_L I_m \cos 2\pi ft$ and $v_C = -X_C I_m \cos 2\pi ft$, so that if $X_L = X_C$ these two voltages will be equal in magnitude but opposite in polarity at every instant and will thus cancel each other out. Note, however, that neither v_C nor v_L is zero. This illustrates the most important point that in a series A.C. circuit the amplitudes of the voltages across the individual elements may not be added, for these voltages will in general not have their maximum values at the same time due to their having different phase angles. It is therefore the instantaneous values that must be added, as was done in obtaining Equation 120. In particular, in the series RLC circuit, the voltage across the inductor is maximum and positive when the voltage across the capacitor is at its maximum negative value. Making the two reactances equal makes these two maxima equal, so that the total voltage across the inductor and capacitor in series is zero. Since as the frequency is increased X_L

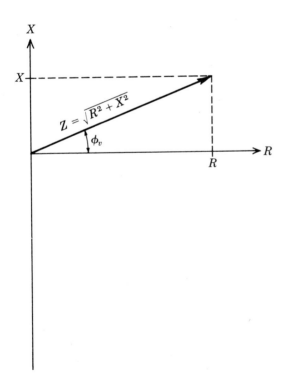
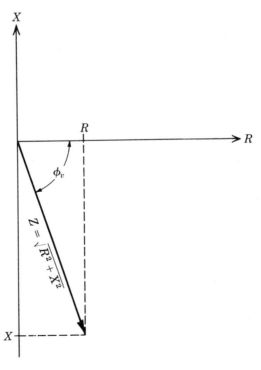

Figure 73 Impedance as a Vector

gets larger while X_C gets smaller, there will surely be a particular frequency f_0 for which they are equal. This is easily found to be

$$f_0 = \frac{1}{2\pi \sqrt{LC}} \qquad (127)$$

● Measurements with the Oscilloscope

The cathode ray oscilloscope is essentially a device for electronically plotting a graph of two interdependent variables. It is built around a cathode ray tube — a long, evacuated glass bulb with an electron-emitting cathode in the small end followed by a series of electrodes used for accelerating the electrons and focusing them into a beam directed at the fluorescent screen which makes up the big end (see Fig. 74). This screen gives off light when struck by electrons, so that a bright spot appears at the point where the beam strikes. This spot may be moved around by deflecting the beam, and two sets of deflection plates mounted at right angles to each other are placed in the tube to do this. When a voltage is impressed across the pair of plates in one of these sets, the electrons in the beam are attracted to the positive plate and repelled by the negative one, and are accordingly deflected in their flight to the screen. A vertical displacement of the spot of light proportional to the voltage across the vertical deflection plates can thus be obtained, and similarly a voltage on the horizontal plates produces a proportional horizontal deflection. Thus, if a varying voltage v_x is connected to the horizontal plates and a voltage v_y, which depends on v_x, according to some function, is put on the vertical plates, the spot will trace out a graph of that function on the screen.

The oscilloscope instrument contains the power supplies necessary for the operation of the cathode ray tube and vertical and horizontal amplifiers whose outputs are connected to the vertical and horizontal deflection plates respectively and which serve to amplify small signal voltages up to the reasonably high levels required for appreciable deflection of the beam. Gain controls, some calibrated by the manufacturer and some requiring calibration by the user, are provided so that the input voltage

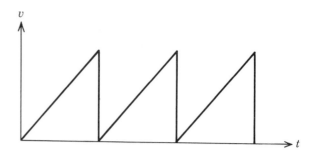

Figure 75   A Sawtooth Voltage

required for a given deflection may be set. The calibration is conveniently given in centimeters per volt. When a voltage that is a function of time is to be studied, the horizontal axis is made a time axis by applying a voltage proportional to t to the horizontal input. Such a voltage is graphed in Fig. 75. It rises linearly with time until it reaches a value corresponding to full beam deflection and then drops quickly back to zero and starts over. It is often called a "sawtooth" voltage, for obvious reasons, and special circuits are provided in the oscilloscope to generate it. When applied to the horizontal deflection plates, it causes the spot to move across the screen at a constant rate and then snaps it back and "sweeps" it across again. During a sweep, the horizontal or x deflection is proportional to time, so that this axis may be calibrated in seconds. If the voltage to be studied is applied to the vertical plates while the spot is moving linearly across the screen, a plot of this voltage as a function of time will be traced out. Furthermore, if this voltage is periodic and the sawtooth is made to have the same period and to start at the same instant that a cycle does, a plot of one complete cycle will be traced over and over again. If the frequency is high enough so that one trace does not have a chance to fade from the screen before the next one starts (period short compared to the screen's *persistence time*), the plot of one cycle will appear to remain stationary and without flickering. To keep the sweep in step with the voltage under study, the oscilloscope includes *synchronizing* circuits, which deliver pulses to the sweep oscillator telling it when to start each sweep. The complexity of these circuits and the elaborateness of the associated controls vary greatly among different oscilloscopes; the details of the instruments provided in your laboratory will be explained by the instructor. In general, however, all oscilloscopes have the following basic controls:

Intensity (or Brightness): Controls the voltage on the cathode ray tube grid and hence the intensity of the electron beam and the brightness of the resulting spot or pattern on the screen.

Focus: Controls the voltage on the first anode and hence the potential difference between it and the second anode, which is at a fixed accelerating voltage. This potential difference has the effect of bringing the electron

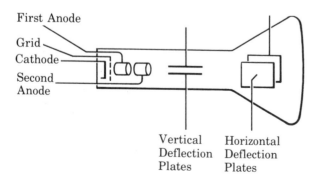

First Anode

Grid

Cathode

Second Anode

Vertical Deflection Plates   Horizontal Deflection Plates

Figure 74   The Cathode Ray Tube

beam to a focus and may thus be adjusted to put the focal point at the screen.

Vertical Centering: Controls the D.C. component of the voltage between the vertical deflecting plates and therefore allows centering of the spot on the screen in the vertical direction. A complete pattern may also be centered to fill the screen evenly in this direction.

Horizontal Centering: Has the same function as vertical centering but in the horizontal direction.

Vertical Gain: Controls the gain of the vertical amplifier and hence the oscilloscope's vertical deflection sensitivity, that is, the number of centimeters of vertical deflection that will be produced by a given input voltage. On some instruments this control is calibrated directly in centimeters per volt, but on the less expensive models you must make your own calibration.

Horizontal Gain: Same as vertical gain, but applies to the horizontal amplifier. Note that when a horizontal time axis is desired, the output of the sweep generator is switched to the horizontal amplifier input, and the horizontal gain then controls the amplitude of the sawtooth applied to the horizontal deflection plates. The gain control then sets the width of the pattern.

Sweep Frequency: Controls the sawtooth repetition rate so that it may be set equal to the repetition rate of a periodic voltage applied to the vertical channel for study. The sawtooth repetition rate may also be set equal to a submultiple of that of the voltage under study, in which case several cycles will appear on the screen. In high-quality oscilloscopes this control is called *sweep speed* and is calibrated directly in seconds for the spot to travel 1 centimeter horizontally. Such an instrument provides a horizontal axis calibrated in time from which the period and hence the frequency of the vertical signal voltage may be read directly. Less expensive instruments do not have this feature and thus do not provide for measurement of the period of the observed alternating current.

Synchronization Controls: These vary in complexity with different oscilloscopes, but in general serve to control the introduction of pulses into the sweep oscillator that tell it when to start each sawtooth. Simple instruments merely have a gain control that sets the level of these pulses. It is advanced from zero until synchronization is obtained but should not be advanced too far, as "oversynchronization" can produce distortion in the pattern by interfering with the linear nature of the sawtooth sweep. There is also a switch to select the source of the pulses. Since the voltage under study and the sweep ordinarily are to have the same period, a very usual source is the voltage under study itself, and the *internal* position of the selector switch makes the necessary connection inside the oscilloscope. For studying voltages derived from the A.C. power line, a *60-hertz* position allows pulses taken from the power line to be introduced into the sweep generator. Finally, an *external* position is provided for cases when synchronization is to be effected by pulses from a separate source. Input terminals are provided on the oscilloscope panel for bringing in such pulses. For further details on the instruments provided in your laboratory, consult the instructor.

● **Lissajous Figures** A very important application of the oscilloscope is the study of the pattern resulting from the application of sinusoidal voltages of the same frequency to both the vertical and horizontal inputs. Such a pattern is called a *lissajous figure.* In general, a lissajous figure is an ellipse and is a stationary pattern on the oscilloscope screen if the two frequencies are indeed equal. Thus, if one signal comes from a calibrated variable-frequency source and the other is unknown, tuning the calibrated source until a stationary ellipse is obtained allows the frequency of the unknown to be read from the calibrated source's dial. The nature of the ellipse can also reveal the phase difference and the ratio of the amplitudes of the two applied sinusoidal signals. Thus, if these signals are represented by

$$v_h = A \sin 2\pi ft \qquad (128)$$

and $$v_v = B \sin (2\pi ft + \phi) \qquad (129)$$

where v_h is the voltage applied to the horizontal input, A is its amplitude, v_v is the vertical input voltage, B is its amplitude, and ϕ is the phase angle between v_h and v_v, then the lissajous figure will be an ellipse whose major and minor axes make an angle θ with the horizontal and vertical axes given by

$$\tan 2\theta = \frac{2 \cos \phi}{\dfrac{A}{B} - \dfrac{B}{A}} \qquad (130)$$

Two interesting special cases deserve particular attention. The first is the very simple one obtained when $A = B$ and $\phi = 0$. Then $v_h = v_v$ and the pattern on the

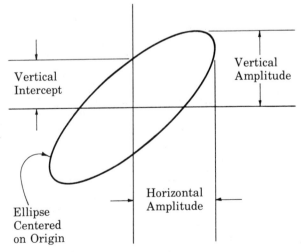

Figure 76 Phase Measurement with a Lissajous Figure

oscilloscope screen is a straight line through the origin at an angle $\theta = 45°$. The second is that of $A = B$ and $\phi = 90°$. In this case $v_v = A \cos 2\pi ft$, and the relation between the vertical and horizontal voltages is therefore

$$v_h{}^2 + v_v{}^2 = A^2 \tag{131}$$

This is the equation of a circle of radius A. In the more general case of $A = B$ and an arbitrary phase angle, an ellipse is obtained from which the phase angle can be determined as follows (see Fig. 76): If the horizontal deflection is given by Equation 128, it will be zero at $t = 0, 1/f, 2/f$, etc., and at these times the vertical deflection, which is given by Equation 129, is $B \sin \phi$. The

vertical deflection at zero horizontal deflection is the vertical intercept marked in Fig. 76 and can thus be read from the oscilloscope pattern. If A is made equal to B, the vertical intercept is $A \sin \phi$ and thus gives $\sin \phi$ when divided by the vertical amplitude. We conclude that the phase angle can be obtained from the relation

$$\sin \phi = \frac{\text{vertical intercept}}{\text{vertical amplitude}} \tag{132}$$

These characteristics of lissajous figures will be useful in our study of the performance of resistors, inductors, and capacitors in A.C. circuits.

APPARATUS

1. Cathode ray oscilloscope
2. Transformer delivering 6.3 volts at 3 amperes
3. Iron-core inductor (about 3 henrys)
4. A board on which are mounted eight $\frac{1}{2}$-microfarad capacitors with individual switches, connected in parallel to two binding posts

5. Unknown resistor (about 2000 ohms)
6. Decade resistance box (0–10,000 ohms)
7. Multimeter with r.m.s. A.C. voltage ranges
8. Assorted hookup wires and connectors

PROCEDURE

1. Connect the output (secondary) of the transformer to the vertical input of the oscilloscope and set the horizontal circuits for a time sweep appropriate for viewing a 60-hertz voltage. Before plugging in the transformer, have the instructor check your arrangement and explain the operation of your particular oscilloscope.

2. Plug in the transformer and adjust the oscilloscope to produce a stationary pattern of one cycle of a sine curve. Learn to use the sweep frequency and synchronizing controls on your instrument by readjusting them to get stationary patterns of two and three cycles.

3. Calibrate the vertical direction on the oscilloscope screen in centimeters per volt at the vertical input terminals. The method appropriate for your oscilloscope will be described by the instructor. Then measure the transformer secondary voltage (which is also the vertical input voltage) with the multimeter. Be sure the meter is switched to the proper range before connecting it. Record the meter reading. Using the vertical calibration just obtained, measure and record the voltage amplitude by converting the amplitude of the sinusoidal pattern displayed on the oscilloscope to volts.

4. Unplug the transformer, disconnect it from the oscilloscope, and wire the transformer, oscilloscope, unknown resistor, and decade box in the circuit diagrammed in Fig. 77. The unknown resistor should be used in the position marked "Element Under Test." Notice that this element and the decade box are in series across the 6.3-volt source (the oscilloscope inputs have very high impedance and thus draw negligible current) so that the same current flows in both and that the voltage across the element under test is applied to the oscilloscope's vertical input and the voltage across the decade resistance box to the horizontal input. These two voltages obviously have the same frequency, and so a stationary lissajous figure will be displayed on the oscilloscope screen. Note, however, that the "high" terminal of the horizontal input is connected to the "low" rather than the "high" terminal of the resistance box. This means that the polarity of the voltage applied to the horizontal input is reversed relative to that applied to the vertical input, so that the pattern will be reversed left-to-right as compared with that shown in Fig. 76. This will have no effect on your measurements, however.

5. Set the oscilloscope to display the lissajous figure, that is, switch the horizontal channel to the input terminals rather than the sweep generator. CAUTION: Be sure you have a pattern rather than just a spot on

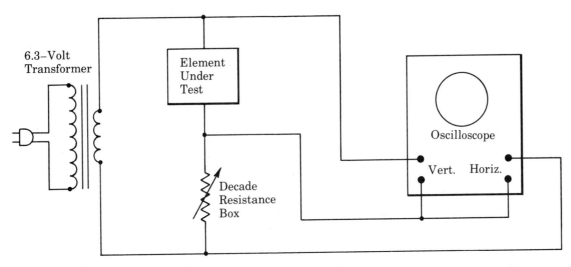

Figure 77   Circuit for Impedance Measurement

the screen. Oscilloscopes should not be left with the beam on but no horizontal or vertical deflection, as the continuous electron bombardment at one point on the screen will soon burn away the fluorescent material there. Set the horizontal and vertical gain controls so that the horizontal and vertical sensitivities are equal and are such as to give a pattern filling most of the screen. To do this, set the resistance box to zero and temporarily shift the "high" horizontal input lead to the "high" vertical input terminal so that both oscilloscope inputs receive the same voltage, namely that across the resistor under test. Then adjust the gain controls so that a straight line extending almost all across the screen and at exactly 45° to the horizontal is obtained. The centering controls should be adjusted until this line passes through the origin, that is, the center of the screen. The angle of 45° is assured if the horizontal extent of the line (its horizontal component) is exactly equal to its vertical extent (the vertical component). Do not touch the centering or gain controls after these adjustments have been made.

6. Disconnect the horizontal input lead from the vertical input terminal and reconnect it to the "low" side of the resistance box so that your circuit is again as shown in Fig. 77. Adjust the resistance box until a line at exactly 45° to the horizontal is displayed. Note that this line will extend up to the left rather than up to the right like the line obtained in Procedure 5 because of the polarity reversal discussed in Procedure 4. Record the final setting of the decade box.

7. Remove the unknown resistor from your circuit and substitute the inductor as the element under test. Because the inductor is not a pure inductance but presents an impedance consisting of a resistance and its inductive reactance in series, the pattern will not be a circle but an ellipse similar to that shown in Fig. 76 except for the left-right reversal. Adjust the resistance box so that the horizontal and vertical pattern amplitudes defined in Fig. 76 are exactly equal. Record the resistance box setting and the values of the vertical amplitude and vertical intercept.

8. Remove the inductor from your circuit and substitute the capacitor board as the element under test. Close four of the eight switches on this board to set up a capacitance of about 2 microfarads between its terminals. Adjust the resistance box so that equal horizontal and vertical amplitudes are obtained. Practical capacitors approach the ideal rather closely, so that their impedance is usually very nearly a pure capacitative reactance. You should therefore get a very good circular pattern. In this case the vertical amplitude and vertical intercept are the same, and Equation 132 gives $\phi = 90°$ as it should for a pure capacitance. Observe this fact, and record the setting of the resistance box. *Note:* The transformer may disturb the sinusoidal nature of the voltage enough so that some distortion may be apparent in the displayed circle, but if a reasonably good transformer is used, this effect will be negligible.

9. Repeat Procedure 8 with all switches closed on the capacitor board to give a capacitance of about 4 microfarads.

10. Rewire your components in the circuit of Fig. 78 in order to observe series resonance. Use the capacitor board at C, the iron core inductor at L, and the resistance box set to 10 ohms at R. Notice that R serves as a current transducer, that is, the current through it produces a proportional voltage across it according to Ohm's law and this voltage is then observed with the oscilloscope. The vertical oscilloscope

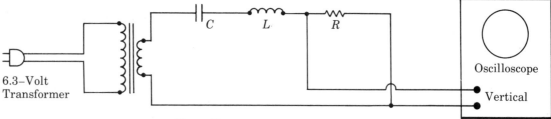

Figure 78 Circuit for Observing Resonance

deflection is thus proportional to the current flowing through the circuit. Set the oscilloscope to observe the 60-hertz alternating current and look for series resonance. To do this, start with all the switches on the capacitor board open except one ($C = 0.5$ microfarad) and close them one at a time until the sine curve displayed on the oscilloscope has maximum amplitude. This corresponds to resonance, for the circuit impedance is a minimum when $X = 0$, hence a fixed voltage source such as the transformer secondary coil will produce maximum current. Record the number of switches closed when maximum current amplitude is observed.

DATA

R.m.s. transformer secondary voltage (from multimeter) _____

Calculated amplitude of transformer secondary voltage _____

Amplitude from oscilloscope _____

Percent error _____

• Resistance Measurement

Resistance box setting _____

Given value of unknown resistor _____

Percent error _____

• Inductance Measurement

Resistance box setting _____

Pattern amplitude _____

Pattern intercept _____

Phase angle ϕ _____

Resistance of inductor _____

Reactance of inductor _____

Inductance of inductor _____

Given value of the inductance _____

Percent discrepancy _____

• Capacitance Measurement

Number of Switches Closed	Resistance Box Setting	Capacitance		Percent Discrepancy
		Measured	Given	
4				
8				

• Series Resonance

Number of switches closed _____

Value of C _____

Value of L _____

Calculated value of the frequency _____

Percent error _____

CALCULATIONS

1. Compute the amplitude of the transformer secondary voltage from the r.m.s. value measured with the multimeter in Procedure 3. Compare this result with the amplitude as determined from the oscilloscope by calculating the percent error.

2. The resistance box setting obtained in Procedure 6 should be equal to the resistance of the unknown resistor. When you have finished the experiment, the instructor will tell you the actual value of this resistor. Compare it with your measured value by calculating the percent error.

3. Calculate the phase angle between the current through and the voltage across the inductor from the data of Procedure 7. The resistance box setting obtained in this step is the inductor's impedance. Using Equations 123 and 124, compute the resistance and reactance of the inductor.

4. From the reactance of the inductor found in Calculation 3 and the known line frequency, compute the inductance of the inductor. Compare your result with the known value by calculating the percent discrepancy. *Note:* The value of the inductance in henrys marked on the inductor is nominal and may differ from its actual value by as much as 20%, so that close agreement between the marked value and the result of your measurement cannot be expected. If a better value of the inductance is available, your instructor will give it to you when you have finished the experiment so that a more meaningful comparison can be made.

5. Calculate the capacitance from your data in Procedure 8 and compare it with the known value by calculating the percent discrepancy. In Procedure 8 four switches were closed, placing four 0.5-microfarad capacitors in parallel for a nominal capacity of 2 microfarads. However, the actual capacity may differ from the nominal value marked on the capacitors by ±10% or more, so that good agreement between nominal and measured values cannot be expected. If your instructor has better values for the capacitors on the capacitor board, he will so inform you when you have finished the experiment so that you can compare your result with a more precisely known capacity.

6. Repeat Calculation 5 using the data of Procedure 9.

7. From the number of switches closed for resonance in Procedure 10, find the resonant value of C. Then using the inductance found in Calculation 4, compute the resonant frequency of your series RLC circuit. Compare your result with the known value of the A.C. line frequency, 60 hertz. *Note:* The value 60 hertz for this frequency is not nominal but very precise (about 1 part in 10,000), since the accuracy of electric clocks depends on it. Thus, if you have good values for L and C, your calculated frequency should be very nearly equal to 60 hertz. Note, however, that you could vary C only in 0.5-microfarad steps, which places a limit on the precision of this experiment quite apart from the precision with which you know the values of the inductance of the inductor and the capacitances of the capacitors on the capacitor board.

QUESTIONS

1. An electric heater is run from the 120-volt A.C. line. What resistance should it have in order for it to heat 200 grams of water from room temperature (20°C) to the boiling point in five minutes? Assume normal atmospheric pressure and neglect the heat capacity of the cup containing the water.

2. Calculate the amplitude of the 120-volt A.C. line voltage.

3. The magnetic field of a very long solenoid (a single-layer cylindrical coil) is, to a good approximation, concentrated in the region inside the solenoid where it is uniform, directed parallel to the solenoid axis, and given by

$$B = \frac{\mu N i}{h}$$

where B is the field strength in teslas, μ is the permeability of the core material (the material on which the solenoid is wound), N is the number of turns in the coil, i is the current through the coil in amperes, and h is the length of the solenoid in meters. Using the definitions of Φ and L, derive an expression for the inductance of a long solenoid.

4. Show that if the henry were written in terms of the dimensions of fundamental quantities in the MKS system, it would be a newton-meter-second²/coulomb².

5. Prove by any method you choose that the average of $(\sin 2\pi ft)(\cos 2\pi ft)$ is zero over one period and hence over any integral number of periods.

6. Derive Equations 125 and 126 as outlined in the theory section.

7. Derive Equation 127.

8. Explain why no vector representing Z can ever be drawn in the left half-plane (the region to the left of the vertical axis) in diagrams such as those shown in Fig. 73 as long as only passive circuit elements (elements like resistors, capacitors, and inductors that do not include a source of power) are considered.

9. Explain why the decade box resistance obtained in Procedure 7 is equal to the impedance of the inductor being tested. Point out in particular why it is equal to the *impedance,* not the resistance or reactance of the inductor.

10. In high-quality oscilloscopes having the sweep frequency control calibrated directly in seconds per centimeter, the horizontal gain control is automatically switched out of the circuit when the sweep system is being used so that the pattern width is permanently preset rather than being adjustable. Why is this so?

11. Show that if there is no phase difference between the two sinusoidal voltages applied to an oscilloscope in a lissajous figure observation, the angle θ between the pattern axis and the horizontal axis is given by

$$tan\theta = \frac{B}{A}$$

where A is the horizontal and B the vertical amplitude. What does the pattern look like under these circumstances? Does Equation 130 agree with the above result when $\phi = 0$?

12. The A.M. broadcast band runs from 540 to 1600 kilohertz (thousands of cycles per second). One common tuning method used in older radio receivers was to have an inductor connected to a variable capacitor driven by the tuning dial. A station was selected by adjusting C so that resonance was obtained with the fixed L at the station frequency. (a) The variable capacitor used usually had a maximum capacity of 365 picofarad (10^{-12} farad). Find the value of the inductance needed to tune in the low-frequency end of the broadcast band. (b) What must the variable capacitor's minimum value be in order that the high-frequency end of the broadcast band may be tuned in?

Vacuum and **35** Semiconductor Diodes

The diode or two-element electronic device allows conduction of an electric current in one direction only. Moreover, the volt-ampere characteristic (how the current through the diode depends on the voltage across it) is not linear even in the forward (current-carrying) direction, and can in some cases be made to serve certain special purposes. These properties make the diode, whether in the form of a vacuum tube or a semiconductor device, a tremendously important circuit element in electronic systems. The purpose of this experiment is to study some diode characteristics and applications.

THEORY

● **The Thermionic Diode** If a metal electrode is placed in a vacuum and heated, a certain number of electrons will break through the potential barrier at the surface and be emitted into the surrounding space. If a second electrode is located near by, and is maintained at a positive potential with respect to the first, the negative electrons will be attracted to it and a current will apparently pass through the vacuum. If, on the other hand, the cold surface is made negative and the hot one positive, no current will flow. The field now wants to draw electrons from the cold to the hot electrode, but the cold one does not emit an appreciable number, and so there are few available to cross the gap. This phenomenon was first observed by Edison and was called the Edison effect. It is the basis of operation of all modern thermionic diodes, which exist to perform the function just described — the conduction of current in one direction only — and which differ only in size and shape to suit various applications in which unidirectional conduction is desired.

The current per unit area carried by electrons emitted through a metal surface is given by

$$J = AT^2 e^{-e\phi/kT} \qquad (133)$$

where J is the current density in amperes per square centimeter, A is a constant having the dimensions of amperes per square centimeter per degree Kelvin squared, e is the electronic charge in coulombs*, ϕ is a constant expressed in volts called the *work function* of the surface, k is Boltzmann's constant in joules per degree Kelvin, and T is the Kelvin or absolute temperature of the surface. Notice that kT is a thermal energy, ϕ is the energy per unit charge required to get charge (in the form of electrons) out of the surface, $e\phi$ is therefore the energy required to get one electron out, and the ratio $e\phi/kT$ is thus the ratio of energy required to extract an electron to the thermal energy available. Equation 133 is known both as Dushman's and as Richardson's equation, and the exponential dependence of emission current on temperature has been well established experimentally. However, the dependence of J on the exponential is so predominant as to mask its dependence on T^2, which has never been decisively verified. Indeed, Richardson originally proposed a relation between J and T in which the T^2 factor was replaced by $\sqrt{T}$. Furthermore, although the Richardson-Dushman equation may be derived very nicely on the basis that the electrons in the metal constitute a perfect Fermi-Dirac gas, the constant A appears as a combination of fundamental constants, and should thus not vary from metal to metal. Experimentally, A is found to differ for different metals, and furthermore has the value 60.2 amperes/cm² – °K² for several materials, whereas the theoretical value is almost exactly twice this. These discrepancies have never been explained satisfactorily, but the Richardson-Dushman equation is nevertheless used, although in a rather empirical manner with experimentally determined values for A and ϕ.

If the positive electrode (usually called either the anode or the plate) is at a sufficiently high potential so that all electrons emitted by the hot electrode (the cathode, or in the case of a simple wire heated by the direct passage of a heating current, the filament) are immediately drawn over to it, the current is given by the Richardson-Dushman equation and is thus a strong function of the temperature. Note that the property of unidirectional conduction depends on this, for if the potential difference is reversed, the anode tries to play the cathode's role, but cannot simply because it is relatively cold and therefore emits far fewer electrons. Under these conditions, the current in the forward direction (anode positive) is almost independent of the anode voltage, as all the emitted electrons are being collected. The anode

---

*This is the e in the exponent. It should not be confused with the same symbol used for the base of natural logarithms.

(or plate) current is thus a function of the cathode temperature only, and is therefore said to be *temperature limited*.

It is possible (and is indeed the usual case in commercial vacuum tubes, which employ cathodes with highly efficient emitting surfaces) to have the plate voltage insufficiently great to draw over immediately all the emitted electrons. There then develops an electron cloud and hence a distribution of negative charge called a *space charge* in the region surrounding the cathode. This space charge forms a virtual cathode that supplies the electrons that flow to the plate and at the same time effectively presents a negative electrode to the cathode surface, thus shielding it from the plate's positive field. Electrons emitted by the cathode then tend to be repelled back to it. In the steady state, the space charge assumes a stable, equilibrium value at which the number of electrons leaving it for the anode is just balanced by the number which join it from the cathode. Under these conditions, the plate current depends on the plate voltage, which determines how many electrons will be drawn out of the space charge per unit time, and not on the cathode temperature, since excess electrons are merely repelled back to the cathode. Therefore, an increase in temperature does not change the net emitted current. The diode's plate current is then said to be *space-charge limited*.

It may be shown that under space-charge-limited conditions, the plate current is proportional to the three-halves power of the plate voltage for all tube geometries, provided only that the electrons may be assumed to be emitted from the cathode with zero initial velocity. The three-halves power law was first derived for the case of plane parallel electrodes by Child and by Langmuir, who found that

$$J = B\frac{V_b^{3/2}}{d^2} \tag{134}$$

where J is the current density as before, V_b is the interelectrode potential (plate voltage) in volts, d is the electrode separation in centimeters, and B is a proportionality constant composed of various fundamental quantities and having the dimensions of amperes per volt$^{3/2}$. This relation is thus often referred to as Child's law or Langmuir's equation. Except for effects due to the initial emission velocities of the electrons (which result in some dependence of J on T), the three-halves power law is well established experimentally.

In rectifier service, that is, in applications where only the diode's property of conducting in one direction is made use of, the three-halves power law is of little interest. In fact, the current should preferably flow through the diode in the forward direction without any attendant voltage drop. Accordingly an *ideal diode* is defined as a circuit element that is an open circuit for one polarity and a short circuit for the other. The vacuum diode is indeed an open circuit when its plate is negative, but the three-halves power law means that in the forward direction it is

not ideal but has a certain voltage drop across it. Therefore it also dissipates some power (namely the voltage across it times the current through it) and thus gives rise to a loss of energy in the system. We shall see below that semiconductor diodes, besides being smaller and requiring no cathode heating, are somewhat better approaches to the ideal and are therefore preferred in almost all present-day rectifier applications.

● **The Semiconductor Diode** It is possible to obtain an essentially unidirectionally conducting device by forming a junction between two types of semiconductor material called P type and N type. Both types are made by "doping" pure semiconductor materials such as silicon or germanium with appropriate impurity elements. Pure silicon and germanium are called *intrinsic semiconductors* and behave in a rather simple way. In both materials the electrons are bound in the crystal lattice and require a certain small amount of energy to break free. This is in contrast to metals, in which large numbers of electrons are free to move. Quantum mechanically speaking, we note that in metals there is a continuum of energy states extending upward (on an energy scale) from the level normally occupied by the electrons so that it is possible to give an electron any energy, however small, and it will immediately accept it and start to move through the lattice with the kinetic energy it has thus acquired. In the semiconductor, on the other hand, there is a range of energies that the electrons in the material cannot have, extending from the levels normally occupied up to a certain level above which is again found a continuum of allowed states. This continuum is called the *conduction band,* while the array of normally occupied states is called the *valence band.* Thus the difference between a metal and a semiconductor is that there is a gap between the valence and conduction bands in a semiconductor whereas there is none in the metal. Hence in a semiconductor, an electron must first be excited with sufficient energy to jump the gap from the valence to the conduction bands before it finds a continuum of available states that will allow it to move about freely, that is, to take part in electrical conduction as a charge carrier. Since few electrons have this energy under ordinary conditions, few are available for conduction, hence the conductivity of an intrinsic semiconductor is relatively low. Note that the conductivity increases as the temperature is raised, for the higher the temperature the more electrons will have sufficient thermal energy to cross the gap into the conduction band.

A P-type semiconductor is made by doping an intrinsic semiconductor with impurity atoms having a valence one less than the intrinsic semiconductor material. The presence of the impurity atoms in the crystal lattice raises the conductivity of the material because there are some energy states available to electrons which are unoccupied, for the impurity atoms bring insufficient electrons with them. Other electrons can thus move into these sites just as in a row of parked cars with one space open: the car behind can move into the space, leaving a space where it formerly stood into which the next car can move, and so forth. Note that if this situation were observed from high in the air, it would look as if a space were moving in the direction opposite from that of the cars. Similarly, the unoccupied sites obtained when an intrinsic semiconductor is doped with a lower valence impurity move in an applied field as if they were positive particles, and indeed these ''holes'' may be regarded as positive electrons and treated as positive charge carriers. Semiconductors that have been doped in this way are thus called P-type because positive charge carriers appear to be responsible for conduction.

Clearly an N-type semiconductor is made by doping the intrinsic material with impurity atoms having a larger valence. These atoms bring in extra electrons, so that again the conductivity of the intrinsic semiconductor is increased, although this time by reason of the presence of negative charge carriers. Conduction therefore proceeds in a manner similar to that in metals.

The modern solid-state rectifier diode is made by forming a junction between pieces of N-type and P-type semiconductor material. At this so-called *PN junction* the holes in the P-type material are brought together with the electrons in the N-type so that they combine and neutralize to form a *depletion layer* in the neighborhood of the junction in which there are few free charge carriers. The flow of holes into the depletion layer from the P-type material leaves a negative charge on that side of the junction, whereas the N-type material becomes positively charged. In this way a potential barrier is set up across the depletion layer that prevents further current flow. Placing across the junction an external potential of such a polarity that the N-type side is made positive may be thought of as drawing the electrons back from the junction in the N-type material and the holes back in the P-type, which is made negative in this situation. The thickness of the depletion layer is thus increased. The external potential may be considered as adding to the internal potential across the depletion layer and thus helping to stop all motion of the *majority* charge carriers (electrons in the N-type material and holes in the P-type). Note, however, that even in N-type material there are a few holes and in P-type there are a few free electrons that had enough thermal energy to reach the conduction band. These are called *minority* charge carriers, and under the conditions just described a small current will flow across the junction because of them. This is called a *reverse current,* and when the N side of the junction is positive the junction is said to be *reverse biased.*

If the N-type material is made negative, however, electrons are driven into the depletion layer, and at the same time holes are driven in from the P-type side. There is a large amount of recombination, the depletion layer is reduced in thickness, and a large current flows. One may consider that the applied potential, by supplying positive charge to the P-type material and negative charge to the N-type, neutralizes the internal potential barrier so that the large current flow can take place. Under these conditions the junction is said to be *forward biased.*

A PN-junction diode is thus an excellent rectifying (unidirectionally conducting) device. Besides being small and requiring no cathode heating, it has the additional advantage over vacuum diodes that in the forward direction the current rises steeply after a volt or so has been applied — much more steeply than the three-halves power function that is found to hold for vacuum diodes. The semiconductor rectifier is thus a closer approximation to the ideal and has accordingly become more and more popular. When reverse-biased, however, a PN junction passes a small current and is thus inferior in this respect to the vacuum diode, which passes no current in the reverse direction no matter what the applied reverse voltage right up to the point where the insulation between the anode and cathode breaks down. The semiconductor diode's reverse current is usually negligible, however, if the temperature is not allowed to rise appreciably. Since the reverse current is due to minority carriers originating from electrons in the crystal occasionally having enough thermal energy to break their bonds and create electron-hole pairs, an increase in temperature results in an increased reverse current. Because of the high reverse voltages often encountered, considerable energy dissipation may be associated with this reverse current, and under some conditions the temperature may continue to rise and a runaway effect occur, ending in the destruction of the diode. Keeping the junction cool is thus an important consideration in semiconductor applications, hence the sometimes odd-looking designs of cases for semiconductor devices, the widespread use of ''heat sinks'' to carry off the heat generated, and the attention paid to temperature in semiconductor rating charts. Moreover, the *peak inverse voltage* (P.I.V.) rating is particularly critical for a semiconductor diode, and good circuit design dictates that a large safety factor be employed. This derating of the diode is especially necessary when large temperature variations may be encountered, since elevated temperatures, because they result in increased numbers of minority charge carriers, tend to reduce the value of the breakdown voltage.

The principal use of diodes is for *rectification,* which means the application of their property of unidirectional conduction to the conversion of alternating current (current that flows first in one direction and then in the other) to direct current (which flows in one direction only). An obvious circuit for doing this is shown in Fig. 79(a). Notice the symbols used. On the left, the circle with the tilde in it represents a voltage source and is marked with the voltage's polarity. In this case of an alternating voltage, the actual sinusoidal dependence of voltage on time is given and the + and − signs mean that if the plus end is treated as the ''high'' side, the observed voltage will be as given, whereas if the polarity is reversed, the ob-

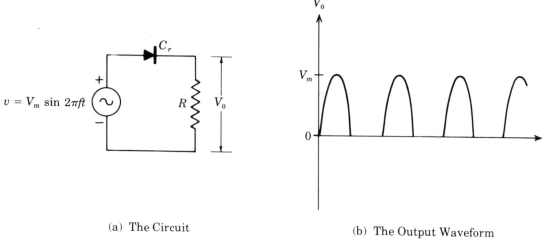

(a) The Circuit   (b) The Output Waveform

Figure 79   Half-Wave Rectification

served voltage will be $-V_m \sin 2\pi ft$, corresponding to a 180° phase reversal. The symbol ▸▸― is universally used to represent a diode, the direction of conduction ("forward" direction) being indicated by the arrow. Note that if this symbol stands for a vacuum diode, the arrow represents the anode or plate. The other end is then the cathode or filament, and the circuit for heating the cathode is understood to be there without being actually shown on the drawing.

If C_r in Fig. 79(a) is ideal, the output voltage V_0 obtained across the load resistor R is as shown in Fig. 79(b), where it is plotted as a function of time. It is direct in the sense that it never goes negative, but it is hardly constant at a particular value as is the D.C. voltage from a battery. In fact, the diode (or *rectifier*, as a diode used in this service is often called) looks like an open circuit during the intervals in which the voltage v from the voltage source is negative, so that V_0 is zero during these intervals. The rectifier has thus "thrown away" all the negative half-cycles and passed all the positive ones. The arrangement of Fig. 79 is consequently called a *half-wave rectifier* circuit.

Despite the variations plotted in Fig. 79(b), the rectified output voltage can be made almost pure direct current (almost constant) by special circuits called *filters*. The purpose of a filter is to distinguish between the A.C. and D.C. components making up the time-dependent voltage of Fig. 79(b), eliminating the A.C. components and retaining the D.C. or constant one. By the D.C. component is meant the average value of V_0, which can be shown to be V_m/π or approximately one third of the amplitude of the A.C. input voltage in the case of the half-wave rectifier. There are many different kinds of filter circuits, but one of the simplest and most effective consists of a large capacitor connected in parallel with the load resistance R, as shown in Fig. 80. Assuming that the diode and A.C. voltage source can supply whatever current is needed, we note that during the positive half-cycles a large charging current will flow into the

capacitor in addition to the current delivered to the load. In fact, the capacitor will ideally receive charge $Q = CV_m$ where C is its capacity in farads and V_m is the amplitude of the A.C. input voltage in volts. During the negative half-cycles, when the diode is an open circuit, the capacitor supplies current to the load, discharging in the process. In fact, as soon as the A.C. voltage v decreases from its maximum value V_m to a value less than the voltage on the capacitor, the diode becomes reverse-biased and hence an open circuit. Thereafter the output voltage is simply the exponentially decreasing voltage obtained when a capacitor discharges through a resistor. The exponential decrease continues until the next positive half-cycle, when v rises above the voltage across C, the diode becomes forward-biased once more, and the capacitor is again charged up to V_m. If the RC time constant is large, the voltage decrease does not get a chance to proceed very far in one period of the A.C. wave, and so the output voltage not only never falls to zero but indeed stays near the value V_m of the A.C. voltage amplitude. Hence this filter does more than just get rid of the A.C. components in the wave of Fig. 79(b). It also raises the average to a value close to V_m. In fact, one form of A.C. voltmeter called a peak-reading voltmeter uses the circuit of Fig. 80 with v as the voltage to be measured and R the meter resistance. Note, however, that both good filtering and making the average approach

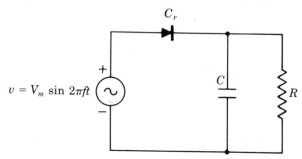

Figure 80   The Half-Wave Rectifier with Capacitor Filter

V_m depend on the time constant RC being very long compared with a period of the alternating current. In the peak-reading voltmeter this is easy to do with reasonable values of C because R, the resistance of the D.C. voltmeter used as the load, can be made very large. In power supplies, however, R represents the load, and in the case of a heavy load (one drawing a lot of current), R is small. Good filtering then requires that C be very large.

The waveform (graph of voltage as a function of time) shown in Fig. 79(b) has a low average and large A.C. component principally because the negative half-cycles of the A.C. input are not used, voltage V_0 being simply zero during those time intervals. This situation would be greatly improved by a rectifier circuit that made use of the negative half-cycles, somehow changing their polarity so that a positive output would be obtained. Such a circuit appears in Fig. 81. Notice that there are two identical voltage sources, each producing $V_m \sin 2\pi ft$, on the left. The upper one is in a circuit with C_{r_1} and R identical to the circuit of Fig. 79(a) and gives rise to a voltage across R identical to that plotted in Fig. 79(b). In addition, the circuit consisting of the lower voltage source, C_{r_2}, and R is also identical to that of Fig. 79(b) except that the polarity of the source is reversed, the "minus" end rather than the "plus" end going to the diode. As a result of this phase reversal, the negative half-cycles appear as positive ones to C_{r_2} and are thus the ones passed by this diode to the upper end of R to fill in the spaces between the positive half-cycles passed by C_{r_1} as illustrated in Fig. 82. Because both positive and negative half-cycles are used, the present scheme is called a *full-wave rectifier*, and because another half-cycle is inserted in each empty space between the half-cycles delivered by the half-wave rectifier, the average output of the full-wave rectifier is twice that of the half-wave circuit, or $2V_m/\pi$. Moreover, positive peaks come at twice the repetition rate of those obtained in the half-wave case, making filtering correspondingly easier.

The disadvantage of the full-wave rectifier circuit of Fig. 81 is the need for two voltage sources. This is less of a problem than it might at first appear because of their series connection. The usual A.C. voltage source supplying a rectifier is the secondary coil of a transformer, and this winding is easily made to give $2V_m \sin 2\pi ft$ (an A.C. voltage having the amplitude of the two sources of Fig. 81 in series) and to have a centertap (a connection to the middle turn of the winding). Then $V_m \sin 2\pi ft$ appears between the centertap and either end, these two voltages are in series with the proper polarity, and in fact the centertapped secondary coil is in every way identical to the arrangement of the two separate sources of Fig. 81. The resulting circuit is shown in Fig. 89. Because of the centertap, it is called the *full-wave centertapped rectifier* system. Note, however, that the transformer secondary coil has to be wound to give a total amplitude of $2V_m$ in order to produce a peak voltage of only V_m at the rectifier output. In other words, a factor of two in

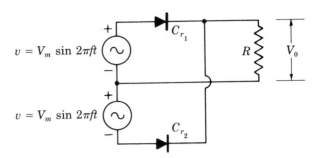

$v = V_m \sin 2\pi ft$

$v = V_m \sin 2\pi ft$

Figure 81 The Full-Wave Rectifier

voltage has been sacrificed. This is somewhat compensated for by the fact that the average current delivered to the load can be twice that for which the transformer is rated, since each half of the secondary coil supplies load current only half the time. However, the fact that the secondary coil has to be wound for twice the load voltage occasionally dictates the use of the different arrangement described below.

The so-called *bridge rectifier* system features full-wave rectification without sacrificing a factor of two in output voltage at the cost of requiring four diodes. To understand the circuit, look first at Fig. 83, which shows two full-wave centertapped rectifier circuits driven from a single pair of voltage sources. Note that in the circuit employing diodes C_{r_2} and C_{r_3}, these diodes are reversed in polarity so that a negative output voltage is obtained at terminal N relative to the common terminal G. Such an arrangement is sometimes used when both a negative and a positive voltage relative to a common ground point are needed. Thus point P is at a positive voltage with respect

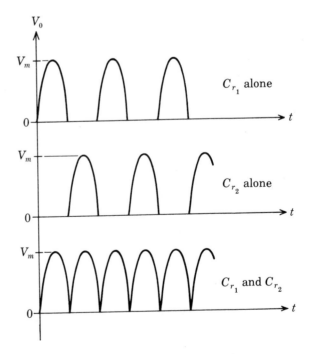

Figure 82 Full-Wave Rectifier Waveforms

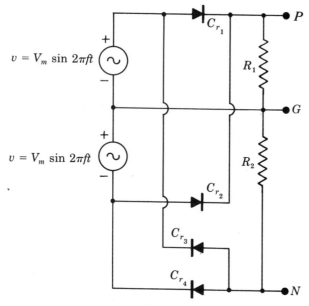

Figure 83   Development of the Bridge Rectifier
Circuit from Two Full-Wave Centertapped
Systems

trated in Fig. 82. An identical voltage appears between N and G except that N is negative with respect to G.

Now, if $R_1 = R_2$, the two currents returned through the common ground line to the centertap (the connection between the two voltage sources in Fig. 83) are equal and opposite, so that the net current in this line is zero. Hence it and the centertap connection can be omitted without changing anything. Point G is then also omitted, and the voltage between P and N is seen to be twice the full-wave centertapped rectifier output, since in effect we have two such systems in series. Thus between P and N the peak voltage is $2V_m$ and hence is equal to the amplitude of the voltage across the two sources in series. The complete circuit is usually drawn as shown in Fig. 84, where the two sources of Fig. 83 (which now have no connection to the line between them) have been combined into a single source giving twice the amplitude, and the load resistors R_1 and R_2 (which are now simply in series) have been combined into an equivalent resistance $R = R_1 + R_2$. You should compare Figs. 83 and 84 and note that they are in fact identical except for the omission of the terminal G and the associated common ground line. Further comparison of Fig. 84 with the circuit of the Wheatstone bridge (Fig. 47) should reveal why the bridge rectifier circuit is so called.

to G, and this voltage has all the characteristics of the output of the full-wave centertapped rectifier as illus-

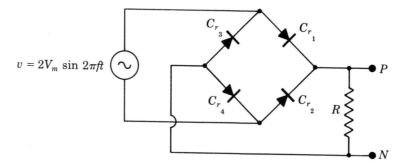

Figure 84   The Bridge Rectifier Circuit

APPARATUS

1. Type 6AU4GT vacuum diode
2. Octal-type radio tube socket, mounted
3. Four type 1N5391 semiconductor diodes
4. Filament transformer, 6.3-volt, 3-ampere centertapped secondary
5. Oscilloscope
6. Electronic voltmeter (EVM)
7. D.C. milliammeter (0–100 milliamperes)
8. 12-volt D.C. power supply or storage battery
9. Tubular rheostat (about 100 ohms)
10. Two carbon resistors (10,000 ohms, 2-watt rating)
11. 68-ohm, 2-watt carbon resistor
12. A board on which are mounted eight $\frac{1}{2}$-microfarad capacitors with individual switches, connected in parallel to two binding posts
13. Electrolytic capacitor (about 10 microfarads, 50-volt rating)
14. Assorted hookup wires and connectors

PROCEDURE

1. Connect the type 6AU4GT diode vacuum tube in the circuit of Fig. 85. Do not plug in or turn on the power supply (or, if you are using a storage battery, do not make the final connections to the battery terminals) until the circuit has been approved by the instructor. The tube, which is used in the horizontal scanning system of many television sets, has a so-called *indirectly heated* cathode. That is, the cathode (or

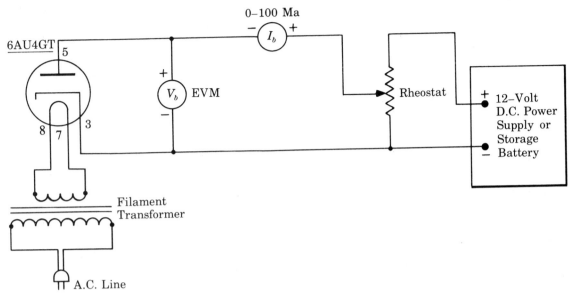

Figure 85   Circuit for Measuring the Vacuum Diode's Volt-Ampere Characteristic

electron emitter) is a coated nickel sleeve that is separately heated by a heater element placed inside it and insulated from it. The heater serves only to heat the cathode and plays no part in the emission of electrons. This separation of the heating and cathode circuits offers certain advantages over the older arrangement of having a filamentary cathode heated by the direct passage of the heating current through the emitting element itself, and almost all modern tubes use it.

The 6AU4GT is equipped with the so-called *octal base* — a very common base style having eight identical, equally spaced connection pins arranged in a circle around a large, central bakelite pin with a key on one side. The key fits into a keyway in the tube socket to determine the one orientation in which the tube can be inserted. The key does not go all the way to the end of the central pin, thus allowing this pin to be started into the socket and the tube rotated until it goes in the rest of the way. The ease with which octal-based tubes may be plugged into sockets located in inaccessible positions is one of the great advantages of this base style.

The connection of the tube elements to the base pins is given in terms of the following standard pin-numbering arrangement: Looking at the tube's *bottom (base) end,* assign no. 1 to the pin immediately to the left of the key on the central pin (with the key in the 6 o'clock position) and number the pins clockwise, ending with no. 8 immediately to the right of the key. *Note:* The clockwise numbering order pertains to the *bottom* view of the tube. If you are looking *down* on the tube from above, the pin numbering will proceed counterclockwise. Provision is made for eight pins on all octal bases, although if some are not needed, they may be omitted. For example, on the 6AU4GT, whose connection diagram is given in Fig. 86, pins 1, 4, and 6 are omitted although there are places for them in the base. Note that when a pin is omitted, its number is omitted from the pin count also. Thus pin 5 on the 6AU4GT remains pin 5 even though there is no pin 4. For convenience, the proper pin numbers have been marked on the wiring diagram of Fig. 85.

All ordinary commercial tubes have oxide-coated cathodes whose very low work function provides copious electron emission with a minimum expenditure of power for cathode heating. Such tubes are not very satisfactory for demonstrations of the temperature limitation of emission current, and no attempt to show this will be made in the present experiment. Instead, the 6AU4GT will be operated at its rated heater voltage of 6.3 volts and the three-halves power law of space-charge limitation investigated. The 6AU4GT works well under these conditions because of its simple geometry. Study the element structure before you plug the tube into its socket, noting in particular the long cylindrical cathode and the plate structure, which presents a concentric cylindrical surface to the cathode.

2. When the instructor has approved your circuit, plug in the filament transformer and allow a couple of minutes for the tube to warm up. Then set the electronic voltmeter (EVM) for a full-scale reading of 10 or 15 volts D.C., make sure the rheostat is set for zero plate voltage (contact all the way to the negative end), and plug in and turn on the power supply or make final connections to the storage battery. Adjust the rheostat for a plate voltage V_b of exactly 1 volt. Read and record the plate current I_b as indicated by the milliammeter.

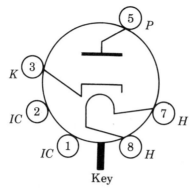

P = Plate
K = Cathode
H = Heater
IC = Internal Connection. Pin may connect
to element support structure and
should not be used.

Figure 86   Base Connection Diagram for the Type 6AU4GT

3. Raise the plate voltage to 2 volts and read and record the plate current. Repeat for plate voltages of 4, 6, 8, and 10 volts.

4. Reset the rheostat to zero volts and turn off the power supply or disconnect the storage battery. Interchange the plate and cathode (pins 5 and 3) connections on the 6AU4GT, apply power, and increase the rheostat setting to see if current will flow through the tube in the reverse direction. Report your observation in the space below.

5. Again set the rheostat to zero volts and shut off the power. Unplug the filament transformer and disconnect the tube from your circuit. Rewire the circuit using the 1N5391 semiconductor diode as shown in Fig. 87. Note that the cathode end of the 1N5391 is marked with a white band. Note also that the circuit of Fig. 87 differs from that of Fig. 85 not only in that no heater power is needed but also in that a 68-ohm

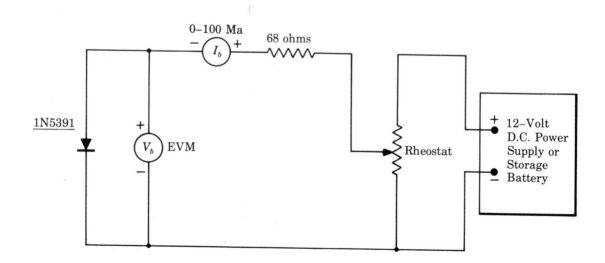

Figure 87   Circuit for Measuring the Semiconductor Diode's Volt-Ampere Characteristic

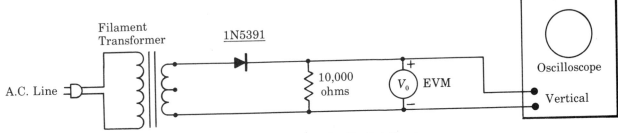

Figure 88   Half-Wave Rectifier Test Circuit

protective resistor is inserted in series with the anode. Have your circuit approved by the instructor.

6. Make sure the rheostat is set for zero voltage before turning on the power supply or connecting the battery. The voltage across the diode will probably not exceed 1 volt, and so the EVM should be switched to its lowest D.C. voltage range. CAUTION: Be very careful when the EVM is set to this range. Do not make any changes in your circuit without first either disconnecting the EVM or switching it to a higher range. When you are sure everything is in order, apply power and slowly advance the rheostat until you just observe some current flow through the diode. Record the voltage at which current flow just begins as the voltage corresponding to zero current. Actually it is the maximum forward voltage before the onset of current through the diode, all lower voltages corresponding to zero current.

7. Advance the rheostat setting until the diode current is 10 milliamperes. Record the diode voltage observed with this current. Repeat for currents of 20, 40, 60, 80, and 100 milliamperes.

8. Reset the rheostat to zero, turn off the power, and interchange the connections to the 1N5391. Reapply power and advance the rheostat to see if a reverse current through the diode is observed. Report your finding in the space below.

9. Turn off and disconnect the power supply, disassemble your circuit, and rewire your 1N5391 diode in the half-wave rectifier circuit shown in Fig. 88. Have your circuit approved by the instructor before you plug in the transformer.

10. Set the EVM to its 10- or 15-volt D.C. range, plug in the transformer, and adjust the oscilloscope to display two cycles of the A.C. voltage wave. Use line synchronization to stabilize the pattern. Note that a D.C. voltmeter reads the *average* of a rapidly varying voltage applied to it. Record the EVM reading as the average output voltage, and read and record the peak output voltage displayed by the oscilloscope. NOTE: To read the peak voltage from the oscilloscope pattern, the vertical axis must be calibrated. You will have to carry out this calibration if your instrument does not have it provided for you.

11. Some oscilloscopes have an "A.C.–D.C." switch on the front panel. This switch does not set the instrument for operation from an A.C. or D.C. power line but controls the response of the vertical amplifier. In the "D.C." position, any voltage, including a constant one, that is connected to the vertical input terminals will cause a vertical deflection of the trace on the screen. Thus, if the oscilloscope is set for a vertical deflection sensitivity of 1 volt per centimeter and you connect a 1.5-volt battery to the vertical input, the trace will shift vertically by 1.5 centimeters and remain in the new position. With the input switch at "A.C.", on the other hand, only A.C. signals will produce a vertical deflection, the oscilloscope's response cutting off when the frequency gets down to 3 or 4 hertz. Connecting the battery as before will produce only a momentary deflection, the trace returning to its zero vertical position at once.

The rectifier output voltage is time-varying but does not have a zero average and may therefore be thought of as the sum of a D.C. component equal to its average value and a "true A.C." component having zero average. With the oscilloscope set for direct current, both components are displayed on the screen. When the oscilloscope is set for alternating current, the D.C. component is blocked. Hence, *switching the vertical input from alternating to direct current displaces the pattern vertically by an amount equal to its average value.* If your instrument has such a switch, make and record a second measurement of the rectifier output's average value in this way.

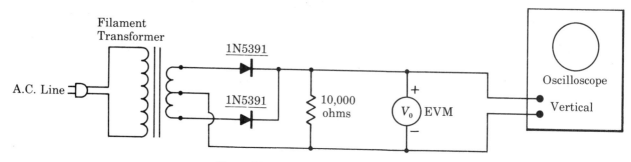

Figure 89   Full-Wave Rectifier Test Circuit

12. Investigate the effect of a capacitor filter by connecting the capacitor board across the output of your rectifier (in parallel with the 10,000-ohm load resistor). Observe the oscilloscope pattern and read and record the average output voltage indicated by the EVM for capacitances of $\frac{1}{2}$, 1, 2, 4, 10, and 14 microfarads. Close the appropriate switches on the capacitor board to obtain these capacitances, and use the 10-microfarad electrolytic capacitor in parallel with the board terminals to obtain the 10- and 14-microfarad values. NOTE: If possible, have the vertical amplifier set for direct current when making these observations. If your oscilloscope permits and the instructor suggests it, measure the average value of the rectifier output voltage by using the A.C.–D.C. switch as described in Procedure 11 for each value of capacitance tried. Note also that the electrolytic capacitor must be connected in accordance with the polarity markings on its jacket.

13. Unplug the transformer and rewire your circuit using two 1N5391's to form the full-wave centertapped rectifier system diagrammed in Fig. 89. Have your wiring checked by the instructor, plug in the transformer, and observe the pattern displayed on the oscilloscope screen. Record the average value of the rectifier output voltage as read on the EVM and determine the peak value from the oscilloscope display. If possible and if the instructor so suggests, make a second measurement of the average output voltage using the oscilloscope's A.C.–D.C. switch as described in Procedure 11.

14. Unplug the transformer, reverse the connections to both diodes, plug the transformer in, and observe the pattern now displayed by the oscilloscope. Notice that the polarity of the EVM must be reversed. Report your observations in the space below.

15. Unplug the transformer and rewire your circuit using four 1N5391's to form the full-wave bridge rectifier of Fig. 90. Have your circuit checked by the instructor and then plug in the transformer. Measure and record the average and peak values of the output voltage as in Procedure 13.

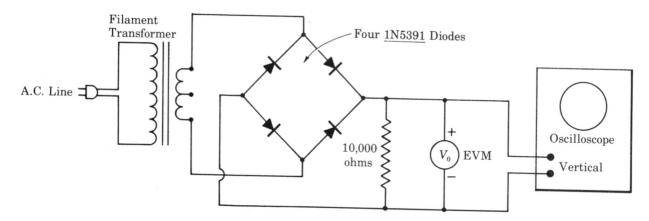

Figure 90   Bridge Rectifier Test Circuit

DATA

Diode	V_b Volts	I_b Milliamperes
6AU4GT	1	
	2	
	4	
	6	
	8	
	10	
1N5391		0
		10
		20
		40
		60
		80
		100

Measured exponent in three-halves power law _____

Percent error _____

Rectifier Circuit	Filter Capacitor, Microfarads	Peak Value of Output Voltage	Average Value of Output Voltage		
			Meter	Oscilloscope	Calculated
Half-wave	0				
	0.5				
	1				
	2				
	4				
	10				
	14				
Full-wave Centertapped	0				
Full-wave Bridge	0				

CALCULATIONS

1. Using the data obtained in Procedures 2 and 3, plot a graph of I_b against V_b on the 1-cycle × 2-cycle logarithmic graph paper. Use the 1-cycle scale for V_b and the 2-cycle scale for I_b. Refer to the instructions on plotting graphs in the Introduction. Draw the best straight line through your points.

2. From the slope of the graph plotted in Calculation 1, find the exponent to which V_b should be raised to make the result proportional to I_b. According to the three-halves power law this exponent should be $\frac{3}{2}$. Compare your result with this theoretical one by finding the percent error.

3. Using the data of Procedures 6 and 7, plot a graph of I_b against V_b on the linear graph paper. Plot the voltage values as abscissas and the current values as ordinates. Arrange your scales to cover as much of the sheet of graph paper as possible.

4. From the peak output voltage values observed in Procedures 10, 13, and 15, calculate the average output voltage expected for each rectifier circuit. Compare the three (two if your oscilloscope has no A.C.–D.C. switch) average values obtained in each case.

QUESTIONS

1. (a) Compare the forward volt-ampere characteristics of the vacuum-tube diode and the semiconductor diode. (b) Which approaches the ideal diode most closely? Explain your answer.

2. (a) Show that the power dissipated in a vacuum diode is proportional to $I_b^{5/2}$. (b) What power is dissipated in an ideal diode? (c) Using your graph from Calculation 3, explain why you would expect the power dissipated in a semiconductor diode to be proportional to the current I_b.

3. What is the purpose of the 68-ohm resistor in the circuit of Fig. 87?

4. When data was taken for the volt-ampere characteristic of the 6AU4GT, the voltage V_b was used as the independent variable, whereas the current was used in the case of the 1N5391. Why?

5. A thin ribbon of tungsten 2 millimeters wide and 5 centimeters long is to be used as the filamentary cathode of a high-power vacuum tube. For tungsten the constant A in Equation 133 has the value 60.2 amperes/cm² °K², and the work function is 4.5 volts. The tube is to have a maximum current rating of 200 milliamperes, but because this is to be under space-charge limited conditions, the cathode must be capable of emitting five times that, or 1 ampere. (a) Remembering that both sides of the tungsten ribbon may be used and that the temperature in Equation 133 is in degrees *Kelvin*, find the temperature in degrees Celsius to which the filament must be heated. (b) Look up the melting point of tungsten and see if the filament can in fact be heated to this temperature. (c) Tungsten filaments are often treated with thorium to reduce their work function. Suppose the present filament had its work function reduced to 3.0 volts in this way. To what temperature would it then have to be heated to yield the same emission?

6. Suppose a sinusoidal A.C. voltage were displayed on an oscilloscope whose vertical amplifier was equipped with an A.C.–D.C. switch. What would you observe when this switch was thrown? Explain your answer.

7. How do the patterns obtained in Procedures 13 and 15 compare? What is the principal difference between them?

8. Using the characteristic of the 1N5391 diode exhibited in the graph of Calculation 3, estimate the amplitude of the A.C. voltage coming from the transformer secondary coil from the peak values of the rectifier output voltage measured in Procedures 10, 13, and 15.

9. A peak-reading A.C. voltmeter is made by measuring the output of a half-wave rectifier with an EVM having an input resistance of 10 megohms (10^7 ohms). What capacitance must be used in parallel with the EVM input so that the time constant will be ten times the period of the 60-hertz power line voltage?

10. A so-called *full-wave voltage doubler* is made by substituting large capacitors for the two lower diodes (C_{r_2} and C_{r_4}) in the circuit of Fig. 84. If R is very large, the resulting circuit delivers a D.C. voltage equal to twice the amplitude of the A.C. input voltage ($2V_m$ in Fig. 84). Explain the action of this circuit.

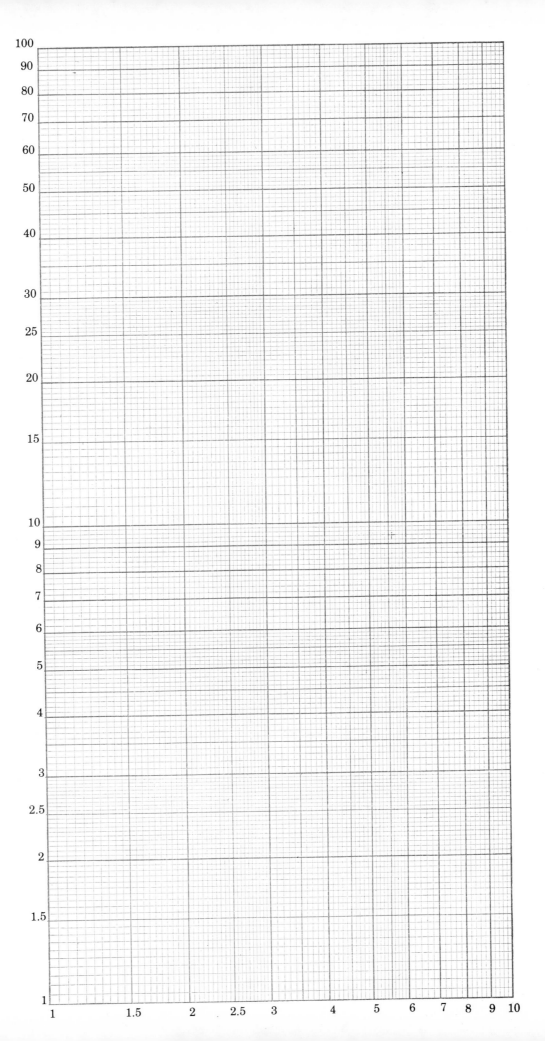

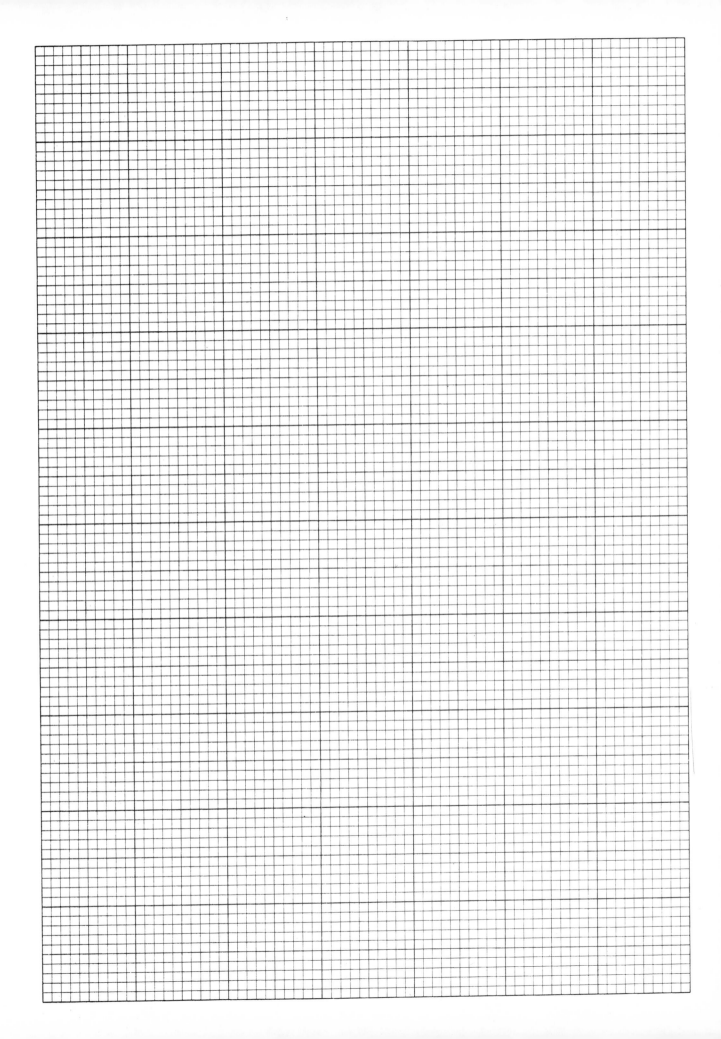

Amplification with Triode and Pentode Vacuum Tubes **36**

The high-vacuum electronic tube has been the mainstay of electronic communications, computing, and control systems of all kinds for the last half-century, and even now it is only partially superseded by semiconductor devices such as the transistor. Its importance as an amplifier, especially in circuits where the voltage is higher than can be readily handled by transistors, entitles it to special study, and the purpose of this experiment is to see how a tube amplifies and to observe some of the characteristics of vacuum-tube amplifier circuits.

THEORY

The construction and operation of a diode (two-element) vacuum tube was described in Experiment 35. It was pointed out that when the cathode is arranged to provide copious electron emission, a negative electron cloud called a space charge forms in its vicinity and the plate current is controlled by the plate voltage according to the three-halves power law. Now, if a third electrode — a grid of fine wire mesh — is interposed between the cathode and the plate, the potential applied to it will have a great effect on the space charge and will thus provide a control over the plate current. In particular, if the grid is made negative with respect to the cathode, it will help the space charge to repel electrons back towards the cathode and will thus cause a decrease in plate current. In fact, if the grid is made very negative, the plate current will be reduced to zero despite the plate's being at a high positive voltage. The tube is then said to be *cut off*, and the grid voltage required to just reduce the plate current to zero with a particular plate voltage is called the *cutoff grid bias* for that plate voltage. Less negative grid voltages allow some plate current to flow, the amount depending on the grid voltage value. Thus the *triode*, or three-electrode (cathode, grid, and plate) vacuum tube is a control device in which the grid voltage controls the plate current. The plate current is affected much more strongly by changes in grid voltage than by changes in plate voltage, hence amplification is possible. Indeed, a quantity called the *amplification factor* μ is defined for any particular triode as the ratio of a small change in plate voltage to the change in grid voltage that would have the same effect on the plate current. Thus, in a triode with an amplification factor of 20, a change of 20 volts in the plate voltage would have the same effect on the plate current as a change of 1 volt in grid voltage. In addition you should note that as long as the grid is negative (the normal case in small amplifiers, for plate current flows over a wide range of negative grid voltages between zero and the cutoff value)

it will not collect any electrons and will therefore draw no current. Thus a negative grid just sits there in the tube looking like an open circuit. Nevertheless, the grid voltage, despite the fact that it has negligible current and negligible power associated with it, controls an appreciable plate current associated with an appreciable plate voltage. Vacuum tubes are therefore capable of large power gains.

In the triode, the plate current depends on both the plate voltage and the grid voltage and is thus a function of two variables. To plot this function completely would therefore require a three-dimensional graph with three mutually perpendicular axes for the plate voltage, plate current, and grid voltage respectively. Since this is impractical, plots of one variable against another for various constant values of the third are published for most commonly available tubes. Of the three possibilities, the most useful and hence the most often given is the so-called *plate characteristic curves* obtained by plotting plate current I_b as a function of plate voltage V_b for various values of grid voltage V_c. Those for a triode-connected type 6AU6 are shown in Fig. 91. Note how each curve goes down to zero plate current at a plate voltage for which the pertinent grid voltage is the cutoff bias.

The type 6AU6 is actually a *pentode* or five-element tube, there being two extra grids placed between the grid already described and the plate. This latter grid, which is nearest the cathode and is the one used to control the plate current, is accordingly called the *no. 1* or *control grid*. When the tube is to be used as a triode, the other two are simply connected to the plate so that the combination of these three electrodes functions as a single electron collector or anode. When the tube is used as a pentode, on the other hand, the second and third grids have special functions that will now be described.

A less-than-ideal feature of the triode is the fact that the plate current, which is supposed to be controlled by

251

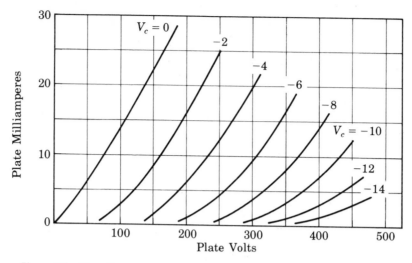

Figure 91 Plate Characteristic Curves for the Type 6AU6 Connected as a Triode

the grid voltage, also depends on the plate voltage. This dependence can be much reduced by the insertion of a second grid between the control grid and the plate, this second grid being maintained at a fixed positive voltage. It therefore looks like the plate as far as the cathode and space charge are concerned; and since its voltage is held constant, the current through the tube is controlled by the control grid voltage only. The plate voltage has little effect on this current because the plate is effectively shielded or screened from the cathode region by the second grid. This grid is therefore called either the *no. 2* or the *screen* grid. A set of plate characteristic curves for the type 6AU6 operated as a pentode with the screen grid at 150 volts is shown in Fig. 92. Notice that in contrast to the curves of Fig. 91, these are very flat, indicating that

there is little change in plate current with plate voltage once this latter is made large enough to collect the electrons. The pentode is said to have a high *dynamic plate resistance* as compared to the triode. The dynamic plate resistance (r_p) is the ratio of a small change in plate voltage to the corresponding change in plate current. It is called "dynamic" because the curves of Figs. 91 and 92 are not linear, so that the tube does not obey Ohm's law and a resistance obtained by dividing plate voltage by plate current has no meaning. Note, however, that the *slopes* of the curves in either Fig. 91 or Fig. 92 give the ratio of a small change in plate current to the corresponding small change in plate voltage and thus represent a *dynamic plate conductance* g_p. The reciprocal of g_p is clearly r_p, and the fact that the pentode characteristics

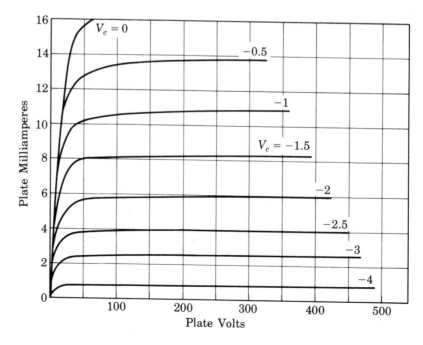

Figure 92 Plate Characteristic Curves for the Type 6AU6 Pentode (Screen Grid Voltage = 150 Volts)

have almost zero slope means that for the pentode, g_p is very small and r_p is correspondingly large.

Because the screen grid is operated at a positive potential, the possibility exists of its attracting electrons and drawing current — an undesirable effect since this current simply passes through the screen grid power supply and represents a power dissipated for no purpose. Screen grid current can be minimized by aligning the screen and control grid wires so that those in the screen are in the shadow of those in grid no. 1. However, secondary electrons coming from the plate can be attracted to the screen and result in a current that both detracts from the plate current and is wasted as noted above. Secondary emission from the plate resulting from electrons in the surface being knocked out by those arriving from the cathode occurs in both the triode and the pentode, but in the triode the secondaries see the negative grid and are immediately repelled back into the plate. In the pentode the secondaries are confronted with the positive screen and can easily be attracted over to it, especially when the plate voltage is low. To prevent this, a third grid is placed between the screen and the plate and operated at a low potential, usually the same as the cathode. Electrons accelerated by the screen have enough energy to get through the third grid and proceed to the plate, but secondary electrons are shielded from the screen's field by this third grid and therefore return to the plate as in the triode. Because the function of the third grid is to suppress secondary emission, it is called the *suppressor grid*.

When a triode is to be used as a *voltage amplifier*, that is, a device for increasing the amplitude of an A.C. signal voltage, it is connected in a circuit such as that drawn schematically in Fig. 93. A power supply supplies voltage V_{bb} to the circuit consisting of the tube (plate and cathode) and the so-called *load resistor* R_L in series, as a result of which current I_b flows around this circuit. Notice the direction of I_b indicated in Fig. 93: it is opposite to the electron flow direction since electrons are

negative but the conventional direction for current is the direction of flow of positive charge.

The purpose of R_L is to convert the varying current I_b (which varies as a result of the control of this current by a signal applied to the grid) into a varying voltage, since a voltage output is wanted. Ohm's law tells us that the potential across R_L is $I_b R_L$, and since this is the difference between the power supply voltage V_{bb} and the plate voltage V_b, we have

$$V_{bb} - V_b = I_b R_L$$

or
$$I_b = \frac{V_{bb}}{R_L} - \frac{V_b}{R_L} \qquad (135)$$

Notice that Equation 135 is a relation between the triode's plate current I_b and its plate voltage V_b and can thus be plotted on the tube's plate characteristics as shown for a typical triode in Fig. 94. The plot is seen to be a straight line with slope $-1/R_L$ and intercept V_{bb}/R_L. Because it is determined by the load resistor, this plot is called the *load line*.

Once the load resistor R_L and the power supply voltage V_{bb} have been established, the relationship between I_b and V_b is fixed by Equation 135, and changes in plate current due to control by the grid voltage will result in the changes in plate voltage dictated by this equation. These changes will be made from some resting or *quiescent* value of plate voltage and current that the tube will assume in the absence of any input signal and that determine a point called the *operating point*, as shown in Fig. 94. The operating point is located at the intersection of the load line and the plate characteristic drawn for the grid voltage applied to the tube in the absence of an input signal, and in fact constitutes the graphic solution of two simultaneous equations, namely Equation 135 and the plate characteristic in question. The operating point is thus chosen by choosing the voltage applied to the grid under quiescent or no-signal conditions. This is the vol-

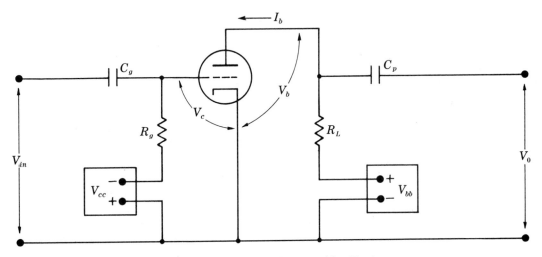

Figure 93   Triode Voltage Amplifier Circuit

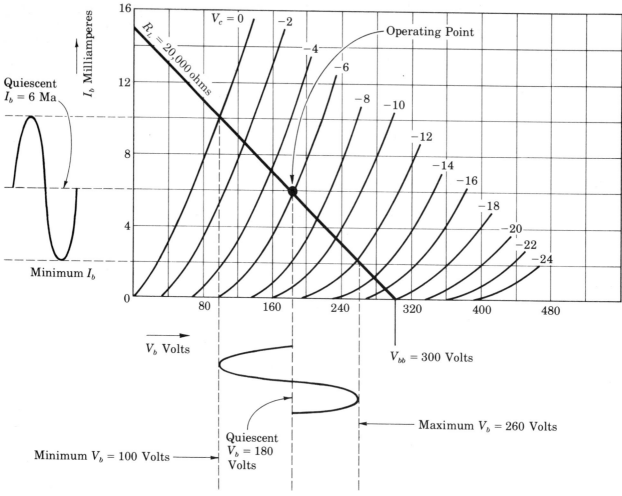

Figure 94 Plate Characteristics and a Sample Load Line for a Typical Triode

tage V_{cc} delivered by the power supply on the left in Fig. 93. Note that V_{cc} and V_c are the same with no input signal, since with no grid current there is no voltage drop across resistor R_g.

The choice of the operating point is made on the basis of allowing the greatest possible linear swing of the plate voltage and current in either direction. Tubes are nonlinear devices as the characteristic curves of Figs. 91 and 92 show, and so if the amplified output voltage is to be a faithful reproduction of the signal voltage at the input, operation must be limited to ranges of plate current and voltage over which the tube's characteristics are reasonably linear. In particular, note that the characteristics become particularly nonlinear at low plate currents, so that operation of the triode of Fig. 94, for example, should be limited to currents above, say, 2 milliamperes. This sets a somewhat arbitrary lower limit on I_b. The upper limit is set by the intersection of the load line with the plate characteristic drawn for zero grid voltage, for in this kind of amplifier the grid must not be made to swing positive. This is because no provision has been made for the flow of grid current, which would result in a voltage drop across the grid resistor R_g. In fact, large values of R_g are chosen on the assumption that there will be no current

through it. The tube's operation with signal can thus be considered to be along the load line from the point where it intersects the low plate current limit to the point where it intersects the characteristic curve for $V_c = 0$. Since the input signal is likely to swing negative as much as positive (this will be exactly true if the input is sinusoidal), the operating point is chosen in the middle of this range as shown in Fig. 94 and the corresponding grid voltage value selected for V_{cc}. If the location so chosen does not happen to fall on one of the given plate characteristic curves, interpolation between the neighboring curves is used to determine the grid bias.

With the above design parameters chosen, the amplifier is ready to accept an input signal. This is applied to the input terminals marked V_{in} in Fig. 93 and passes through capacitor C_g. The purpose of C_g is to pass the A.C. signal but to block the D.C. grid bias voltage from getting back to the signal source. The signal voltage develops across R_g so that the grid voltage V_c is the sum of the D.C. component V_{cc}, which sets the operating point, and the A.C. signal voltage V_{in}, which makes V_c swing above and below its quiescent value. The tube's plate current varies with these swings, and the corresponding plate voltage variations shown in Fig. 94 appear across

R_L. C_p stops the D.C. component of plate voltage while passing the variations on to the output terminals. Because C_g and C_p block D.C. voltages from appearing where they are not wanted, these capacitors are often called *blocking capacitors*. They are also called *coupling capacitors* because they couple the signal voltages from one circuit to the next.

The gain of a voltage amplifier may be determined graphically by a construction such as that shown in Fig. 94. Here the maximum allowed swing in grid voltage (and hence the maximum acceptable input signal amplitude) is 6 volts above and below the chosen grid bias $V_{cc} = -6$ volts. The corresponding plate voltage swing may be read off the voltage axis. The quiescent plate voltage is seen to be 180 volts, this being the D.C. component that must be blocked from the output by C_p, and the swing is up to about 260 volts and down to about 100 volts for an output swing in either direction of 80 volts. This represents the maximum signal-handling capability of the amplifier, and under these conditions the gain is 80/6 or 13.3. Note that an input signal with an amplitude larger than 6 volts is not permissible. Indeed, for minimum distortion, input signal levels should be kept appreciably less than this. For this reason circuits like that of Fig. 93 are often called *small-signal amplifiers*. Note also that because an increase in plate current causes a decrease in plate voltage in consequence of the action of R_L, a rising grid voltage produces a falling plate voltage. This type of amplifier therefore reverses the phase of the input signal so that the output voltage, in addition to being larger than the input, is the negative of it. This feature of voltage amplifiers of the sort diagrammed in Fig. 93 must often be taken into consideration, and such circuits are sometimes put into a system for the express purpose of reversing the phase of the signal. An amplifier so used is called a *phase inverter*.

The gain of a voltage amplifier of the sort here considered is always less than the amplification factor μ of the tube, but it can be made to approach μ by making the value of R_L large. This can be seen from the fact that the larger R_L is, the smaller the variations in plate current needed to produce a given output voltage. Thus, as I_b comes closer to being constant, the change in plate voltage becomes more nearly equal to μ times the change in grid voltage. One method of measuring the amplification factor of a triode is therefore to see how the gain it delivers in the circuit of Fig. 93 varies as R_L is increased. The amplification factor will be the limit that this gain approaches.

Pentodes can also be used as voltage amplifiers in circuits similar to that of Fig. 93, except that a source must be provided for the screen grid voltage. This may be the same as the plate voltage supply, but since the screen is usually operated at a potential considerably less than V_{bb}, means must be provided for dropping the voltage to a suitable level while at the same time maintaining that level constant. A simple series resistor chosen so that the current drawn by the screen will drop the voltage to the proper value is often used, with a capacitor connected from this grid to the cathode to remove voltage variations that would result from variations in screen current. However, a better method that also serves to introduce the gas-discharge voltage-regulator tube will be used in the present work.

The type 0A2 is a cold-cathode gas-discharge diode consisting of two elements — a cathode and an anode, neither of which is heated — enclosed in a glass envelope containing a gas at a low pressure. When some voltage in excess of the so-called *breakdown* value is applied to this diode, the gas atoms become ionized (lose one or more of their electrons) and the gas becomes a conductor due to the presence of large numbers of electrons and positive ions, all of which (although the heavier ions play a relatively minor role) serve as charge carriers. An interesting and important property of a conducting gas discharge is that the voltage across it is very nearly constant and independent of the current it is carrying over a considerable current range. Thus, in contrast to a resistor, whose volt-ampere characteristic is Ohm's law, the gas discharge diode has a volt-ampere characteristic in which the voltage is fixed at some value characteristic of the gas and the gas pressure regardless of the current. Such a device can be used as a voltage regulator in the circuit of Fig. 95. Here an unregulated voltage higher than the tube's breakdown voltage (which is somewhat higher than its operating voltage) is applied at the input. The tube breaks down and current flows through it and R_t in series. Notice that the current automatically takes the value that makes the drop across R_t equal to the difference between the tube's operating vol-

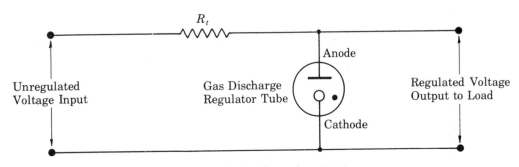

Figure 95   Gas Diode Regulator Circuit

tage and the input voltage, which means that the input voltage must be higher than the desired regulated output voltage. This is characteristic of all voltage regulators, which operate by automatically adjusting the input-output difference to maintain the output voltage constant despite variations in the input. In the present case, the automatic adjustment results from the constant-voltage characteristic of the gas discharge. If the input voltage goes up or down, the current through R_t goes up or down as needed to make the drop across this resistor equal to the difference between whatever value the input voltage may assume and the fixed voltage across the tube. If the load draws more current, the tube draws less to compensate and *vice-versa*. Thus, as long as the tube is not required to pass a current outside its range (5 to 30 milliamperes for the type 0A2), its operating voltage, which is the regulated output voltage, remains essentially constant. The value of this voltage is characteristic of the tube and cannot be varied except by changing tubes. For this reason, several gas-discharge regulator tubes differing only in their operating voltages are available commercially. The type 0A2 has been chosen here because its operating voltage of 150 volts is a recommended value for the screen grid of the type 6AU6.

Because there is very little dependence of plate current on plate voltage (very high plate resistance) in a pentode, almost the entire swing in grid voltage is effective in producing a plate current swing and hence an output voltage across R_L. If the effect of changing plate voltage can be disregarded, the change in plate current is equal to the change in grid voltage multiplied by a quantity g_m called the *transconductance*. This parameter is the real measure of how good the tube is at controlling the cur-

rent through its plate-cathode circuit by means of the voltage on its grid. Being the ratio of a change in plate *current* to a corresponding change in grid *voltage*, it is a *conductance* and the prefix *trans-* expresses the fact that it is the ratio of a current in one circuit (the plate) to the voltage in another (the grid). Transconductance was formerly called *mutual conductance*, which is the reason for the subscript m in g_m, the standard symbol for this parameter.

In view of the above considerations, the variations in the plate current of a pentode will be g_m times the variations in grid voltage (the input signal), and the output voltage will be the plate current changes times R_L, as in the triode. Hence the voltage gain, which is the output signal divided by the input signal, should be equal to $g_m R_L$. This will be approximately true in a pentode amplifier provided R_L is not made too large. One purpose of this experiment will be to test this conclusion. However, although μ for a triode remains almost constant regardless of the choice of operating point, g_m for a pentode does not remain constant and will therefore change when different load resistors and hence load lines and operating points are chosen. The proportionality between gain and R_L is therefore not perfect, although it is reasonable enough for the 6AU6 so that a good average value of this tube's transconductance can be obtained.

Because of their large r_p, pentodes offer considerably higher gain than do triodes. They are, however, less linear, so that signal voltage levels must be strictly limited if excessive distortion is to be avoided. Pentodes are therefore used to best advantage in applications where a small signal is to be amplified as much as possible by a single tube.

APPARATUS

1. Type 6AU6 tube
2. Type 0A2 tube
3. Two 7-pin miniature tube sockets, mounted
4. Filament transformer, 6.3-volts, 3-amperes
5. Power supply, 300 volts D.C. and 6.3 volts A.C.
6. Oscilloscope
7. Electronic voltmeter
8. Decade resistance box (10,000 ohms)
9. Decade resistance box (1000 ohms)
10. Dry battery (9-volts)
11. Four carbon, 1-watt, 5%-tolerance resistors (470,000; 100,000; 47,000; and 27,000 ohms).
12. 10,000-ohm, 10-watt wirewound resistor
13. Volume-control-type potentiometer (5000 ohms, wirewound)
14. Three paper capacitors (0.25-microfarad, 400-volt rating)
15. Assorted hookup wires and connectors

PROCEDURE

1. On the triode plate characteristics for the type 6AU6 (Fig. 91), draw load lines for a plate supply voltage of 300 volts and load resistances of 27,000; 47,000; and 100,000 ohms. Assuming a minimum permissible plate current of 0.5 milliampere, choose a proper operating point on each load line. Record the values of plate current, plate voltage, and grid voltage at each of your operating points.

2. Assemble the triode voltage amplifier circuit using the type 6AU6 as shown schematically in Fig. 96. Use the 10,000-ohm decade resistance box for R_1, the 1000-ohm box for R_2, and the 27,000-ohm resistor for R_L.

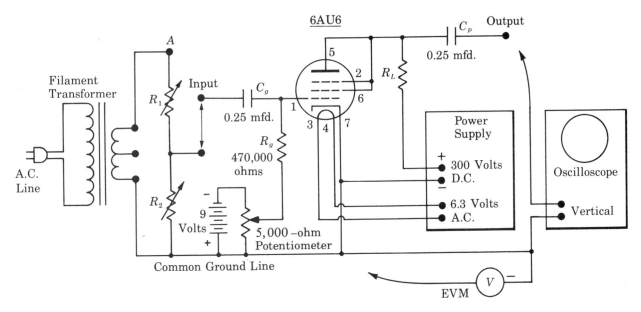

Figure 96   Triode Amplifier Circuit

Notice that the tube's heater is operated from a source of 6.3 volts A.C. included in the power supply, while the filament transformer serves as the source of a 60-hertz test signal to be applied to the amplifier input. The electronic voltmeter's common ("low" side) lead should be connected to the tube's cathode (the common ground line). The EVM's probe is then available to test voltages at various points in the circuit. The "low" side of the oscilloscope's vertical input is also connected to the common ground line, and the "high" side is provided with a lead that may be connected either to the amplifier output terminal or to point A to observe the 60-hertz A.C. voltage across the filament transformer secondary coil. A lead is provided to connect the amplifier input to the junction of R_1 and R_2 but should be omitted in this step.

The type 6AU6 is a so-called "miniature" tube and has a standard 7-pin miniature base. This base has seven wire connecting pins arranged in a circle and spaced as if there were supposed to be eight. The eighth pin position is blank and serves to determine the one way in which the tube may be plugged into its socket. The pins on miniature tubes are easily bent and care must be taken when inserting the 6AU6 to avoid damaging them. Note that the pins are numbered *clockwise when viewed from the base end,* starting with pin no. 1 on one side of the blank space and proceeding in order until pin no. 7 is reached on the other. Remember that if the tube is viewed from above (base down), the pin numbering appears to be counterclockwise. The base diagrams for both the type 6AU6 and the type 0A2 (which is also a 7-pin miniature) are given in Fig. 97, and for convenience the pin numbers are marked on the connections to the respective tube elements in the schematic diagram of Fig. 96.

When your circuit is completely wired, call the instructor. Do not plug in or turn on the power supply until the instructor has checked your work. CAUTION: The power supply output voltage is high enough to give you a painful shock. Be very careful when working around your circuit and do not make any changes with the power on. When measuring voltages with the EVM, be sure you have the meter switched to the right range and, when touching the probe to a particular circuit point, use care not to short-circuit it to any adjacent circuit points.

3. Adjust the 5000-ohm potentiometer to give the grid bias voltage corresponding to the operating point you chose on the 27,000-ohm load line in Procedure 1. *Note:* This so-called potentiometer is really a rheostat with connections made to both ends of its resistance element as well as to the movable contact arm, but such a device is commonly called a potentiometer in electronic work because of its resemblance to the slide-wire arrangement in Experiment 29. Measure V_{cc} at the movable contact arm (middle terminal) of the potentiometer. Do not attempt to measure the voltage at the tube grid, for even though the EVM has an input resistance of about 10 megohms, it will draw enough current when connected to the grid to cause a noticeable voltage drop in the 470,000-ohm grid resistor R_g.

4. Plug in and turn on the power supply. Allow a minute or two for the tube to warm up, and then adjust V_{bb} to exactly 300 volts as measured by the EVM. Measure and record the 6AU6 plate voltage V_b.

5. Set R_1 to 1000 ohms, R_2 to zero, and connect the lead between the junction of R_1 and R_2 and the

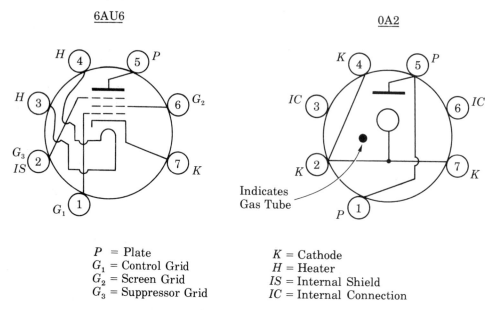

Figure 97 Tube Base Connection Diagrams

P = Plate
G_1 = Control Grid
G_2 = Screen Grid
G_3 = Suppressor Grid

K = Cathode
H = Heater
IS = Internal Shield
IC = Internal Connection

amplifier input. Plug in the filament transformer, connect the oscilloscope vertical input to point A (Fig. 96), and set the oscilloscope to display two cycles of the transformer secondary voltage. Use line (not internal) synchronization to stabilize the pattern. Adjust the vertical gain so that the pattern has a convenient amplitude. Note that a vertical calibration is not necessary for this experiment, but an easily reproducible amplitude should be chosen. This means that the vertical gain should be set to bring the sine curve peaks to easily remembered marks on the oscilloscope screen so that this amplitude can be duplicated later. Do not change the vertical gain setting after this has been done.

6. Disconnect the oscilloscope vertical input from point A and connect it to the amplifier output. Increase the value of R_2 until the displayed sine curve has the same amplitude as that established in Procedure 5. Record the value of R_2 needed to do this.

7. Compare the sine curves displayed on the oscilloscope in Procedures 5 and 6 and report your observation in the space below.

8. Turn off the power supply and substitute the 47,000-ohm resistor for the 27,000-ohm resistor at R_L. Then repeat Procedures 3–6 for $R_L = 47,000$ ohms.

9. Repeat Procedure 8 with $R_L = 100,000$ ohms.

10. Turn off the power supply and unplug both it and the filament transformer. Rewire your circuit to use the 6AU6 as a pentode, as shown in Fig. 98. Make $R_L = 27,000$ ohms and temporarily omit the lead between the junction of R_1 and R_2 and the amplifier input. Notice that both anode connections (pin 1 and pin 5) on the 0A2 are used. This is a safety measure found in all commercial circuits using gas-discharge regulators. Its purpose is to prevent voltage from being applied to the regulator load (in this case the 6AU6 screen grid) in case the regulator tube is inadvertently left out of its socket. Observe that if the 10,000-ohm resistor and the screen grid were connected to a single 0A2 anode pin and the 0A2 were not plugged in, the resistor would run directly to the screen from the high voltage power supply and the screen voltage would be much too high. Capacitor C_s bypasses to the common ground line any A.C. voltages that might develop across the regulator tube and thus assures that the 6AU6 screen voltage will be pure direct current. For this reason C_s is called a *bypass capacitor*. Other parts of the pentode circuit are very similar to those for the triode. Do not plug in either the power supply or the filament transformer until your circuit has been checked by the instructor.

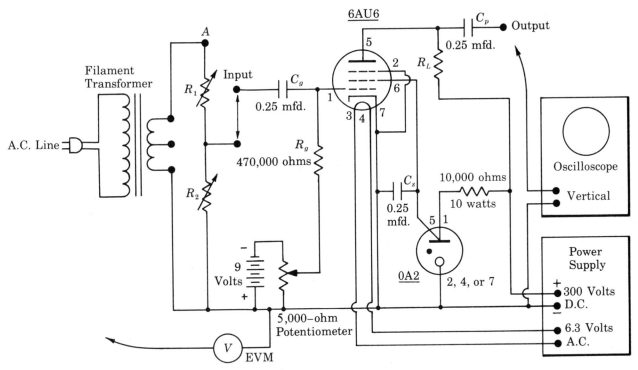

Figure 98   Pentode Amplifier Circuit

11. Before turning on your pentode amplifier, repeat Procedure 1 using the pentode plate characteristics of Fig. 92. In choosing operating points on your three load lines, again assume a minimum I_b of 0.5 milliamperes, but this time notice that before these lines intersect the characteristic curve for $V_c = 0$, they go through a region where the pentode plate characteristics become very nonlinear. Hence, for the pentode a minimum plate voltage rather than intersection with the curve for $V_c = 0$ must be specified to determine the high plate current operating limit. Use 50 volts for the 6AU6. Choose your operating points half way between the minimum plate voltage and minimum plate current points on each load line and record the corresponding values of plate current, plate voltage, and grid voltage.

12. Repeat Procedures 3–9 for your pentode amplifier. After turning on the power, note the glow of the gas discharge in the 0A2 and check the 6AU6 screen voltage. The rated operating voltage for the 0A2 is 150 volts, but this varies by a few volts either way with different samples of the type. Any voltage between 145 and 155 volts is satisfactory. Report the observations made when repeating Procedure 7 in the space below. Be sure to shut off all power when you have finished the experiment.

DATA

• Triode

Load Resistor R_L, Ohms	Calculated from Curves			Measured			
	Grid Bias V_{cc}	Plate Voltage V_b	Plate Current I_b	Plate Voltage V_b	Plate Current I_b	Value of R_2	Gain
27,000							
47,000							
100,000							

Estimated value of the amplification factor μ _____

Value of μ listed by the manufacturer _____

• Pentode

Load Resistor R_L, Ohms	Calculated from Curves			Measured			
	Grid Bias V_{cc}	Plate Voltage V_b	Plate Current I_b	Plate Voltage V_b	Plate Current I_b	Value of R_2	Gain
27,000							
47,000							
100,000							

Average value of the transconductance g_m _____

Value of g_m listed by the manufacturer _____

CALCULATIONS

1. From Equation 135, calculate and record I_b for each value of R_L used in both the triode and pentode amplifiers. For the triode, $V_{bb} = 300$ volts and the necessary values of V_b were measured in Procedures 4, 8, and 9. For the pentode, V_{bb} is again 300 volts and V_b was measured for each load resistance in Procedure 12. Compare each measured value of I_b and V_b with those predicted from the plate characteristics.

2. The combination of the resistances R_1 and R_2 forms a so-called *voltage divider* whose purpose is to produce a voltage V_{in} across R_2, which is a known fraction of the voltage V_s impressed across R_1 and R_2 in series. This fraction is given by

$$\frac{V_{in}}{V_s} = \frac{R_2}{R_1 + R_2} \tag{136}$$

Notice that V_{in} is the amplifier input signal and that R_2 was adjusted in Procedures 6, 8, 9, and 12 until the output voltage V_0 was equal to V_s in each case. Hence the amplifier was amplifying the input signal back up to the same level V_s from which the voltage divider cut it down. Therefore the gain is $(R_1 + R_2)/R_2$. Use this conclusion to compute the gain obtained with each load resistor in both the triode and pentode amplifiers.

3. Plot your values of the triode amplifier gain against the corresponding load resistances using the values of R_L as abscissas and the gain values as ordinates. Choose your scales so as to cover as much of the sheet of graph paper as possible but allow for some extrapolation of your curve to higher R_L. *Note:* If R_L were zero there would be no output voltage and therefore no voltage gain, so that the origin is a point on your graph. Draw a smooth curve through the four points you have plotted.

4. Extrapolate the graph of Calculation 3 to higher R_L and estimate the limit that the gain approaches as R_L gets very large. This should be the amplification factor of the type 6AU6 operated as a triode. Look up the rated value in a tube manual and compare it with your estimate.

5. Plot your values of the pentode amplifier gain against the corresponding load resistances as in Calculation 3 but without making provision for extrapolation. Draw the best straight line through your four points.

6. From the slope of your graph in Calculation 5, find the transconductance of the type 6AU6 operated as a pentode and compare it with a value obtained from a tube manual. *Note:* The manual may list several values for different operating conditions. Take the one for the conditions most nearly approaching those in effect in this experiment.

QUESTIONS

1. Define plate resistance and explain why the gain of a triode amplifier should approach the tube's amplification factor when the load resistance is made larger than the plate resistance.

2. The relation $\mu = g_m r_p$ holds for all triodes. Give a qualitative derivation of this relation.

3. Line synchronization of the oscilloscope pattern was used in Procedure 5 so that an important difference between the patterns displayed in Procedures 5 and 6 could be observed in Procedure 7. What was this difference and why did line rather than internal synchronization have to be used to observe it?

4. Using the plate characteristics given in Fig. 94, find by any method you wish the amplification factor μ of the triode for which these curves were drawn. Describe the method you use.

5. Why does no one talk about the amplification factor of a pentode?

6. Under what conditions does the approximation $g_m R_L$ for the gain of a pentode amplifier fail? Why?

7. Derive Equation 136.

8. How much current could the screen grid of your pentode amplifier draw from the 0A2 regulator before the regulation would fail? *Hint:* The minimum current that must flow through the 0A2 in order to maintain the discharge is 5 milliamperes.

9. Ideally, capacitor C_g in Fig. 93 should be a short circuit for the A.C. signal voltage so that V_{in} will be passed on undiminished to the tube grid. This will be essentially the case for signal frequencies high enough to make the reactance of C_g small compared to R_g. (a) Derive an expression in terms of C_g and R_g for the frequency at which the signal voltage at the tube grid, and hence the gain of the amplifier as a whole, is reduced by a factor of $1/\sqrt{2}$ below the value in effect if C_g were a short circuit. (b) Calculate this frequency for your actual amplifiers (Figs. 96 and 98).

10. (a) The frequency discussed in Question 9 is called the *lower half-power point*. Explain why. (b) What do you think might cause the amplifier gain to fall off for high-frequency signals and thus give rise to an upper half-power point?

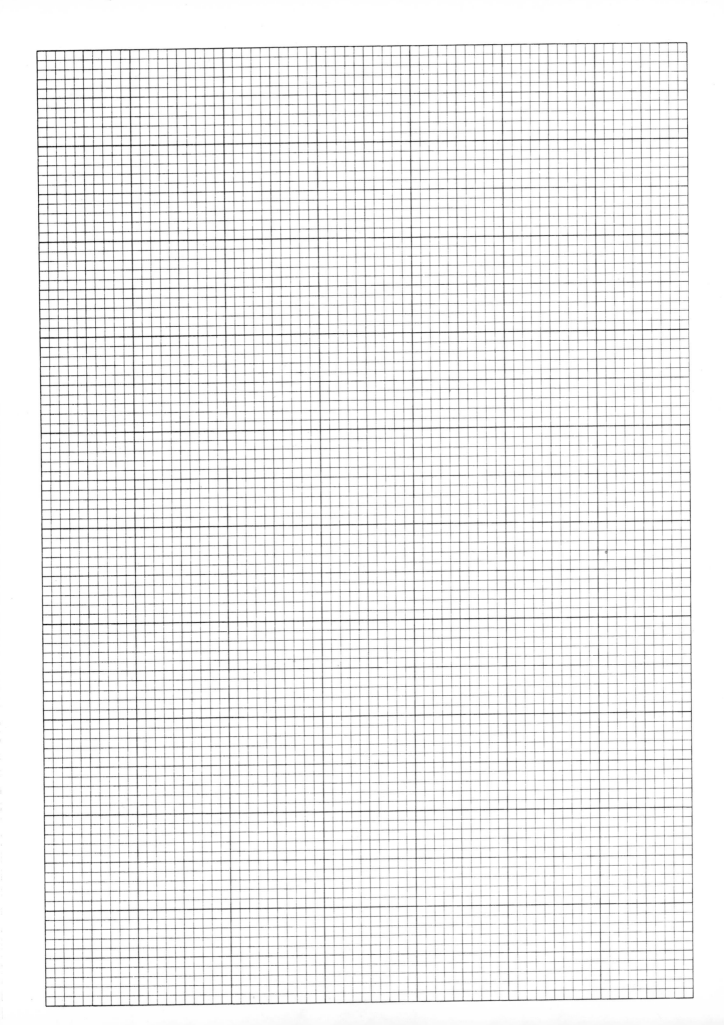

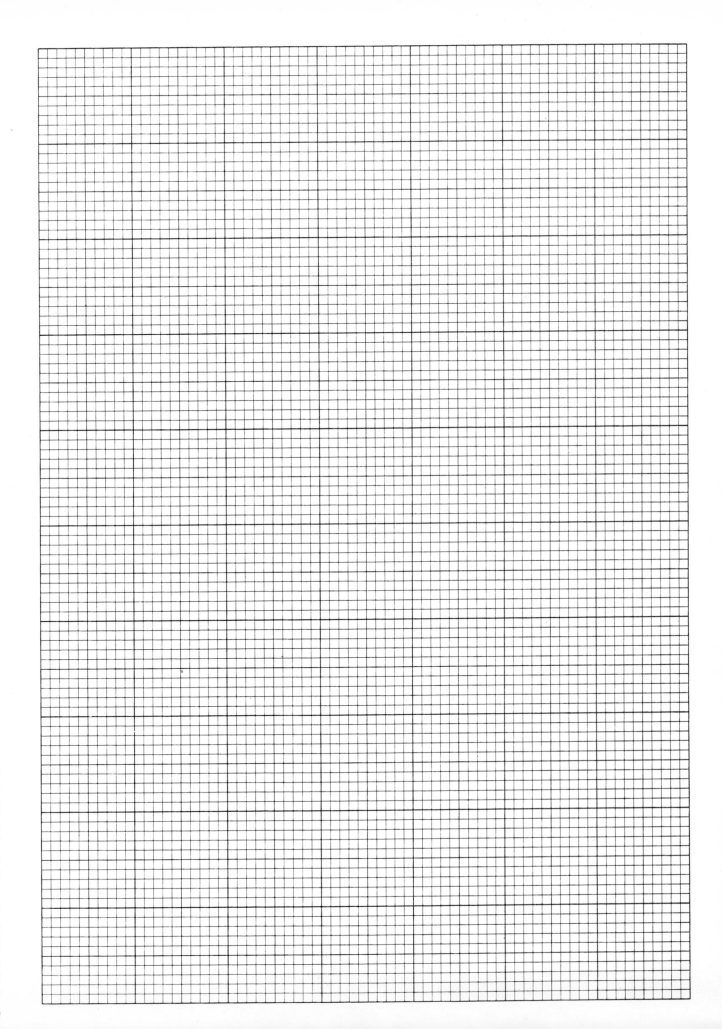

Amplification with Transistors 37

In the last two decades, the solid-state control device called the transistor has developed to the point where it has supplanted vacuum tubes in almost all low-power applications. Its advantages include smallness and the absence of a hot cathode with its attendant need for heating power and possibility of burning out. In operation, the transistor differs from the vacuum triode (or, more correctly, the vacuum pentode, since transistor and pentode characteristic curves look very similar) in that it operates at lower voltage and higher current levels and that the controlling signal is basically a current rather than a voltage. It is, however, a control and amplifying device similar to a vacuum tube, and many of the considerations discussed in Experiment 36 apply to it. A review of that experiment (or a reading of the theory section if Experiment 36 has not been done) is suggested in preparation for the present work, whose purpose will be to study the characteristics of a typical NPN transistor. A review of the section on semiconductor diodes and the PN junction in Experiment 35 may also prove helpful.

THEORY

In Experiment 35 we saw how a PN junction could behave as a unidirectional conductor. It is also possible to make a semiconductor triode with amplifying properties, that is, control of current in an output circuit by a relatively small amount of power in an input circuit. Such a device is called a *transistor*. In its modern form it consists of two PN junctions back to back and may be thought of as two diodes connected in series opposing (cathode-to-cathode or anode-to-anode), although if two semiconductor diodes were actually connected in this way, they would not work as a transistor because junction transistor action depends on actual physical proximity of the two junctions.

A PNP transistor is made by sandwiching a piece of N-type semiconductor material between two pieces of P-type, in which case two PN junctions with their N sides connected together are formed. A piece of P-type material can also be sandwiched between two N-type pieces, thus forming two PN junctions with their P sides connected together. Such a transistor is designated NPN. The two kinds have very similar properties, the principal difference being the reversal of the polarity of all voltages for one relative to the other. Figure 99 illustrates the NPN case. The material in the middle is always called the *base,* that on one end the *emitter,* and that on the other the *collector.* The terms "emitter" and "collector" have some meaning in the light of present-day transistor operation since they respectively emit and collect charge carriers, but the name "base" stems from the first point-contact transistors, in which the emitter and collector were point contacts on a "base" of semiconductor material and which are now obsolete. The term has persisted, however, and is now defined as referring to the material in the middle of the sandwich.

The diode formed by the emitter-base junction is shown forward-biased in Fig. 99. It therefore offers little resistance to the passage of current. On the other hand, the base-collector junction is reverse-biased and would pass negligible current if it were isolated. However, the emitter-base junction conducts by reason of the electrons in the N-type emitter being driven across the junction into the base. There they take some time to recombine with the holes, so that if the base is made very thin, the probability of these electrons drifting across the base-collector junction can be made quite high. It is in fact possible to "collect" from the base most of the electrons "emitted" into it by the emitter, so that most of the emitter current goes to the collector and the base current is comparatively small. Nevertheless, the presence of the emitter-base bias and the resulting base current is essential to the existence of any appreciable collector current at all. A large collector current is thus controlled by a small base current, so that the transistor exhibits amplifying properties. Note, however, that the collector (output) current is controlled by the base (input) *current,* whereas in the vacuum tube the plate (output) current is controlled

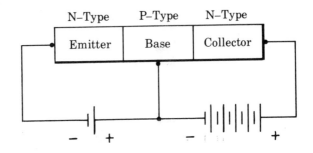

Figure 99   Transistor Arrangement and Biasing

265

by the grid (input) *voltage*. A fundamental difference between the tube and the transistor is that the tube is a voltage-controlled device whereas the transistor is current-controlled.

Nevertheless, the transistor is similar to the tube of Experiment 36 in that the collector is comparable to the plate, the base to the grid, and the emitter to the cathode: the emitter and cathode both emit charge carriers, the collector and plate both collect them, and the base and grid both control how many get through. The NPN transistor is particularly close to the tube since the charge carriers are electrons in both cases. While there is a PNP transistor, there is no practical high-vacuum tube through which the current is carried by positive charge carriers. In this experiment, the operation of the type 2N2219 NPN transistor will be studied, and the similarity to much of the work in Experiment 36 should be noted.

To begin with, a collector family of characteristic curves may be drawn for a transistor just as a plate family can for a tube. In the case of the transistor, it is the collector current that is plotted as a function of collector-to-emitter voltage *for various values of base current*. Note carefully that the curves correspond to different values of base *current,* not voltage. As noted above, it is a current rather than a voltage that is the control variable in a transistor. One purpose of this experiment will be to plot a set of collector characteristics for the 2N2219. You should find that, except for the magnitude of the voltage and current scales and the fact that your curves will be labeled with values of base current rather than grid voltage, your set will look very much like that given for the pentode in Fig. 92.

When a transistor is to be used as a voltage amplifier, it is connected in a circuit such as that of Fig. 100. Notice the standard symbol for a transistor shown in this figure. The arrowhead on the emitter lead indicates the direction of current flow and is drawn in this figure for an NPN transistor. The same diagram could be used for a PNP transistor if the polarity of V_{CC}, the direction of I_C, and the direction of the arrowhead on the transistor symbol were all reversed. A battery or other power supply supplies voltage V_{CC} to the circuit consisting of the collector-emitter path through the transistor and the load resistor R_L in series, as a result of which current I_C flows around this circuit. R_L converts variations in I_C into voltage variations, which are passed by blocking capacitor C_C to the output terminals, where they appear as the output signal voltage V_0. As in the case of vacuum tubes, the drop across R_L is the difference between the supply and collector voltages, so that

$$V_{CC} - V_{CE} = I_C R_L$$

or

$$I_C = \frac{V_{CC}}{R_L} - \frac{V_{CE}}{R_L} \qquad (137)$$

in analogy with Equation 135 for the vacuum-tube plate current. Thus a load line can be drawn on the collector characteristics of a transistor just as it can on the plate characteristics of a tube, and an operating point can be chosen in a similar manner. Since the transistor curves resemble those for a pentode, a minimum collector voltage and a minimum collector current are chosen to limit the operating range as was done in Procedure 11 of Experiment 36. The choice of operating points determines the input bias as in the vacuum-tube case, but now it is a *base bias current* rather than a grid bias voltage that must be provided. Figure 100 shows that this is done by supplying a current I_B to the base through resistor R_B from the V_{CC} supply. If the base-emitter junction were a perfect short circuit, this base current would simply be given by

$$I_B = \frac{V_{CC}}{R_B} \qquad (138)$$

In fact this approximation is extremely good for the 2N2219 in the circuit of Fig. 100, where V_{CC} is always large compared to the base-emitter voltage V_{BE} and the input resistance (resistance of the base-emitter junction) is small compared to the values required for R_B. Equation 138 can therefore be used to calculate R_B so that the base current needed to establish the chosen operating point will be obtained.

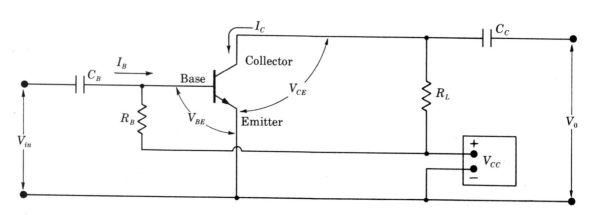

Figure 100   Transistor Amplifier Circuit

Variations in I_B result in variations in I_C, and since (like the dynamic plate resistance of a pentode) the dynamic collector resistance of a transistor is high, the attendant collector voltage variations due to the action of R_L have little effect. The ratio of the change in I_C due to a change in I_B is called the transistor's *forward current transfer ratio* and is designated either β or h_{fe}. It is analogous to the transconductance g_m of a pentode. However, note that because the controlling signal in a tube is a voltage whereas that in a transistor is a current, g_m is a conductance, being the ratio of a current change to a voltage change, but β is dimensionless, being the ratio of two current changes. For the pentode, variations in the grid voltage constitute the input signal V_{in}, and the voltage gain V_0/V_{in} can be approximated by $g_m R_L$ as described in Experiment 36. For the transistor, however, we can only say that $V_0 \approx \beta R_L \Delta I_B$, where ΔI_B represents variations in base current due to the input signal. If this signal is a voltage (as it will be if we are trying to make a voltage amplifier), it can cause variations in I_B by being applied to the base as Fig. 100 indicates. In this case V_{in} develops across the transistor's *dynamic input resistance* h_{ie}, so that $\Delta I_B = V_{in}/h_{ie}$. The voltage gain of the transistor amplifier is then given approximately by

$$A_V \equiv \frac{V_0}{V_{in}} \approx \frac{\beta R_L}{h_{ie}} \qquad (139)$$

where A_V is the gain factor (output signal divided by input signal) and V_{in} is the input signal voltage applied as shown in Fig. 100. Equation 139 assumes that variations in the collector-to-emitter voltage V_{CE} have no effect on I_C owing to the high dynamic collector resistance of the transistor, and this assumption is fully justified for most transistors as long as R_L is not made too large. Thus Equation 139 predicts that A_V will be proportional to R_L

over at least some range of load resistor values. This turns out not to be correct, however, because h_{ie}, instead of being constant, is highly dependent on the quiescent (no-signal) value of I_C and is therefore different for the different operating points that must be chosen when R_L is changed. In fact, h_{ie} is roughly proportional to $1/I_C$. If Equation 137 is written as $I_C = (V_{CC} - V_{CE})/R_L$ and we assume

$$h_{ie} \approx \frac{a}{I_C} \qquad (140)$$

where a is some proportionality constant, h_{ie} becomes proportional to R_L according to the relation

$$h_{ie} \approx \frac{a R_L}{V_{CC} - V_{CE}} \qquad (141)$$

where the quiescent collector voltage V_{CE} (the collector voltage corresponding to the operating point) is assumed not to change much when the operating point is changed as needed to suit different values of R_L. It turns out that, for reasonably large load resistances, the operating points all come at about the same collector voltage, although at different collector currents I_C. Equation 141 is then roughly correct. Substitution in Equation 139 yields

$$A_V \approx \frac{(V_{CC} - V_{CE})\beta}{a} \qquad (142)$$

which shows that over a certain range of R_L, A_V becomes constant and independent of the load resistor value. Equations 141 and 142 can be used to estimate the value of h_{ie} at the operating point chosen for a load resistor in the range where the proportionality between h_{ie} and R_L is valid.

APPARATUS

1. Type 2N2219 transistor
2. Filament transformer, 6.3-volts, 3-amperes
3. Small variable-line transformer ("Variac" or "Powerstat")
4. 12-volt power supply or storage battery
5. Oscilloscope
6. Electronic voltmeter
7. D.C. milliammeter (0–50 milliamperes)
8. Tubular rheostat (about 100-ohms)
9. Decade resistance box (0–10,000 ohms)
10. Decade resistance box (0–1000 ohms)

11. Four deposited carbon or other 1%-tolerance resistors (100,000, 200,000, and two 300,000 ohms)
12. Six carbon 1-watt, 5%-tolerance resistors (4700, 2200, 1500, 1000, 680, and 470 ohms)
13. Volume-control-type rheostat (200,000 ohms)
14. Two electrolytic capacitors (about 500 microfarads, 35-volt rating)
15. Assorted hookup wires and connectors

PROCEDURE

1. Assemble the circuit shown in Fig. 101 for measuring the collector characteristics of the type 2N2219 transistor. A bottom view of the 2N2219 showing the connection arrangement is given at the left in this figure. Use the tubular rheostat at R, and begin with a resistance of 600,000 ohms at R_B. This resistance is obtained by connecting the two 300,000-ohm resistors in series. Have your circuit checked by the instructor

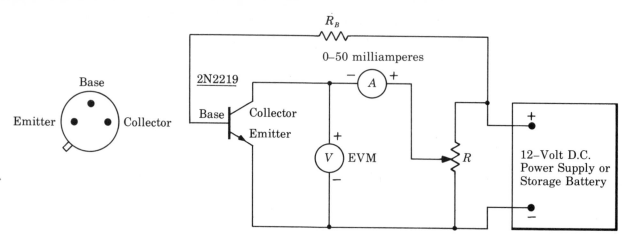

Figure 101   Circuit for Measuring Transistor Characteristics

before plugging in or turning on the power supply (or, if a storage battery is used, before connecting the battery leads). *Note:* The base current will be set by choosing the value of R_B on the assumption that the power supply or battery is producing 12 volts. The instructor will check that the power supply is set for 12 volts or that the battery voltage is correct while inspecting your wiring.

2. Start with the rheostat sliding contact all the way to the "low" end ($V_{CE} = 0$), and make sure that the EVM is switched to the 15-volt D.C. range before applying power. Plug in and turn on the power supply or connect the leads to the storage battery. Note that with $R_B = 600,000$ ohms, $I_B = 12/600,000 = 2 \times 10^{-5}$ amperes or 20 microamperes. Take data for the collector characteristic curve for this base current by setting V_{CE} at 1, 2, 4, 6, 8, 10, and 12 volts in turn and reading and recording the corresponding collector currents. *Note:* At the higher collector voltage and current settings, the transistor will warm up and the observed collector current will drift upward. Allow a few seconds for the current to stabilize after setting the voltage, but do not attempt to wait for it to be completely steady. If the drift is very pronounced, a better set of data can be obtained by starting with $V_{CE} = 12$ volts and working downward.

3. Repeat Procedure 2 with base currents of 40, 60, 80, 100, 120, 140, 160, and 180 microamperes. These currents are set by substituting resistances of 300,000; 200,000; 150,000; 120,000; 100,000; 85,714; 75,000; and 66,667 ohms in turn at R_B. All of these resistances can be made up from series and parallel combinations of the 100,000-, 200,000-, and 300,000-ohm resistors supplied as part of your apparatus. Thus 150,000 ohms results from connecting the two 300,000-ohm resistors in parallel, 120,000 ohms from connecting one of the 300,000-ohm resistors in parallel with the 200,000-ohm resistor, and 85,714 (600,000/7) ohms by adding the other 300,000-ohm resistor in parallel. All the necessary combinations are shown in the data table. Be sure to reduce V_{CE} to zero with the rheostat whenever you make a change in R_B, and turn off the power supply or disconnect the storage battery when you are through taking data.

4. Plot the collector characteristics of your 2N2219 using the data just obtained in Procedures 2 and 3. Plot the collector voltage values as abscissas along the long side of the graph sheet and the collector currents as ordinates along the short side. Choose scales so as to spread your curves over as much of the graph sheet as possible. Then, on the same graph sheet, plot load lines for load resistors of 4700, 2200, 1500, 1000, 680, and 470 ohms.

5. Choose operating points on each load line as described in Experiment 36. Assume a minimum permissible collector current of 0.5 milliampere and a minimum permissible collector-to-emitter voltage of 1 volt, and choose each operating point midway between the points where the pertinent load line intersects these limits. Note and record the base current corresponding to each operating point. Interpolate between neighboring characteristic curves if necessary to find the base current required for an operating point that doesn't fall on one of the characteristic curves you have drawn. Also record the collector current and collector voltage determined by each of your operating points.

6. Wire the transistor voltage amplifier circuit given in Fig. 102. Use the 10,000-ohm decade resistance box for R_1, the 1000-ohm box for R_2, and the 470-ohm resistor for R_L. The filament transformer supplies a 60-hertz test signal that is reduced by the voltage divider (made up of R_1 and R_2) to a level suitable for the amplifier input. However, we will find that an amplitude of only about 1 volt is desirable at point A, and so connections to the transformer secondary coil are made to the center tap and one of the other terminals so that

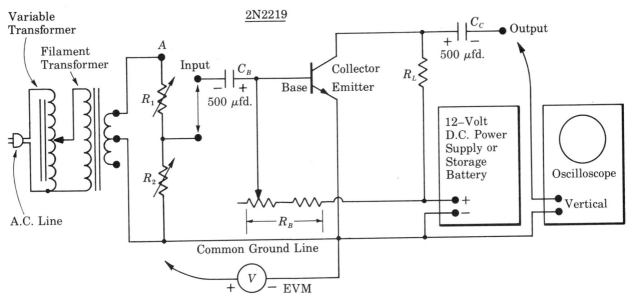

Figure 102   Amplifier Test Circuit for the 2N2219

only half of this winding is used and hence only half the voltage delivered. In addition, a variable transformer is connected between the filament transformer and the A.C. line as shown. This allows the filament transformer primary voltage, and hence the output from its secondary coil, to be set at any desired value. It will be used to adjust the signal level at point A as required.

The electronic voltmeter's common ("low" side) lead is connected to the 2N2219's emitter (the common ground line), and the EVM probe is then available to test voltages at various points in the circuit. The "low" side of the oscilloscope's vertical input is also connected to the common ground line, and the "high" side is provided with a lead that may be connected either to the amplifier output terminal or to point A to observe the A.C. voltage across the voltage divider. A lead is provided to connect the amplifier input to the junction of R_1 and R_2, but should be omitted for the present. The combination of resistors required for R_B is discussed below and should also be temporarily omitted.

7. Resistance R_B is made up of a suitable combination of the 100,000-, 200,000-, and 300,000-ohm resistors in series with the 200,000-ohm rheostat. Calculate and record the value of R_B needed to establish the base bias current found in Procedure 5 for the 470-ohm load line. Assume 100,000 ohms (the half-scale value) for the rheostat, and determine a suitable combination of the fixed resistors to be connected in series with it to give the needed value of R_B as closely as possible. Assemble the combination so determined and measure its resistance with the EVM switched to an appropriate ohmmeter range. Set the resistance to the exact value required by adjusting the 200,000-ohm rheostat. Then connect the combination in your amplifier circuit without disturbing this adjustment. Have your circuit approved by the instructor before applying power to it.

8. Plug in and turn on the power supply or make final connections to the storage battery. Measure and record the collector voltage V_{CE}.

9. Set R_1 to 1000 ohms, R_2 to zero, and connect the lead between the junction of R_1 and R_2 and the amplifier input. Plug in the variable transformer and set it so that the voltage at point A (Fig. 102) has an amplitude of about 1 volt. This will be obtained at about 20% of the transformer dial's full-scale reading. Connect the oscilloscope vertical input to point A, and set the controls to display two cycles of the 60-hertz test signal. Use line (not internal) synchronization to stabilize the pattern. Adjust the vertical gain so that the pattern has a convenient amplitude. If your oscilloscope has a vertical calibration, you can use it to check the voltage at point A and readjust the variable transformer accordingly, but this is not essential for the experiment. However, the final setting of the vertical gain should in any case be such that the displayed sine curve covers most of the oscilloscope screen and falls between easily remembered marks so that its amplitude may be duplicated later. Do not change the vertical gain once this setting has been made.

10. Disconnect the oscilloscope vertical input from point A and connect it to the amplifier output. Increase the value of R_2 until the displayed sine curve has the same amplitude as that established in Procedure 9. Record the value of R_2 needed to do this.

11. Compare the sine curves displayed in Procedures 9 and 10 and report your observation in the space below.

12. Turn off the power supply or disconnect the storage battery and substitute the 680-ohm resistor for the 470-ohm resistor at R_L. Then repeat Procedures 7–10 for $R_L = 680$ ohms.

13. Repeat Procedure 12 with $R_L = 1000, 1500, 2200,$ and 4700 ohms in turn. Be sure to turn off the power supply or disconnect the storage battery when you have finished the experiment.

CALCULATIONS

1. From Equation 137, calculate and record I_C for each value of R_L. Note that $V_{CC} = 12$ volts and the necessary values of V_{CE} were measured in Procedure 8. Compare these measured values of I_C and V_{CE} with those predicted from the characteristic curves.

2. The combination of the resistances R_1 and R_2 forms a so-called *voltage divider* whose purpose is to produce a voltage V_{in} across R_2 which is a known fraction of the voltage V_s impressed across R_1 and R_2 in series. This fraction is given by

$$\frac{V_{in}}{V_s} = \frac{R_2}{R_1 + R_2} \tag{136}$$

Notice that V_{in} is the amplifier input signal voltage and V_s is the voltage at point A. In Procedures 10, 12, and 13, R_2 was adjusted until the output voltage V_o was equal to V_s, so that the amplifier was amplifying V_{in} back up to the same level V_s from which the voltage divider cut it down. In other words, the amplifier was simply undoing what the voltage divider had done, so that the gain is

$$A_V \equiv \frac{V_o}{V_{in}} = \frac{V_s}{V_{in}} = \frac{R_1 + R_2}{R_2}$$

Use this conclusion to compute the gain obtained with each load resistor.

DATA

Collector Current I_C, Milliamperes

Collector-to-Emitter Voltage V_{CE}, Volts	$I_B = 20\mu a$ $R_B = 600K\Omega$ (300KΩ 300KΩ)	$I_B = 40\mu a$ $R_B = 300K\Omega$ (300KΩ)	$I_B = 60\mu a$ $R_B = 200K\Omega$ (200KΩ)	$I_B = 80\mu a$ $R_B = 150K\Omega$ (300KΩ 300KΩ)	$I_B = 100\mu a$ $R_B = 120K\Omega$ (200KΩ 300KΩ)	$I_B = 120\mu a$ $R_B = 100K\Omega$ (100KΩ)	$I_B = 140\mu a$ $R_B = 85,714\Omega$ (200KΩ 300KΩ 300KΩ)	$I_B = 160\mu a$ $R_B = 75,000\Omega$ (100KΩ 300KΩ)	$I_B = 180\mu a$ $R_B = 66,667\Omega$ (100KΩ 200KΩ)
1									
2									
4									
6									
8									
10									
12									

Calculated from Curves / Measured

Load Resistor R_L, Ohms	Base Bias I_B, Ma	Value of R_B, Ohms	Collector Current I_C, Ma	Collector Voltage V_{CE}, Volts	Collector Voltage V_{CE}, Volts	Collector Current I_C, Ma	Collector Voltage V_{CE}, Volts	Value of R_2, Ohms	Voltage Gain A_V
470									
680									
1000									
1500									
2200									
4700									

h_{ie} at the operating point on the 4700-ohm load line _____

Gain at high values of R_L _____

β from the characteristic curves _____

3. Plot gain against load resistance using your values of R_L as abscissas and the gain values found in Calculation 2 as ordinates. Choose your scales so as to cover as much of the sheet of graph paper as possible, but allow for some extrapolation of your curve to higher R_L. *Note:* If R_L were zero there would be no output voltage and therefore no voltage gain, so that the origin is a point on your graph. Draw a smooth curve through your seven plotted points.

·4. Note whether the curve of Calculation 3 seems to approach a constant value of the gain for large R_L, and by careful extrapolation estimate and record this value.

5. Calculate the forward current transfer ratio β in effect near the operating point on the 4700-ohm load line. To do this, draw a vertical (constant V_{CE}) line through this operating point on your 2N2219 characteristics and note the collector currents determined by the intersections of this line with the curves drawn for base currents of 20 and 40 microamperes. Subtract the lower I_C from the higher and note that *this is the change in collector current caused by changing I_B from 20 to 40 microamperes with V_{CE} constant at its operating point value.* Hence, by definition, β at the operating point for R_L = 4700 ohms is this change in collector current divided by the 20-microampere change in base current. Record your result.

6. Subtract V_{CE} at the 4700-ohm operating point from V_{CC} (12 volts) to obtain $V_{CC} - V_{CE}$ for use in Equation 142. Also substitute your values of A_V and β from Calculations 4 and 5 in this equation and solve it for the constant a. Then use this result in either Equation 140 or 141 to find h_{ie} for your transistor as operated with R_L = 4700 ohms.

QUESTIONS

1. Why was it necessary to limit V_s to an amplitude of about 1 volt? What would have happened if the filament transformer had been plugged directly into the A.C. line and the full secondary voltage used as V_s?

2. The collector conductance of a transistor (analogous to the plate conductance g_p of a vacuum tube) is the ratio of a small change in collector current to the corresponding small change in collector-to-emitter voltage with no change in base current. This parameter, usually designated h_{oe}, is thus the slope of the collector characteristic passing through the point where h_{oe} is to be evaluated. Find h_{oe} for your transistor at $I_B = 60$ microamperes and $V_{CE} = 6$ volts. How does its reciprocal (analogous to the tube's r_p) compare with the values of R_L used in this experiment?

3. Line synchronization of the oscilloscope patterns was used in Procedure 9 so that an important difference between the patterns displayed in Procedures 9 and 10 could be observed in Procedure 11. What was this difference and why did line rather than internal synchronization have to be used to observe it?

4. How does the input resistance of a transistor amplifier compare with that of an amplifier using a vacuum tube? How will this affect its operation?

5. In the derivation of Equation 139, the signal voltage V_{in} was assumed to be applied across only the transistor input (base-emitter) resistance h_{ie} so that ΔI_B was equal to V_{in}/h_{ie}. Resistor R_B is also connected to the transistor base, so that V_{in} is actually applied across h_{ie} and R_B in parallel. Why could R_B be neglected in the determination of ΔI_B?

6. A type 2N3394 transistor has $\beta = 100$ and $h_{ie} = 2750$ ohms when $I_C = 1$ milliampere and $V_{CE} = 10$ volts. If the operating point fixed by these values of I_C and V_{CE} is to be used and a 15-volt power supply is available for V_{CC}, find (a) the necessary load and base resistances R_L and R_B and (b) the gain of the resulting amplifier. *Note:* To find the base current required at the operating point, assume β to be a constant right down to zero collector current, that is, assume I_C and I_B to be strictly proportional over their entire range.

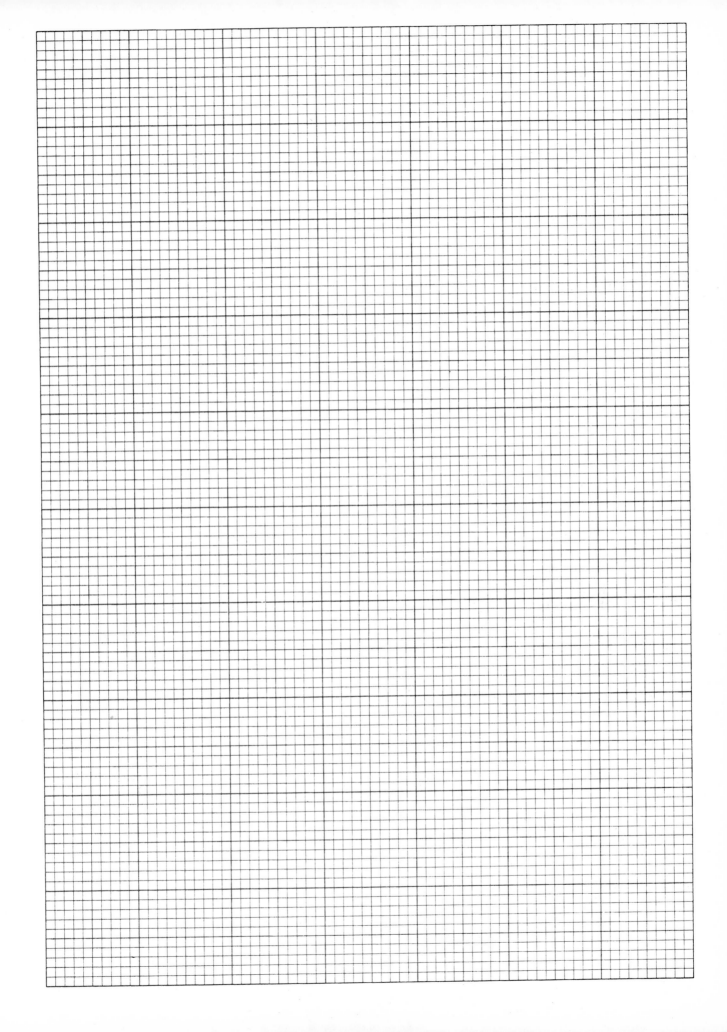

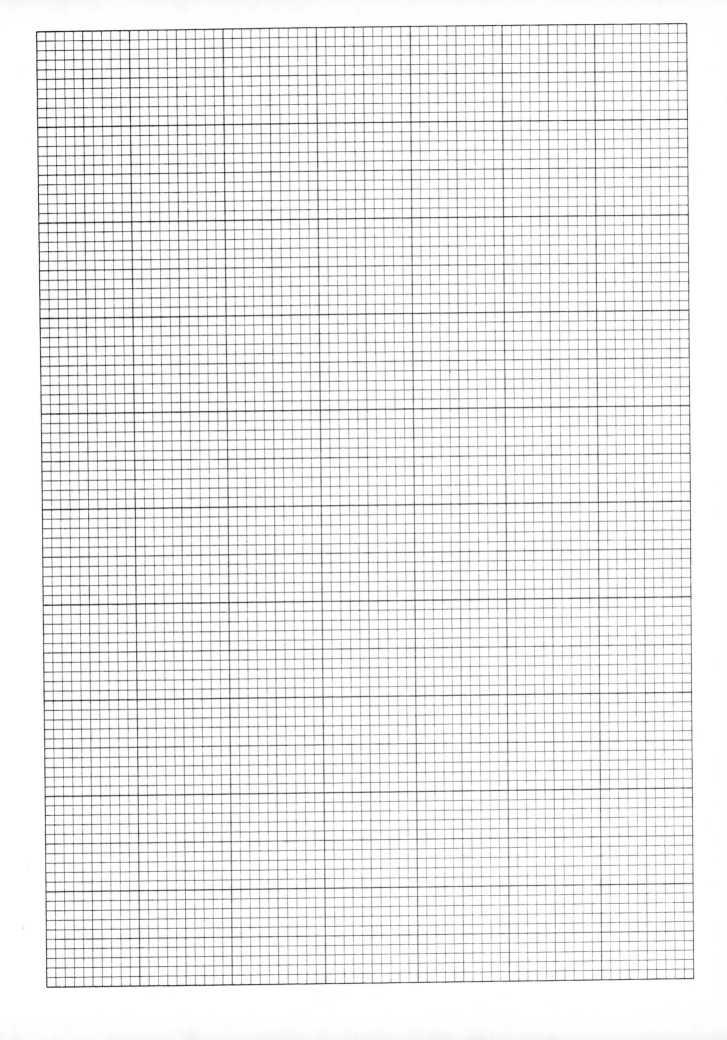

Operational Amplifiers 38

An operational amplifier is one characterized by high gain (20,000 to a million), a reasonably low output impedance, and a frequency response that extends from some fairly high value all the way down to zero (*i.e.*, it is a so-called "D.C. amplifier"). The name comes from the fact that operational amplifiers are often used to perform the electrical analogue of various mathematical operations such as addition, multiplication, integration, and their inverses. Today operational amplifiers are all solid-state units built with integrated-circuit techniques and completely encapsulated in small, sealed packages of which there are several standard styles. Connections for input, output, and supply power are brought out as "leads," *i.e.*, metal tabs suitable for connection in printed circuits. Operational amplifiers can also be built using vacuum tubes, and many such were produced before the advent of transistors. However, because tubes are relatively large, fragile, and characterized by much higher power requirements than the solid state units, they are now considered obsolete. The purpose of this experiment is to see how a typical modern operational amplifier works in some standard circuits and to observe its performance.

THEORY

Operational amplifiers may include several stages of amplification and much complex circuitry, but the details of their construction will not concern us. We shall be interested only in the fact that we have a unit with an input and an output characterized by a large voltage gain A_v, a reasonably low output impedance so that the output may supply a certain amount of current and thus power to a load, a reasonably high input impedance to avoid loading the source of the input signal, and a frequency response extending from zero (D.C.) up to a value sufficiently high to allow needed response speeds to be realized. The schematic symbol for an operational amplifier is an isosceles triangle pointing in the direction of amplification, *i.e.*, with the output lead shown coming out of the apex and the input lead or leads going into the base. Fig. 103(a) is the schematic diagram for a simple operational amplifier, defined as one having a single input. Input voltage V_{in} is applied between this input and common ground, and output voltage V_o appears between the output terminal and common ground. The amplifier exhibits voltage gain A_v, which means that as long as the input signal has a frequency within the amplifier's response range and the output is lightly loaded, V_o is A_v times V_{in}. Because operational amplifiers are, as we shall see, usually used in negative-feedback circuits, the amplifier is designed to invert, *i.e.*, to have an increase in V_{in} result in a decrease in V_o. Thus

$$V_o = -A_v V_{in} \qquad (143)$$

Modern operational amplifiers are almost all of the differential type, which can be represented schematically as shown in Fig. 103(b). Here there are two inputs, and the input voltage to be amplified is applied between

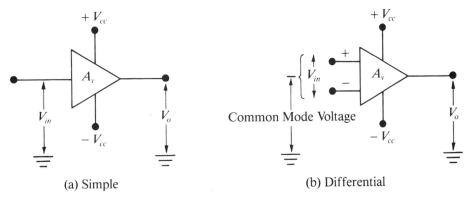

(a) Simple

(b) Differential

Figure 103 Operational Amplifier Schematic Diagrams

them. They are labeled + and − to indicate that if the + input is made increasingly positive with respect to the − input, the output voltage V_o will rise (become more positive) with respect to ground. Consequently, the + input is often called the *non-inverting input,* whereas the − input is the *inverting* one. A differential operational amplifier may be turned into a simple one by grounding one input, in which case it inverts if the + input is grounded and does not if we ground the − input. In addition, V_o is supposed to be affected only by a voltage applied *between* the two inputs and not at all by a voltage applied between the paralleled inputs and ground. Thus if the two inputs are connected together and a voltage applied between ground and the two of them in common, the output voltage should remain zero and the amplifier's operation should be unaffected no matter what this applied voltage may be (provided it stays in a range quoted in the amplifier's specification sheet). Such a voltage is called a *common mode voltage,* and the amplifier's ability to remain unaffected by it is called its *common mode rejection* (CMR). Actually, V_o will, in practice, be slightly affected by the application of a common mode voltage, and a common mode gain, *i.e.,* the ratio of the change in output voltage to the change in common mode voltage that produced it, can be quoted. Usually the ratio of A_v to the common mode gain, a quantity called the *common mode rejection ratio* (CMRR), is what is listed in operational amplifier data sheets. Because the common mode gain is likely to be less than one and the amplifier's voltage gain A_v likely to be on the order of 100,000, the CMRR is often very large indeed. Thus it may be given in decibels, in which case the quoted number is 20 times the common logarithm of the actual gain ratio. Differential operational amplifiers have the advantages of (1) not requiring that the input voltage be referred to ground and (2) being either inverting or non-inverting. Because the differential or "push–pull" amplifier circuit is better from the point of view of circuit balance and is no harder to build than the simple or "single-ended" type, almost all modern operational amplifiers have differential inputs.

Fig. 103 shows the power supply connections explicitly, although in many schematic diagrams the fact that power must be supplied to the amplifiers is taken for granted, and these connections are consequently not shown. Notice that because the amplifier's output voltage may swing both above and below ground, both a positive and a negative power supply are required. Most solid-state operational amplifiers are designed for a standard power supply voltage V_{cc} of 15 volts but will work perfectly well with a V_{cc} of 12 volts, so that standard storage batteries may be used. The junction of the positive and negative power supplies is the common ground point.

Fig. 104 shows the basic circuit for an operational amplifier employing negative feedback. On the assumption that the amplifier's input impedance is so high that negligible current flows into the input, a current

$$i = \frac{V_1 - V_o}{R_1 + R_f}$$

flows around the circuit composed of $V_1, R_1, R_f,$ and V_o in series. The amplifier's input voltage V_{in}, which because the + input is grounded is the voltage between the − input and ground, is then seen to be given by

$$V_{in} = V_1 - iR_1 = V_1 - \frac{V_1 - V_o}{R_1 + R_f} R_1$$

But the output voltage must be given by Equation 143, so that substitution for V_{in} yields

$$- \frac{V_o}{A_v} = V_1 - \frac{R_1}{R_1 + R_f} (V_1 - V_o)$$

which, after an appropriate rearrangement of terms, becomes

$$V_o \left(\frac{R_1}{R_1 + R_f} + \frac{1}{A_v} \right) = - V_1 \left(\frac{R_f}{R_1 + R_f} \right) \qquad (144)$$

Now the voltage gain A_v is very large for all frequencies in which we are interested; thus $1/A_v$ is very small. In particular, $1/A_v \ll R_1/(R_1 + R_f)$ and may therefore be neglected in comparison. This can be appreciated by noting that A_v will be at least 20,000 whereas R_f will be at most ten times R_1, so that we are comparing 1/20,000 or less with 1/11 or more. Hence to a very good approximation, Equation 144 may be reduced to

$$V_o \left(\frac{R_1}{R_1 + R_f} \right) = - V_1 \left(\frac{R_f}{R_1 + R_f} \right)$$

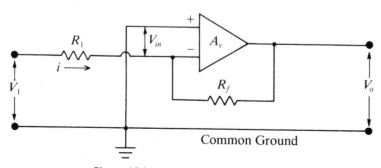

Figure 104   The Basic Inverting Multiplier

and finally to

$$V_o = -V_1 \frac{R_f}{R_1} \qquad (145)$$

This most important result tells us that the output voltage V_o is the negative of the input voltage V_1 multiplied by the fixed factor R_f/R_1. We can say that we have a device that multiplies V_1 by $-R_f/R_1$ or that our complete circuit is an inverting amplifier having a gain G with feedback (as distinguished from the operational amplifier's unfedback gain A_v) of R_f/R_1. The important point is that G is fixed by the values of a couple of resistors (which can be made very precise and stable) and *does not depend on the gain of the amplifier at all*. In fact, *as long as A_v is large enough so that the approximation made in deriving Equation 145 remains valid, the actual value of A_v can vary all over the place as the amplifier ages and supply voltages change without affecting the relation between V_o and V_1.* Because of the precision with which G can be maintained, the circuit in Fig. 104 is usually thought of as an analog multiplier rather than an amplifier of gain G.

There is a much simpler way of looking at the circuit of Fig. 104 that lets us understand why Equation 145 should be true without going through a lengthy derivation. The crucial point is that the gain A_v of the operational amplifier itself is high. This means that only a very tiny input voltage will produce an appreciable output voltage. Thus suppose we have an operational amplifier whose V_o may swing over a range of -10 to $+10$ volts. This is typical of solid-state amplifiers designed to operate from a pair of 15-volt power supplies. If $A_v = 100,000$ (a typical value), *only a ten-thousandth of a volt at the input drives the output over its full range.* But this will not be allowed to happen, for as soon as V_o gets away from zero, the feedback through R_f will apply a correcting signal at the input. Thus although the $-$ input is not grounded, it is maintained at almost exactly ground potential by the action of the circuit and is called a *virtual ground.* If the voltage at this point can really

be taken to be zero, we can then write

$$i = \frac{V_1}{R_1} = -\frac{V_o}{R_f}$$

from which Equation 145 immediately follows.

Fig. 104 illustrates the basic circuit in which negative feedback is applied to an operational amplifier to produce a system capable of performing analog operations. There are many possible modifications for adapting the circuit to particular conditions or setting it up to perform operations other than simple multiplication by a fixed factor. A typical adder circuit is shown in Fig. 105. It is the same as Fig. 104 except that more than one input has been provided. If the amplifier really takes no current at its input, the sum of the currents at their junction at the $-$ input must be zero. That is,

$$i_1 + i_2 + i_3 + i_f = 0 \qquad (146)$$

Because the $+$ input is grounded, the amplifier's input voltage is that between the $-$ input and ground, and if this is negligible (the $-$ input is a virtual ground), we have $i_1 = V_1/R_1$, $i_2 = V_2/R_2$, $i_3 = V_3/R_3$, and $i_f = V_o/R_f$. Substitution in Equation 146 yields

$$\frac{V_1}{R_1} + \frac{V_2}{R_2} + \frac{V_3}{R_3} = -\frac{V_o}{R_f}$$

or $\quad V_o = -\left(\dfrac{R_f}{R_1}V_1 + \dfrac{R_f}{R_2}V_2 + \dfrac{R_f}{R_3}V_3\right) \quad (147)$

Equation 147 shows that the negative of the output voltage V_o is equal to the sum of the voltages V_1, V_2, and V_3, each multiplied by a factor that we are free to choose by choosing the resistances R_1, R_2, R_3, and R_f. In particular, if we want $-V_o$ to be just the sum of $V_1 + V_2 + V_3$, we have only to make all the resistances equal. Note that Equation 146 states a fundamental rule of circuit theory, namely that the algebraic sum of all the currents entering a junction must be zero. Because of this, the $-$ input in the adder circuit is sometimes called the *summing point.*

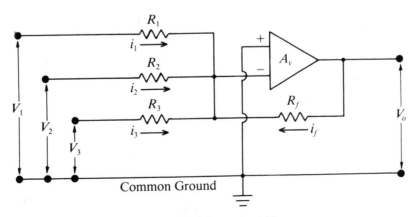

Figure 105 The Analog Adder

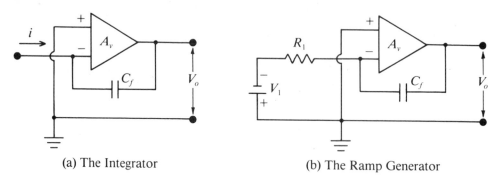

(a) The Integrator   (b) The Ramp Generator

Figure 106   Analog Integrator Circuits

An operational amplifier may be used to perform integration by substituting a capacitor C_f for the feedback resistor R_f appearing in our earlier circuits. Fig. 106(a) shows this arrangement. If the $-$ input is a virtual ground, the voltage across the capacitor is the output voltage V_o, and the capacitor's charge q is given by

$$q = C_f V_o$$

But again, if no current flows into the amplifier input, current i must all flow into the capacitor. Therefore q will be simply i times the time, or if i varies with time

$$q = - \int i\,dt$$

where dt is a time interval so small that i may be considered constant over it, and the minus sign comes from the fact that i flows into the side of C_f opposite to that connected to V_o. Putting these results together yields

$$V_o = - \frac{1}{C_f} \int i\,dt \qquad (148)$$

so that the output voltage is proportional to the time integral of the input current. If we now look at the circuit of Fig. 106(b), we see that, provided the $-$ input remains a virtual ground, i has the constant value V_1/R_1. Consequently Equation 148 now becomes

$$V_o = - \frac{V_1}{R_1 C_f} t \qquad (149)$$

where t is the time in seconds measured from an initial instant at which the charge on C_f and hence the output V_o was zero. Note in Fig. 106(b) that the battery supplying V_1 has been connected to make this voltage negative to remove the negative sign in Equation 149 and make the output voltage positive. This voltage will thus *increase exactly linearly with time*, the slope of a graph of V_o against t being the constant $V_1/R_1 C_f$, which we can choose at will. Because this graph, which is a straight line sloping up to the right, looks like a ramp, the circuit of Fig. 106(b) is often called a *ramp generator*. According to Equation 149, the ramp could go on forever, but of course a limit is imposed by the maximum output

voltage of which the operational amplifier is capable (usually about 10 volts). The circuit may be reset and V_o returned to zero by momentarily short-circuiting the capacitor.

One important application of operational amplifiers that does not employ feedback is service as a so-called *Schmitt trigger*. The original Schmitt trigger is a bistable circuit characterized by an output voltage that can assume one of two values. Circuits of this general type are called ''flip-flops'' because they ''flip'' or switch their output voltage from one possible value to the other practically instantaneously without this voltage remaining at any intermediate value. The Schmitt trigger has two inputs, one of which is maintained at some reference voltage V_{ref} while a signal voltage of some sort is applied to the other. When the signal voltage is above the reference voltage, the output voltage has one of its two values, and when the signal voltage is below V_{ref}, V_o has its other value. Switching takes place as the signal voltage passes through the reference value. An actual Schmitt trigger employs two transistors in a special circuit incorporating a large amount of positive feedback, but an operational amplifier with no feedback can be arranged to provide essentially the same performance. The important point once again is that the operational amplifier's unfedback voltage gain A_v is large. Taking 100,000 as a representative value, we again note that an input voltage V_{in} of only one ten-thousandth of a volt is enough to drive the output voltage V_o to its maximum value of 10 volts. In practice there are always enough noise voltages or other stray voltages around so that even with the amplifier inputs connected together and grounded, V_o will not stay around zero but will ''hang up'' at either its maximum positive or its maximum negative value. The operational amplifier with no feedback thus exhibits two, and only two, stable output voltage values, just as the Schmitt trigger does.

Now let us assume that the $-$ input is maintained at a positive reference voltage V_{ref}, which can, for example, be selected by a rheostat as shown in Fig. 107. A signal voltage V_1 is applied between the $+$ input and ground. We note at once that if $V_1 = 0$, $V_{in} = - V_{ref}$

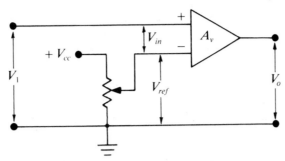

Figure 107   The Schmitt Trigger Application

and the amplifier is driven far into saturation with V_o "hung up" at its maximum negative value. This situation continues as V_1 is increased until suddenly, as V_1 gets very close (of the order of a ten-thousandth of a volt) to V_{ref}, the amplifier will suddenly find itself driven the other way and V_o will switch to its maximum positive value. This changeover takes place extremely rapidly, the switching speed being a characteristic of the amplifier and its load. The speed with which V_o can switch from one value to another in response to an instantaneous change in V_{in} is called the amplifier's *slewing rate* and is quoted in volts/second in data sheets supplied with commercial units.

The operational amplifier chosen for this experiment is the type 741, a very popular general-purpose commercial unit. It is made with integrated circuit techniques that allow 20 transistors and their associated circuitry to be packaged in a tiny plastic encapsulation having a standard size and connection tab arrangement called an "8-pin DIP." A top view of this package, drawn four times the actual size and showing the pin connections for the type 741, is given in Fig. 108. Note that this is a *top view* and that to observe the pin numbering correctly the orienting mark must be to your left. The connections for − input, + input, the negative supply voltage − V_{cc}, the positive supply voltage + V_{cc}, and the output voltage V_o are clearly shown. In addition there is no connection to pin 8, and we will not use the offset null pins 1 and 5. The term "offset" refers to the fact that the slight imbalances in practical differential amplifier circuits result in a zero output voltage corresponding not to a zero input voltage but to a V_{in} slightly "offset" from zero. The type 741 provides for connection to an external circuit that allows the offset to be "nulled," *i.e.*, adjusted to exactly zero. This is needed only for certain very critical applications and will not concern us here.

TYPE 741 CHARACTERISTICS

Supply voltage V_{cc}	± 12 to ± 15 volts
Range of output voltage V_o	± (V_{cc}−2) volts
Voltage gain A_v	Approximately 200,000
Input resistance	Approximately 2×10^6 ohms
CMRR	90 decibels
Common mode input voltage range	± 13 volts
Maximum short-circuit output current	± 25 milliamperes

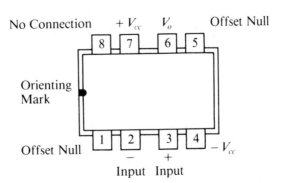

Figure 108   Type 741 Connection Diagram (Top View)

APPARATUS

1. Two type 741 operational amplifiers (8-pin DIP package)
2. Two 12 to 15-volt power supplies or storage batteries
3. Two good 1.5-volt dry cells
4. D.C. voltmeter (0–15 volts)
5. Single-pole, double-throw relay (12-volt D.C. coil)
6. Type 2N2219 transistor
7. Two single-pole, single-throw switches
8. Decade resistance box (10,000 ohms)
9. Decade resistance box (1000 ohms)
10. Four carbon 1-watt 5% tolerance resistors (470,000, 4700, 1500, and 1000 ohms)
11. Volume-control-type rheostat (5000 ohms, wirewound)
12. Oil or paper dielectric capacitor (2.0 micro-farad)
13. Electrolytic capacitor (about 500 microfarads, 35-volt rating)
14. Assorted hookup wires and connectors
15. Stop watch or stop clock

PROCEDURE

1. Wire the analog multiplier circuit shown in Fig. 109. Use the 10,000-ohm decade box for the feedback resistor R_f and the 1000-ohm box for the input resistor R_1. Notice that the dry cell is connected with its positive side to ground so that the amplifier output will be positive. Fig. 109 shows the metal decade-box cases connected to the common ground line, and you should do this if your boxes have terminals for this purpose. Common ground is established as the junction of the two power supplies, and the + input is connected to this line to fix the common mode voltage at zero. Set R_f at zero and R_1 to 1000 ohms. Do not plug in or turn on the power supplies or, if you are using storage batteries, do not make final connections to their terminals until your wiring has been checked by the instructor.

2. When your wiring has been checked and approved, turn on the power supplies or connect the storage batteries. If you are using power supplies with variable output voltage, start with the controls all the way down. Temporarily disconnect the 15-volt voltmeter from the amplifier output and use it to measure the two power-supply voltages. Set each to between 12 and 15 volts, making sure the two voltages have the same magnitude ($| + V_{cc} | = | - V_{cc} |$). Note that the supply for $- V_{cc}$ has its positive side grounded, which means that the voltmeter connections must be temporarily interchanged to allow measurement of this voltage. Then reconnect the voltmeter to the amplifier output and check that it reads zero. If it does not, see whether R_f has been set to zero and R_1 to 1000 ohms. If the voltmeter still does not read zero, shut off power to the circuit and recheck your wiring. If trouble persists, call your instructor.

3. When you are sure your circuit is working properly, set R_f to 1000 ohms and read the voltmeter. Record this reading in the data table. Repeat this procedure with $R_f =$ 2000, 3000, 4000, 5000, and 6000 ohms.

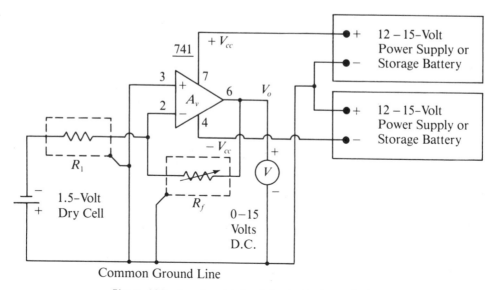

Figure 109 Experimental Circuit for the Analog Multiplier

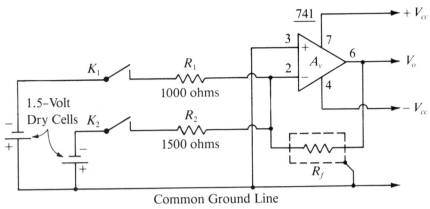

Figure 110 Circuit Changes for the Analog Adder

4. Shut off the power to your circuit by unplugging the power supplies or disconnecting the storage batteries. Then make your setup into a simple analog adder by installing the modification diagrammed in Fig. 110. This consists of removing the 1000-ohm decade box and connecting R_1, R_2, K_1, K_2, and the two dry cells to the − input as shown. Note that R_1 and R_2 are 1-watt carbon resistors. Set the decade box at R_f to 3000 ohms and make sure switches K_1 and K_2 are both open. Have your wiring checked by the instructor before applying power to the circuit.

5. Turn on the power by plugging in the power supplies or making final connections to the storage batteries. If necessary, check the supply voltages $+ V_{cc}$ and $- V_{cc}$ with the voltmeter as described in Procedure 2. When the voltmeter is reconnected to the amplifier output, it should read zero.

6. Close switch K_1 and read and record V_0. Open K_1 and close K_2, and again read and record V_0. Repeat with K_1 and K_2 both closed.

7. Shut off power to your circuit and rewire it as an analog integrator as shown in Fig. 111. R_1 is a 470,000-ohm, 1-watt carbon resistor and C_f is a 2-mfd capacitor with oil or paper dielectric. Use the same dry cell you used in the analog multiplier and note that again its positive terminal is grounded so that you will get a positive output voltage. Close K_1 to short-circuit C_f and have your wiring checked by the instructor before applying power to the circuit.

8. Turn on the power by plugging in the power supplies or making final connections to the storage batteries. If necessary, check $+ V_{cc}$ and $- V_{cc}$ with the voltmeter as before. Reconnect the voltmeter to measure V_0. It should read zero with K_1 closed.

9. Open K_1 and simultaneously start the stop watch (or stop clock). You should observe a steady rise in V_0 as read on the voltmeter. Stop the stop watch when the voltmeter reads exactly 10 volts. Reset V_0 to zero by closing K_1. Record the stop watch reading.

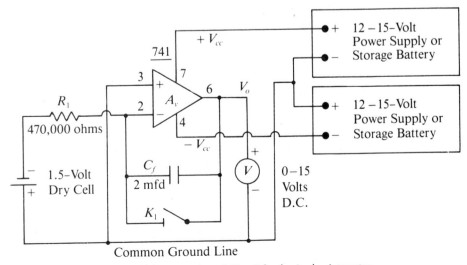

Figure 111 Experimental Circuit for the Analog Integrator

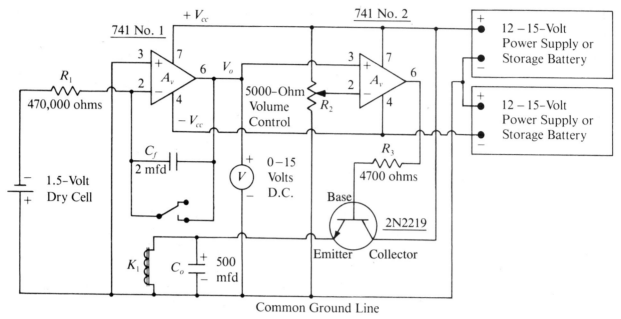

Figure 112   The Sawtooth Generator

10. Repeat Procedure 9 twice to obtain a total of three measurements of the time required for V_o to rise from 0 to 10 volts.

11. An operational amplifier can be used as a Schmitt trigger to reset the integrator–ramp generator to zero by short-circuiting C_f automatically whenever V_o rises above a preset value. Shut off power to your present circuit and add the second type 741 to make the complete circuit shown in Fig. 112. In this circuit the type 2N2219 transistor is normally "turned off" by having a negative voltage applied to its base from the output of the No. 2 operational amplifier. At the start of the operation, this amplifier's output is hung up at its maximum negative voltage because its + input is at zero while its − input is at some positive voltage V_{ref} selected by the volume-control-type rheostat R_2. Operational amplifier No. 1 is connected as an integrator–ramp generator as in Procedures 7 through 10, R_1 and the dry cell being the source of constant current that makes the integrator output V_o be a constantly rising voltage or "ramp." When V_o reaches V_{ref}, amplifier No. 2 switches, sending a large base current to the 2N2219 which is thereby "turned on" so that current passes through its collector–emitter circuit from the $+V_{cc}$ power supply to the coil of relay K_1. The relay then closes, short-circuiting C_f and resetting the integrator so that V_o is returned to zero. Amplifier No. 2 then flips back to a negative output, turning off the transistor and deenergizing the relay. With the relay contacts open, C_f can recharge and the cycle repeats. C_o is a large electrolytic capacitor that keeps current flowing in the relay coil for a brief interval after the transistor is turned off. This allows the relay contacts to stay closed long enough to completely discharge C_f. R_3 limits the base current to the transistor and thus limits the collector–emitter current, protecting the transistor against an excessive current surge when it is turned on and starts charging C_o. A plot of the resulting V_o as a function of time appears in Fig. 75 (see Exp. 34), which illustrates why this wave is called a "sawtooth" and the circuit that produces it a "sawtooth generator."

When wiring your circuit, be careful to avoid mistakes in making connections to the operational amplifiers and the transistor. Fig. 108 is the connection diagram for the Type 741 and Fig. 101 (in Exp. 37) for the 2N2219. If you are using power supplies, leave them unplugged and with their switches off until your wiring has been approved by the instructor. If you are using storage batteries, do not make final connections to their terminals until the instructor tells you to go ahead.

12. When your wiring has been checked, plug in and turn on the power supplies or make the final connections to the storage battery terminals. Temporarily disconnect the voltmeter from the output of amplifier No. 1 and, if you are using power supplies, use it to check and set the supply voltages $+V_{cc}$ and $-V_{cc}$ as described in Procedure 2. Then measure V_{ref} and set it to + 4 volts with R_2. Return the voltmeter to the No. 1 amplifier output as shown in Fig. 112. You should see V_o rise linearly until it gets to 4 volts, at which point it will suddenly return to zero and start up again. If the circuit appears to be operating properly, time five complete cycles of the sawtooth wave. Do this by starting the stop watch just as V_o returns to zero and counting five such returns before stopping the watch. Record your result.

13. Shift the voltmeter from the output of amplifier No. 1 to the − input of No. 2 to measure V_{ref}. Set this voltage to + 6 volts. Then return the voltmeter to the amplifier No. 1 output and again time five complete cycles of the observed wave. Record your result.

14. Repeat Procedure 13 with V_{ref} set to + 8 volts and again with V_{ref} = + 10 volts.

DATA

Multiplier

R_f, Ohms	V_o, Volts	R_f, Ohms	V_o, Volts
1000		4000	
2000		5000	
3000		6000	

Slope of graph _____   Dry cell voltage from graph _____

Adder

K_1	K_2	V_o, Volts		Percent Error
		Measured	Predicted	
Open	Open			
Closed	Open			
Open	Closed			
Closed	Closed			

Integrator–Ramp Generator

Time for 0–10 volt rise:

Trial 1 _____ Trial 2 _____ Trial 3 _____ Average _____

Calculated Value _____ Percent error _____

Sawtooth Generator

V_{ref}, Volts	Time for 5 Cycles, Seconds	Period, Seconds		Percent Error
		Measured	Calculated	
4				
6				
8				
10				

CALCULATIONS

1. Plot the analog multiplier output voltage against R_f using the data obtained in Procedures 2 and 3. Draw the best straight line through the plotted points.

2. Find the slope of the graph plotted in Calculation 1. From this slope, and assuming that R_f is exactly 1000 ohms, compute the dry-cell voltage.

3. Calculate what the analog adder output voltage V_o should be for each condition of switches K_1 and K_2. Use Equation 147 with $R_1 = 1000$ ohms, $R_2 = 1500$ ohms, $R_f = 3000$ ohms, and $V_1 = V_2 =$ the dry-cell voltage obtained in Calculation 2. There is, of course, no V_3.

4. Compare your predicted values of the analog adder output with the measured values by finding the percent error.

5. Calculate the average of your three measurements of the time for a 10-volt rise in the integrator–ramp generator output.

6. Calculate the theoretical value of the time for a 10-volt rise from Equation 149 using $R_1 = 470,000$ ohms, $C_f = 2.0$ mfd, and $V_1 = $ the dry-cell voltage obtained in Calculation 2.

7. Compare the theoretical value of Calculation 6 with the measured value of Calculation 5 by finding the percent error.

8. Divide each of the time measurements made in Procedures 12 through 14 by 5 to obtain the sawtooth period corresponding to each of your four values of V_{ref}.

9. Calculate the sawtooth period for each V_{ref} from Equation 149, in which we note that t is the period when $V_o = V_{ref}$.

10. Compare your four measured values of the sawtooth period with the corresponding predicted values obtained in Calculation 9 by finding the percent error.

QUESTIONS

1. Explain how the analog multiplier circuit of Fig. 104 can function as a voltage regulator for a power supply. *Hint:* Large operational amplifiers can supply considerable output current, and remember that V_o is exactly R_f/R_1 times V_1.

2. Discuss the errors found in Calculation 4 for your results with the analog adder. Note in particular whether these errors are consistent with the resistance tolerances on R_1 and R_2 and your assumptions about the dry-cell voltages.

3. Suggest how the analog integrator circuit could be modified to make it perform differentiation instead of integration.

4. Explain how a Schmitt trigger can be used to generate a square (or rectangular) wave from a sine-wave input. In particular, point out what determines the repetition rate and pulse length of the rectangular wave.

5. In the sawtooth circuit of Fig. 112, suppose the inductance of the relay coil can be neglected so that the coil can be represented by its resistance alone (about 200 ohms). (a) What is the time constant of the relay coil and capacitor C_o? (b) If the relay contacts open when the coil voltage drops to 60 percent of its value at the instant the transistor turns off, how long after turn-off do the contacts remain closed? (c) What is the effect of making C_o too large?

6. The sawtooth period should be given by Equation 149 with t set equal to the period T, and V_o to V_{ref}, so that

$$T = \frac{R_1 C_f V_{ref}}{-V_1}$$

Show that your results from Calculation 8 are in accord with this linear relation between T and V_{ref}.

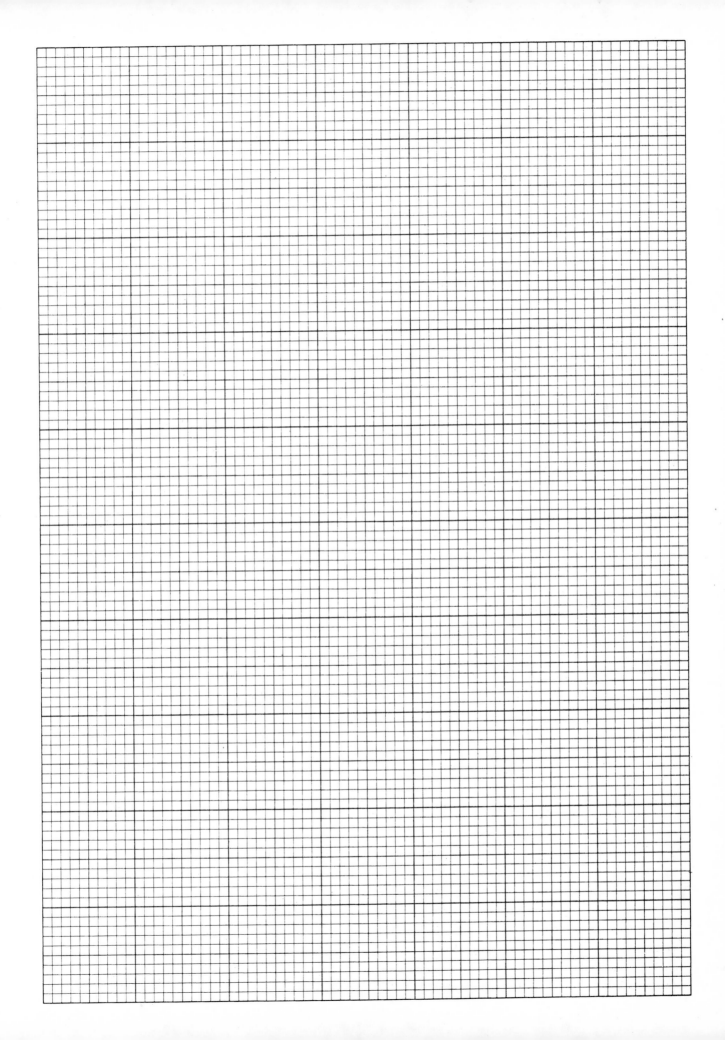

Geometrical Optics **39**

Although light is an electromagnetic wave, its wavelength is so short that when dealing with it in equipment whose dimensions are large by comparison (such as lenses, mirrors, etc.), we may ignore its wave nature. Light can then be discussed in terms of beams or rays that travel in straight lines (the law of rectilinear propagation) unless bent by reflection or refraction. When a light ray strikes the boundary between two media of different densities, all or part of it is reflected according to the law of reflection. If the medium into which the incident ray is directed is transparent, most of the light enters this second medium, but the ray is bent where it passes through the boundary if it does not strike normally (*i.e.*, at right angles to the boundary surface). This bending of light rays is called *refraction* and follows a third law. The purpose of this experiment is to study these three laws by tracing the pattern of light rays. We shall find that a great deal of this work is geometrical in nature, hence the study of light rays in experimental setups in which their wave nature may be ignored is called *geometrical optics* and the laws of reflection, refraction, and rectilinear propagation are known as the laws of geometrical optics.

THEORY

A light ray travels in a straight line if it is in a transparent medium of uniform density. This simple fact bears the somewhat grandiose name of *the law of rectilinear propagation*. If, however, a light ray strikes a boundary surface separating two transparent media of different density, some of the light will be reflected and some will proceed into the new medium but in an altered direction. The situation is illustrated in Fig. 113, which shows a ray in Medium 1 incident on a boundary separating Mediums 1 and 2 and producing a reflected ray coming back in Medium 1 and a refracted ray proceeding into Medium 2. Regular reflection takes place at a smooth surface such as the boundary shown in this figure. Such a surface might be the surface of a mirror (the boundary between glass and air) or of a lake (a boundary between water and air). The law of reflection states that a single ray of light will proceed, after reflection, in a direction such that the angle of incidence equals the angle of reflection. These angles are the angles made by the respective rays with the *normal*, *i.e.*, the perpendicular drawn to the boundary surface at the point of incidence. In addition, the incident ray, the normal, and the reflected ray must all lie in the same plane. Fig. 113 shows a ray incident on a boundary between two media, the normal, the angle of incidence, i, the reflected ray, and the angle of reflection s. The law of reflection requires $i = s$.

The inter-medium boundary does not, of course, have to be plane. Consider, for example, a spherical mirror, *i.e.*, a mirror whose surface is a portion of a sphere. It can be shown fairly easily that if a concave spherical mirror is held facing a beam of parallel light rays, the reflected rays will (to a close approximation) converge to a point. This point is called the mirror's *principal focus, F* (Fig. 114). Another important point is the *center of curvature, C*, which is the center of the spherical surface of which the mirror is a part. The *principal axis* of a mirror is a straight line drawn from the center of curvature through the focus to the mirror's surface. The distance from the center of curvature to the mirror along the principal axis is obviously a radius R of the reflecting surface. The distance from the focus to the mirror along the principal axis is the mirror's *focal length, f,* and can be shown to be $\frac{1}{2}R$.

Refraction is the bending of light rays when they pass obliquely from one medium to another (see Fig. 113). The direction of a particular ray after refraction is given

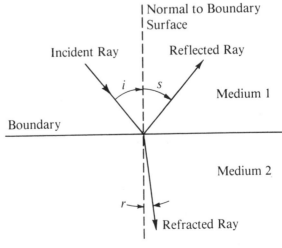

Figure 113 Reflection and Refraction at a Plane Surface

289

by the law of refraction, often called Snell's law after its discoverer, Willebrord Snell (1591–1626). This law states that the ratio of the sine of the angle of incidence to the sine of the angle of refraction is equal to the ratio of the velocity of light in the first medium (the medium of incidence) to that in the second. As in the case of reflection, the incident ray, the refracted ray, and the normal to the boundary all lie in the same plane, and the angle of refraction is the angle between the refracted ray and the normal as shown in Fig. 113. Snell's law may be stated mathematically by the expression

$$\frac{\sin i}{\sin r} = \frac{v_1}{v_2} = \frac{n_2}{n_1} \tag{150}$$

where i = angle of incidence
r = angle of refraction
v_1 = velocity of light in Medium 1
v_2 = velocity of light in Medium 2
n_1 = index of refraction of Medium 1
n_2 = index of refraction of Medium 2

The index of refraction of a particular medium is the ratio of the velocity of light in vacuum to its velocity in the medium. Thus $n_1 = c/v_1$ and $n_2 = c/v_2$, where c represents the velocity of light in vacuum. Hence $v_1/v_2 = n_2/n_1$, as indicated in Equation 150. Note that the velocity of light in air is very nearly the same as its velocity in vacuum, so that the index of refraction of air is very nearly 1. It may be taken as equal to 1 for most purposes.

A lens is a piece of glass or other transparent material bounded in its simplest form by spherical surfaces which is used to converge or diverge rays of light passing through it. It is similar to a spherical mirror in this respect but is used to converge or diverge *transmitted* rays rather than rays that are reflected back on themselves, and it does this by refraction rather than reflection. Fig. 115 illustrates this situation for a converging lens bounded by convex spherical surfaces. These surfaces have centers of curvature (not shown in the figure), and the principal axis of the lens is the line joining these

points. The principal focus F is the point on the left of the lens to which rays coming in from the right parallel to the principal axis are converged by the lens's action. F lies on the principal axis at a distance f, the focal length, from the lens. As long as the same medium (for example, air) lies on both sides of the lens, it makes no difference which way the light passes through it, and parallel rays incident from the left will be converged to focal point F' located on the principal axis at the same focal length f to the right of the lens that F was to the left. The focal length should properly be measured from the center of the lens, but we commonly deal with so-called "thin" lenses, whose thickness is so small that measuring f from the point where the principal axis intersects the lens surface introduces negligible error.

Snell's law shows that if a ray of light passes from one medium into another whose index of refraction is less than that of the first (*e.g.*, a ray from an underwater source passing from water into air), the angle of refraction r is greater than the angle of incidence i and the ray is bent away from the normal. As the angle of incidence is increased, eventually the refracted ray will graze the boundary surface. In this case $r = 90°$ and Equation 150 becomes $\sin i_c = n_2/n_1$, where i_c, the angle of incidence for which the refracted ray grazes the boundary, is called the *critical angle*. If the angle of incidence is made greater than the critical value, no light proceeds into the new medium and total reflection occurs at the bounding surface.

A prism is a solid piece of transparent material having flat surfaces. When a ray of light impinges on a prism it is refracted on entry into the prism material and again when it exits through one of the other surfaces. Prisms of various shapes are used to deflect light rays in predetermined desired ways, but a common form is the triangular prism whose principal use is in optical spectrometers. Used this way it deviates the incident beam through an angle D, called the *angle of deviation*, as shown in Fig. 116. D depends on the index of refraction of the prism material, and because the velocity of light in transparent media depends on the light's wavelength, the index of refraction and hence D will be different for light of different colors. A prism spectrometer can thus be used to separate a beam of light into its component wavelengths. For light of a fixed wavelength, D depends on the direc-

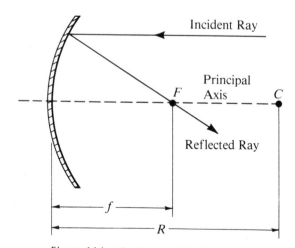

Figure 114   The Concave Spherical Mirror

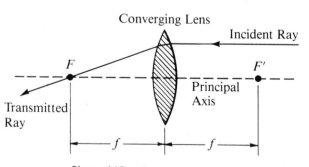

Figure 115   The Converging Lens

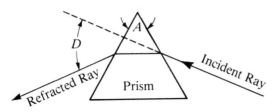

Figure 116   Deviation by a Triangular Prism

curs when the path of the ray inside the prism is symmetrical. This means that the initial angle of incidence is equal to the final angle of refraction, and the path of the ray inside the prism is parallel to the face opposite the prism's apex angle A (Fig. 116). Under these conditions the index of refraction of the prism material is given by

$$n = \frac{\sin \frac{1}{2}(A + D)}{\sin \frac{1}{2}A} \qquad (151)$$

This constitutes one method of measuring index of refraction.

tion of the incident ray (the initial angle of incidence), but for one particular direction of incidence the angle of deviation becomes a minimum. Minimum deviation oc-

APPARATUS

1. Low-power helium–neon laser
2. Drawing board
3. Plane mirror
4. Concave mirror
5. Convex mirror
6. Plano-convex lens
7. Semicircular lens

8. 60° triangular prism
9. Meter stick
10. Protractor
11. Drawing compass
12. White paper sheets for drawing board
13. White cards, masking tape, pins

PROCEDURE

1. Tape a sheet of white paper to the drawing board. Place the plane mirror on the sheet near the center with the reflecting surface facing your right. Plug in and turn on the laser and allow a few minutes for it to warm up. *Note:* There are two important features of laser light. First, a laser emits a narrow beam which spreads out very little and is thus ideal for use as a "ray" in this experiment. Second, laser light is highly monochromatic, meaning that it contains, to a close approximation, just one wavelength. For a helium–neon laser this wavelength is 6328 angstrom units. An angstrom unit (abbreviated Å) is 10^{-10} meters or 10 nanometers; thus the wavelength of the helium–neon laser may also be quoted as 6.328×10^{-7} meters, 6.328×10^{-5} centimeters, or 632.8 nanometers. Although knowledge of the wavelength of the light used in this experiment is not necessary because we are doing geometrical optics, you should realize that the refractive index you will encounter for the lens and prism material will have the value corresponding to 6328 Å.

2. Place the laser on the laboratory bench just beyond the right edge of the drawing board and direct it so that the beam passes over the board just above the paper surface and strikes the mirror at an angle of incidence of about 30°. CAUTION: Although the low-power laser is used in this experiment so that the beam will not hurt your skin, *be very careful not to look into the beam.* Keep the beam down on the bench and take care not to shine it in anyone's face, including your own.

3. Using the mirror as a straight edge, draw a heavy line with a sharp pencil to represent the reflecting surface on the paper. Make a mark on the paper directly under the point where the laser beam strikes the mirror. Also make marks near the right-hand edge of the paper directly under the incoming (incident) beam and under the outgoing (reflected) beam. *Note:* In clear air the laser beam will be invisible because no part of the light travels to your eye. It can be made visible by blowing chalk dust or cigarette smoke into the beam path, in which case the small particles will scatter some of the light to your eye so that you can see it. There will be enough scattering at the mirror surface so that you will be able to see a red spot where the beam strikes and so be able to mark the paper immediately below it. To place a mark under the incident and reflected rays, hold your pencil vertical with the point down and move it sideways through the beam. You will see the beam spot on the pencil and will easily be able to tell when the pencil is centered in the beam path. You can then lower the point onto the paper to make the desired marks.

4. Remove the mirror from the paper and point the laser toward the wall or other backstop behind the

bench to stop the beam. Do this whenever you are not actually using the laser. With the drawing compass and meter stick, construct the normal to the line representing the mirror at the point of incidence. Draw this normal as a broken line. Draw solid lines representing the incident and reflected rays. Use the meter stick as a straight edge and connect the point you marked under the incoming beam in Procedure 3 and the point of incidence to represent the incident ray. Connect the point under the reflected ray and the point of incidence to represent the reflected ray. Place arrowheads on these lines to indicate direction of propagation. With the protractor, measure and record the angles of incidence and reflection.

5. Remove the sheet of paper from the drawing board and set it aside for inclusion in your report. Put down a new sheet of paper and secure it with tape. Place the concave mirror on this sheet near its left edge and facing to the right. Note that for the two-dimensional work we are doing here the mirror is cylindrical rather than spherical. Trace the reflecting surface on the paper with a sharp pencil so that you get a heavy curved line representing the mirror. Now remove the mirror and locate its center of curvature on the paper as follows: draw any two chords to your curved line and construct their perpendicular bisectors. The intersection of these two lines is the mirror's center of curvature. Pick a point near the middle of the curved line and, using the meter stick as a straight edge, draw a straight line from this point through the center of curvature. This line will serve as the mirror's axis. Now construct two lines, one on each side of the axis, parallel to it, and separated from it by two inches. These lines should extend from the right-hand edge of the paper to the curved line representing the mirror and will represent incoming rays. Put arrowheads on them to show that they are directed toward the mirror.

6. Replace the concave mirror on the paper so that its edge matches the curved line traced from it in Procedure 5, thus ensuring that the mirror is back in its original position. Direct the laser beam at the mirror so that it lies right above one of the incoming ray lines drawn in Procedure 5. Do this by making the spot where the beam strikes the mirror come right above the point where the ray line intersects the curved mirror tracing, while at the same time a pencil held vertically in the beam path near the right edge of the paper points to the line below it. Observe the reflected ray and mark a point directly under it near the right-hand edge of the paper as described in Procedure 3.

7. Repeat Procedure 6 with the laser beam directed along the other ray line.

8. Remove the mirror from the paper and direct the laser beam at the wall to stop it. Draw lines representing the reflected rays from the respective points of incidence on the mirror through the points marked in Procedures 6 and 7. These lines should intersect on the axis, the point of intersection being the mirror's principal focus. If they do not, mark the best average position on the axis and call this the focus. With the meter stick, measure and record the focal length and radius of curvature of the concave mirror. Put arrowheads on your reflected ray lines to indicate that they are directed away from the mirror.

9. Repeat Procedures 5 through 8 for the convex mirror. Note, however, that this mirror diverges the reflected rays so that, instead of *coming to* a focus, they appear to *come from* a focus behind the mirror. To handle this situation, modify your procedure as follows:

a. Place the mirror near the center of the paper rather than near the left edge to leave room to draw projections of the rays and axis behind it.

b. Extend the axis line across the paper behind the mirror.

c. To locate the principal focus, draw the projections of the reflected rays as broken lines behind the mirror. These projections are called *virtual rays*, since there is really no light there. They should intersect at a point on the axis. This is the principal focus. Because it lies behind the mirror, the focal length (as well as the radius of curvature) is considered to be negative.

10. Remove the sheet of paper on the board for inclusion in your report and replace it with a clean sheet. Place the plano-convex lens on the paper near its center with the convex side facing to the right. With a sharp pencil trace the outline of the lens. Then remove the lens and construct the perpendicular bisector to the straight line representing the lens' flat side. Extend this perpendicular bisector all the way across the paper. It will represent the lens axis. Also construct two lines, one on each side of the axis, parallel to it and spaced away from it by two inches. These lines should be drawn from the right edge of the paper to the convex surface of the lens and will represent incoming rays. Put arrowheads on them to show that they are directed toward the lens.

11. Replace the lens on the paper, centering it carefully in the tracing to ensure its return to its original position. Direct the laser beam at the lens, adjusting the laser so that the beam lies above one of the incoming ray lines drawn in Procedure 10. Do this as described in Procedure 6. Observe the refracted ray issuing from

the left (plane) side of the lens. Mark the point where this ray crosses the axis. Also mark a point on the paper directly below the point where the beam exits the lens.

12. Repeat Procedure 11 with the laser beam directed along the other ray line.

13. Remove the lens from the paper and point the laser beam at the wall to stop it. Draw lines representing the refracted rays through the points marked in Procedures 11 and 12. Note that the points where these rays intersect the axis should coincide, this being the lens' principal focus. If they do not, mark the best average position on the axis and call this the focus. Establish a point on the axis in the center of the lens tracing and measure the focal length as the distance from this point to the focus. Record your result.

14. Replace the sheet of paper on the board with a new sheet. Lay the semicircular lens on this sheet near its center with the flat side facing to the left. Aim the laser beam in from the right so that it strikes "normally" in the center of the curved surface of the lens, passes through the lens material without being refracted, and exits normally through the middle of the flat surface, again without refraction because the ray is normal to the boundary between the two media involved.

15. Stick a pin into the drawing board at the midpoint of the flat surface of the lens, to serve as a pivot about which the lens may be conveniently rotated. Note that because this is a semicircular lens, the pin is at the center of the curved surface. Carefully rotate the lens about the pin as an axis, noting that in this way the laser beam is kept normal to the curved surface at the point of incidence so that no refraction occurs as the beam enters the lens material. At the point of *exit,* however, the angle of incidence is being increased from zero, and the beam is bent *away* from the normal because it is passing from a *denser* medium (the lens material) to a *less dense* medium (air). Observe the refracted beam by holding a small white card in its path and noting the spot where it strikes. Continue rotating the lens until the refracted beam just grazes the flat face. You will notice two interesting effects at this point: just as you get to an angle of refraction of 90°, the refracted beam will disappear (the spot will vanish from the card) and a strong reflected beam emerging normally from the curved side will appear and can be detected by observing the spot on your card when it is held in the appropriate position. This is total reflection occurring as you set the angle of incidence of the ray *inside the lens* on the flat surface to its critical value. With the lens in this critical position, trace its outline on the paper with a sharp pencil. Also mark points under the incoming beam where it comes over the right edge of the paper, where it strikes the curved surface of the lens, under the totally reflected beam where it exits the curved surface, and where it passes over the paper's right edge.

16. Remove the lens from the drawing board and direct the laser beam toward the wall to stop it. Construct the perpendicular to the line representing the flat face of the lens at the point of incidence. This point should be the center of this line and should be marked for you by the pin, which may now be removed. The perpendicular should be drawn as a broken line and represents the normal to the lens's flat surface. Also draw the incident and totally reflected rays through the appropriate points marked in Procedure 15 and place arrowheads on them to show their directions. Note that these rays are radii of the curved lens surface inside the lens. Measure and record the angles of incidence and reflection at the flat surface.

17. Remove the sheet of paper on the board for inclusion in your report and replace it with a fresh sheet. Place the triangular prism on the paper near the center with one side toward you. Direct the laser beam at the prism as illustrated in Fig. 116 and observe the refracted beam by holding the white card in its path and noting the spot where it strikes. Carefully rotate the prism and watch how the spot moves on the card. You should be able to find an arrangement in which rotation of the prism in one direction causes the spot to move to an extreme position and then start to move back. Set the prism so that the spot is in its extreme position. This is minimum deviation.

18. Mark points under the incident beam where it comes over the right-hand edge of the paper on the drawing board and where it strikes the prism. Similarly, mark points under the refracted beam where it emerges from the prism and where it passes over the left edge of the paper. With a sharp pencil, and being careful not to let the prism move, trace the prism's outline on the paper.

19. Remove the prism and shut off the laser. Draw the incident and refracted rays through the points established in Procedure 18 and place arrowheads on them to show their directions. Draw a line from the point of incidence of the incoming ray on the prism to the point at which the refracted ray exits the prism to represent the ray inside the prism material. See if this ray is parallel to the prism base (the face opposite the apex angle). Extend the incident ray line through the prism as a broken line so that you can measure the angle of deviation D as shown in Fig. 116. With the protractor, measure and record this angle and the prism's apex angle A.

DATA

Plane Mirror
Angle of incidence   _____   Percent error   _____
Angle of reflection   _____

Curved Mirrors

Mirror	Focal Length	Radius of Curvature	½ Radius of Curvature	Percent Error
Concave				
Convex				

Lens
Focal length   _____

Critical Angle
Angle of incidence   _____   Index of refraction of
Angle of reflection   _____   lens material   _____

Triangular Prism
Apex Angle A _____   Index of refraction of
Angle of deviation D _____   prism material   _____

CALCULATIONS

1. Compare the angles of incidence and reflection measured in Procedure 4 by finding the difference between them, treating this difference as an error, and calculating the percent error.

2. From your data in Procedure 8, calculate one-half the radius of curvature of the concave mirror. This should be the mirror's focal length. Compare it with your measured focal length by finding the percent error.

3. Repeat Calculation 2 for the convex mirror using your data from Procedure 9.

4. Assuming the index of refraction of air to be 1, calculate the index of refraction of the semicircular lens material from the value of the critical angle measured in Procedure 16.

5. Use Equation 151 to obtain the index of refraction of the triangular prism material using your data from Procedure 19.

QUESTIONS

1. (a) What is meant by the spherical aberration of a mirror? (b) Why are the concave mirrors used in automobile headlights parabolic instead of spherical?

2. What is meant by the chromatic aberration of a lens?

3. Under what conditions is the angle of refraction greater than the angle of incidence?

4. What is wrong with a lens that has astigmatism?

5. A ray of light shines from under water through the surface into the air above. Find the critical angle of incidence of this ray on the water–air interface. Take $n = \frac{4}{3}$ for water and 1 for air.

6. Show that the angle of deviation D produced by a triangular prism is minimum when the prism is positioned so that the ray inside it is parallel to the prism's base.

7. (a) Compute the velocity of light in the triangular prism from your value of the index of refraction for this prism and the known value of the velocity of light in vacuum. (b) Repeat for the semicircular lens.

8. A ray of light impinges at an angle of incidence i on the surface of a transparent slab of thickness d. The material of the slab has index of refraction n. (a) Show that the ray that emerges on the other side of the slab is parallel to the incident ray. (b) Derive an expression giving the lateral displacement of the emergent ray from the incident one in terms of i, d, and n.

9. A ray is incident on the plane surface of a transparent medium having index of refraction n. Derive an expression for the angle of incidence that will cause the reflected ray and the refracted ray (the ray transmitted into the medium) to be perpendicular to each other. The incident and reflected rays are in air, whose index of refraction can be taken to be 1. *Note:* The angle of incidence found in this question is called *Brewster's angle*.

10. Show that, for a spherical concave mirror that is small enough so that the angle of incidence of any ray coming in parallel to the mirror's axis and intercepted by the mirror is small, the rays are converged to a point (the principal focus) that is one-half the radius of curvature of the mirror.

11. A straight stick is held so that it extends through the surface and down into a pool of clear, still water. Explain why the stick appears bent at the surface to an observer viewing it from above. *Hint:* Remember that the human eye knows nothing about reflection or refraction but judges an object's location on the assumption that rays coming to it from the object travel in straight lines.

12. A six-foot man's eyes are 4 inches below the top of his head. (a) How tall must a flat mirror mounted on the wall be so that the man can just see all of himself when he looks in it? (b) How high above the floor should the bottom edge of this mirror be placed?

13. A woman looks at herself in a plane mirror. (a) Why does her image appear to be behind the mirror? (b) Is this a virtual or real image? (c) Why is it reversed left-to-right but not top-to-bottom?

Measurement of the **40** Focal Length of Lenses

The formation of images by lenses is one of the most important studies in geometrical optics. The purpose of this experiment is to observe the real images formed by various lenses and to verify the lens equation; in particular, to measure the focal length of some positive and negative lenses and the equivalent focal length of a combination of thin lenses.

THEORY

When a beam of rays parallel to the principal axis of a lens impinges upon a converging lens, it is brought together at a point called the principal focus of the lens. The distance from the principal focus to the center of the lens is the focal length of the lens; the focal length is positive for a converging lens and negative for a diverging lens. The relation between the object distance p, the image distance q, and the focal length f of a thin lens is given by the lens equation

$$\frac{1}{p} + \frac{1}{q} = \frac{1}{f} \qquad (152)$$

The magnification produced by a lens (the linear magnification) is defined as the ratio of the length of the image to the length of the object. This is also equal to the ratio of the image distance to the object distance. Therefore

$$\text{Magnification} = \frac{q}{p} \qquad (153)$$

The principal focal length of a converging lens may be determined by forming an image of a very distant object on a screen and measuring the distance from the screen to the lens. This distance will be the focal length, since the rays of light from a very distant object are very nearly parallel. A more accurate method of determining the focal length of a positive lens is to measure the image distance corresponding to a suitable and known object distance, and to calculate the focal length from the lens equation, Equation 152.

Another method, which demonstrates the meaning of this equation very clearly, makes use of the following arrangement: An object and a screen on which an image of the object is to be focused are placed a known distance D apart, and a convergent lens of focal length f is placed between them. Since q is the distance from the lens to the image (*i.e.*, the screen), the sum $p + q$ must be equal to D. Hence $q = D - p$, and substitution in Equation 152 to eliminate q produces a quadratic equation in p whose solutions are

$$p_1 = \frac{D}{2}\left(1 + \sqrt{1 - 4\frac{f}{D}}\right) \qquad (154)$$

and

$$p_2 = \frac{D}{2}\left(1 - \sqrt{1 - 4\frac{f}{D}}\right) \qquad (155)$$

There are thus two positions of the lens between the object and the screen for which an image on the screen will be in focus *provided D is at least four times the lens's focal length*. If D is just equal to $4f$, the square root in Equations 154 and 155 disappears and $p_1 = p_2 = D/2$, that is, there is only one lens position, namely the midpoint of the object-screen distance, for which a focused image is obtained. This condition is easily observed and the corresponding value of D measured. The focal length of the lens is then found by dividing this result by 4.

When two thin lenses are in contact, the equivalent focal length of the combination may be measured experimentally by one of the above methods. It may also be calculated in terms of the individual focal lengths. Thus

$$\frac{1}{f} = \frac{1}{f_1} + \frac{1}{f_2} \qquad (156)$$

where f is the equivalent focal length of the lens combination, f_1 is the focal length of the first lens, and f_2 is the focal length of the second lens.

A concave lens by itself cannot form a real image, since it is a diverging lens, hence a different method must be used for measuring its focal length. This is done by placing the negative lens in contact with a positive lens of shorter and known focal length, measuring the equivalent focal length of the combination experimentally, and then using Equation 156 to solve for the focal length of the negative lens.

299

APPARATUS

1. Optical bench
2. Illuminated object
3. Lens holders
4. Screen

5. Two convex lenses, A and B (of about 20-cm and 10-cm focal length respectively)
6. One concave lens, C (of about −20-cm focal length)
7. Metric ruler

PROCEDURE

1. Measure the focal length of lens A directly by obtaining the image of a very distant object on the screen and measuring the image distance. The object may be a tree or a house about a block away.

2. Repeat Procedure 1 for lens B.

3. Determine the focal length of lens A by the use of the lens equation. Place the illuminated object at one end of the optical bench, then place the screen at a distance of about five times the focal length of the lens. With the object and screen fixed, find the position of the lens for which a sharp, enlarged image is produced on the screen. Make sure that the object, lens, and screen all lie along the same straight line (the principal axis of the lens) and that they are all perpendicular to the axis. Record the position of the object, the lens, and the screen; record the measurements to 1 millimeter. Measure the size of the object and the size of the image; record these measurements to 0.5 millimeter.

4. Using the arrangement of Procedure 3, move the lens back and forth to find two positions for which the image on the screen is in focus. Move the screen a few centimeters closer to the object and again observe that a focused image is obtained for two different lens positions. Notice that these positions are closer together than they were before the screen was moved. Continue to move the screen closer to the object and to observe that there are two lens positions that give a focused image until these two positions coincide at the midpoint between the object and the screen. Measure and record the value of the object-screen distance D corresponding to this condition.

5. Repeat Procedure 3 using lens B.

6. Repeat Procedure 4 using lens B.

7. Repeat Procedure 3 using the combination of lenses A and B in contact.

8. Repeat Procedure 3 using the combination of lenses B and C in contact.

DATA

Lenses	Focal Length Measured Directly	Object Distance p	Image Distance q	Distance D for Single Lens Position	Size of Object	Size of Image
A						
B						
A and B						
B and C						

Lenses	Focal Length From Lens Equation	Focal Length From Equation 156	Focal Length From D for Single Lens Position	Magnification	
				Size of Image / Size of Object	$\dfrac{q}{p}$
A					
B					
A and B					
B and C					
C					

CALCULATIONS

1. Calculate the focal length of lens A from the data of Procedure 3, and the focal length of lens B from the data of Procedure 5. Compute the magnification produced in each case from the ratio of the size of the image to the size of the object and from the ratio of the image distance to the object distance.

2. Calculate the focal lengths of lenses A and B from the values of the object-screen distance D found in Procedures 4 and 6 respectively.

3. Calculate the focal length of the combination of lenses A and B from the data of Procedure 7. Compare the result with the value obtained from theory, using Equation 156.

4. Calculate the focal length of the combination of lenses B and C from the data of Procedure 8. By using this value of the focal length of the combination, compute the focal length of the negative lens C, using Equation 156.

QUESTIONS

1. Draw the ray diagram for the optical arrangement of parts in Procedure 3. Represent the object by a small arrow.

2. Derive Equation 153 from the geometry of your diagram in Question 1.

3. (a) What is meant by the spherical aberration of a lens? (b) How can this aberration be reduced in a simple lens?

4. What is a real image?

5. What is a virtual image?

6. In a film projection apparatus it is desired to produce pictures 12 feet wide on a screen 50 feet from the lens; the size of the picture on the film is 1 inch wide. What must be the focal length of the projection lens used?

7. Substitute $q = D - p$ in Equation 152 and show how the solutions 154 and 155 are obtained.

8. (a) Under what conditions will a convex (converging) lens produce a virtual image from a real object? (b) A magnifying glass is just a simple converging lens. Explain how it works in view of your answer to part (a).

The Telescope and Microscope 41

Combinations of lenses are used extensively in many kinds of optical instruments. It is instructive to see how such a combination operates in actual practice. As exam-ples, this experiment presents a method of constructing an astronomical telescope, an opera glass, and a microscope, and of determining the magnifying power of each.

THEORY

There are three main types of telescopes: the astronomical telescope, the terrestrial telescope, and the opera glass.

The astronomical telescope, which is the simplest type, consists of two lenses, the objective lens and the eyepiece. The objective lens is a positive lens of long focal length, which forms a real inverted image of a distant object in its focal plane. The eyepiece is a positive lens of short focal length used to magnify the image produced by the objective lens. The eyepiece is used as a magnifying glass, by means of which an enlarged virtual image is formed of an object placed just inside its focus. In this instance the object used is the image formed by the objective lens. The eye looks directly at the enlarged virtual image produced by the eyepiece.

In the terrestrial telescope the image produced by the objective lens is made erect by means of another positive lens. This lens is placed beyond the first image and at such a distance as to produce a real inverted image of it, so that the second image is erect. The eyepiece is used to magnify this image, as described above.

The opera glass is a telescope in which the eyepiece is a divergent lens. This lens is so placed that the rays from the objective lens strike it before they converge to form an image. Thus the image that would have been formed by the objective lens serves as the virtual object for the negative lens. After leaving this lens, the rays diverge as if they came from an enlarged virtual image. The distance between the two lenses is made equal to the difference of their focal lengths. This type of telescope has a small field of view, but it has the advantage that it is shorter than the other forms of telescope and gives an erect image of the distant object.

Like the astronomical telescope, the microscope uses two convergent lenses, one as an objective and one as an eyepiece, but in this case the objective is the lens with the shorter focal length. The object is placed at a distance only slightly greater than the focal length of this lens, so that a highly enlarged real image is formed. The longer focal length lens is used as the eyepiece and is placed as in the telescope so that the real image produced by the objective falls just inside its focal length. A further magnified virtual image is thus obtained. The microscope gives a much greater magnification than does the eyepiece used alone as a magnifier because in the microscope the eyepiece is used to view the enlarged real image that the objective forms from the object, rather than viewing the object directly.

The magnifying power of an optical instrument is defined as the ratio of the angle subtended at the eye by the image of an object viewed through the instrument to the angle subtended at the eye by the object viewed directly.

APPARATUS

1. Optical bench
2. One long-focus convergent lens (about 25 cm)
3. One short-focus convergent lens (about 10 cm)
4. One short-focus divergent lens (about 18 cm)
5. Lens holders
6. Screen
7. Illuminated object
8. Vertically mounted metric scale — for example, scale and support stand for optical lever

applications (see Experiments 9 or 23)
9. Telescope magnification scale. This scale is made up of a strip of white paper 2 feet long with a series of thick black lines drawn horizontally across it at regular intervals of about 2 inches. The lines are numbered to facilitate counting.
10. Vernier caliper
11. Desk lamp

PROCEDURE

1. Measure the focal length of the two convergent lenses directly by obtaining the image of a very distant object on the screen and measuring the image distance. The object may be a tree or a house about a block away. Record each focal length.

2. The focal length of a divergent lens cannot be measured directly, since a divergent lens by itself cannot form a real image. Hence a different method must be used. The negative lens is placed in contact with a positive lens of shorter and known focal length, the equivalent focal length of the combination is then measured experimentally, and finally the focal length of the negative lens is calculated from Equation 156 (see Experiment 40).

Place the divergent lens in contact with the short-focus convergent lens. Measure the focal length of this combination directly by obtaining the image of a very distant object on the screen and measuring the image distance. Finally, calculate the focal length of the negative lens by using Equation 156. The focal length of the short-focus convergent lens was found in Procedure 1.

3. Construct a simple astronomical telescope using the long-focus convergent lens as the objective and the short-focus convergent lens as the eyepiece. Special care must be taken to have the various optical parts at the same height and well lined up. In addition, the lenses must be perpendicular to the optic axis of the system. Mount the eyepiece near one end of the optical bench and the screen in front of it at a distance of its focal length away. Place the illuminated object at the far end of the laboratory table and point the optical bench at it. Mount the objective lens on the optical bench on the other side of the screen from the eyepiece and move it until it focuses a real inverted image of the object on the screen. Then remove the screen and adjust the eyepiece so that on looking through it you can see a sharp inverted image of the illuminated object.

4. In order to measure the magnifying power of your telescope, point it in the direction of the telescope scale. Illuminate the scale with the desk lamp if necessary. Look through the telescope with one eye and directly at the scale with the other. The magnified image of the scale will be superimposed on the unmagnified scale. Count the number of divisions of the scale as viewed directly that cover exactly one division of the magnified scale. This number gives the magnifying power of the telescope.

5. Construct an opera glass using the same lens for the objective as before and the short-focus divergent lens as the eyepiece. This time place the screen at one end of the optical bench, point the optical bench at the illuminated object at the far end of the laboratory table, and mount the objective lens so that a real inverted image of the object is formed on the screen. Mount the eyepiece on the same side of the screen as the objective lens and spaced from this lens by a distance equal to the difference of their focal lengths. Remove the screen and adjust the eyepiece so that on looking through it you can see the largest clear and erect image that can be formed.

6. Determine the magnifying power of the opera glass as in Procedure 4.

7. Construct a microscope using the short-focus convergent lens as the objective and the long-focus convergent lens as the eyepiece. Place the illuminated object at one end of the optical bench and mount the short-focus lens in front of it at a distance of a little more than this lens's focal length. Mount the screen on

the other side of this lens and slide it along the optical bench until a real inverted image is sharply focused on it. Then mount the eyepiece on the other side of the screen from the objective and at a distance from the screen equal to the eyepiece's focal length. Finally remove the screen and adjust the eyepiece so that when you look through it you see a sharp inverted image of the illuminated object.

8. Replace the illuminated object with the vertical metric scale, positioning this latter so that it is the same distance from your microscope objective that the illuminated object was. Illuminate the scale with the desk lamp. Look through your microscope and make small adjustments to the eyepiece position to bring the scale divisions into sharp focus. Place the vernier caliper against the eyepiece and adjust the jaws until they span exactly five millimeter divisions as seen in the microscope. Remove both lenses and mount the caliper on the eyepiece support without disturbing the setting of its jaws. Look at the scale directly through the space between the jaws, keeping your eye in the same position it was in when looking through the microscope, and note the length of scale bracketed by these jaws. You will have trouble distinguishing the millimeter divisions at this distance, but you should be able to see the centimeter divisions quite clearly. Read the length of the scale seen between the caliper jaws, estimating to 0.1 centimeter. This is the length subtended at the eye by the same angle that subtended 5 millimeters (0.5 centimeter) in the image. This length in centimeters divided by 0.5 centimeter is therefore the magnifying power of the microscope.

DATA

Focal length of the long-focus convergent lens   _____

Focal length of the short-focus convergent lens   _____

Focal length of the short-focus divergent lens   _____

Magnifying power of the astronomical
telescope   _____

Magnifying power of the opera glass   _____

Magnifying power of the microscope   _____

QUESTIONS

1. If a telescope is accurately focused on a distant object, in what direction must the eyepiece be moved to focus on a near object? Explain.

2. (a) How does the terrestrial type of telescope differ from the astronomical telescope? (b) Explain how the change is accomplished.

3. What is the basic difference between the astronomical telescope and the microscope?

4. Draw a ray diagram, approximately to scale, showing the illuminated object, the real image formed by the objective, and the virtual image formed by the eyepiece of your astronomical telescope. Represent the object by a small arrow.

5. The magnifying power of an astronomical telescope 24 inches long is equal to 15. Determine the focal length of the objective lens and that of the eyepiece. Assume that the telescope is focused on a distant object.

6. When two lenses are used in tandem (with the light passing first through one and then through the other, as in all the optical systems studied in this experiment), the position of the final image may be determined by applying Equation 152 twice, once to find the position of the image formed by the first lens and again for the second lens on the assumption that the image formed by the first in the absence of the second may be treated as the second lens's object. Derive Equation 156 in this way. Assume that the two lenses in contact are so thin that the combination is thin also. *Note:* Normally the object is placed on the left of the lens. In a case where the object is on the right, the object distance p is considered negative.

7. Explain how the method of Procedure 8 measures the magnifying power of the microscope in accordance with the definition of magnifying power.

8. For both the astronomical telescope and the opera glass, the distance between the lenses is approximately equal to the *algebraic* sum of their focal lengths. Explain why this is so, remembering that the focal length of a diverging lens is negative, so that in the opera glass the lens separation is the *difference* between the *values* of the focal lengths.

Analysis of Light **42** by a Prism Spectrometer

Light is made up of many different colors, each of which has a definite wavelength. When a beam of light is passed through a glass prism, it is broken up into a spectrum. Each chemical element produces its own characteristic spectrum, so that by analyzing the light given off by a substance it is possible to identify the elements it contains. The object of this experiment is to introduce the student to the prism spectrometer and to use this important optical instrument to study the bright-line and continuous spectra obtained from certain light sources.

THEORY

In general, light is made up of many component wavelengths, each of which corresponds to a definite color. When a beam of light is passed through a glass prism, refraction takes place when the beam enters the prism and again when it leaves, with the result that the beam is bent through an angle D, called the angle of deviation. This angle is a function of the index of refraction n of the glass (see Equation 151 in Experiment 39), and since n is different for different wavelengths (because they propagate with different velocities in transparent materials such as glass), each component wavelength in the incident beam is bent through a different angle. The incident beam is thus separated into beams of its component colors (wavelengths) which may then be displayed as a *spectrum*. We say that the incoming light has been *dispersed* into a spectrum, and the property exhibited by the prism of separating this light into its constituents is called *dispersion*.

Spectrum analysis is the decomposition of a beam of light into its constituent wavelengths and the examination of the image so formed. A spectrometer (or spectroscope, as instruments not provided with means for recording the analyzed light are often called) is an optical instrument for producing and analyzing spectra. A diagram of a prism spectrometer is shown in Fig. 117. It consists of three essential parts: a collimator, a prism, and a telescope. The collimator is a tube with a converging lens at one end and a fine slit of adjustable width at the other. The length of the tube is equal to the focal length of the lens so that the collimator renders the rays of light from the slit parallel. The prism deviates these rays and disperses them into a spectrum. The telescope contains an objective lens, which brings the rays of light to a focus in its focal plane, and an eyepiece, through which the image of the spectrum is viewed. If the light is all of one color, or monochromatic, a single image of the slit appears. If a source contains several colors, several images of different color will appear side by side, each being an image of the slit formed by one component of the light. The telescope can be rotated about the prism so that light emerging at different angles can be viewed, and the angular position of each slit image can be read off from the divided circle under the telescope mount.

There are three main types of spectra: continuous, bright-line, and absorption. A continuous spectrum is produced by light from an incandescent solid or liquid. It contains all wavelengths and hence appears as a continuous gradation of colors, merging with one another with

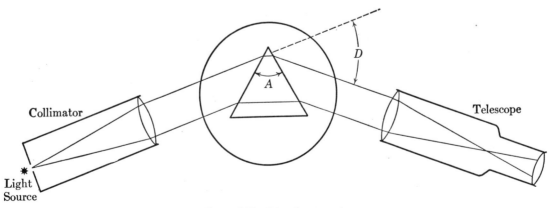

Figure 117 Prism Spectrometer

TABLE 42-1 HELIUM LINES

Wavelength in Angstrom Units	Color
7065	red
6678	red
5876	yellow
5048	green
5015	green
4922	blue-green
4713	blue
4471	deep blue
4388	blue-violet
4026	violet
3964	violet
3889	violet

TABLE 42-2 MERCURY LINES

Wavelength in Angstrom Units	Color
6907	red
5790	yellow
5770	yellow
5461	green
4916	blue-green
4358	blue
4078	violet
4047	violet

very gradual changes from red to violet. The bright-line spectrum is produced by exciting an element in the gaseous state by means of an electric discharge or by heating. Such a spectrum consists of bright lines corresponding to light of definite colors, separated by dark spaces, the number and relative positions of the lines depending on the element used and the method of excitation employed. An absorption spectrum is produced when light from a source furnishing a continuous spectrum is passed through some cooler absorbing medium. The resulting spectrum is crossed by dark spaces or lines that form the absorption spectrum of the material.

Wavelengths of visible light are usually expressed in angstrom units (abbreviated Å) or, in more recent literature, in nanometers. An angstrom unit is defined as 10^{-10} meters and thus equals 10^{-8} centimeters. A nanometer is 10^{-9} meters. Hence a nanometer is 10Å.

The principal lines in the helium spectrum are given in Table 42-1. The principal lines in the mercury spectrum are given in Table 42-2. In these spectra other fainter lines are also present, but the ones listed are the brightest and easiest to find.

APPARATUS

1. Spectroscope
2. Mercury discharge tube
3. Helium discharge tube
4. Hydrogen discharge tube
5. Sodium light source
6. Desk lamp
7. Discharge-tube power supply
8. Stand and clamp for holding the discharge tubes (unless part of the power supply)

PROCEDURE

1. The spectrometer should be already adjusted with the collimator and telescope properly focused and aligned and the prism correctly mounted and leveled. CAUTION: Do not handle the spectrometer until the instructor has demonstrated its operation to you. This instrument is easily thrown out of adjustment, and needless delay or even permanent damage can be caused by careless handling. Be extremely careful when using it.

2. Mount the helium discharge tube in its holder and place it immediately in front of the slit at the end of the spectrometer's collimator. The spectrometer should be positioned so that the slit faces directly into the center of the discharge tube. Connect the leads from the power supply to the discharge-tube terminals and turn the supply on. CAUTION: The discharge tube operates at a high voltage and its power supply can deliver enough current to give a painful shock. Do *not* touch either the tube or the wires leading to it while the power supply is in operation.

3. Adjust the angular position of the telescope so that the helium spectrum is visible when you look into the eyepiece. Note that the spectrum consists of a number of vertical lines of different colors. Each of these lines is an image of the collimator slit formed by the pertinent wavelength, but because most optical spectrometers use such slits, the word "line" has come to mean a particular wavelength in a spectrum. Now adjust the spectrometer for minimum deviation. To do this, move the telescope so that the bright yellow line in the spectrum at 5876 Å is roughly in the center of the field of view. Slowly turn the prism table in such a direction as to diminish the angle of deviation while watching this line. It should move first one way and then

the other as the prism is rotated. If it moves out of the field of view, the telescope position must be readjusted to follow it. Lock the prism table in the position corresponding to the extreme position of the yellow line. This is the orientation giving minimum deviation for light of this wavelength. The prism also gives maximum dispersion when set this way, which is why it is so adjusted for use in a spectrometer. *Note:* When you rotate the prism table, every line in the spectrum will move through a minimum deviation position, but the table setting for minimum deviation will be slightly different for different lines, *i.e.*, different wavelengths. The yellow line is chosen because it is at about the middle of the visible spectrum. Therefore, the prism setting for minimum deviation of this line represents a good average for the entire spectrum. Do not change the prism table position after the minimum deviation setting just described has been found.

4. Adjust the position of the telescope so that the first red line of the helium spectrum falls exactly on the intersection of the eyepiece cross hairs. Note that directly under the arm supporting the telescope is a clamping device that rotates with the arm and permits the telescope to be clamped to the base by means of a radial screw. A tangent screw on the arm allows a fine adjustment of the telescope setting; however, the tangent screw can be used only after the telescope has been clamped with the radial screw. Rotate the telescope until the first red line is near the vertical cross hair and then clamp it by means of the radial screw. Use the tangent screw to set the intersection of the cross hairs exactly on the red line. Record the angular setting of the telescope to the nearest minute of arc, reading the divided circle with the help of the vernier. Also record the first red line's color and known wavelength (see Table 42-1).

5. Repeat Procedure 4 for the other lines of the helium spectrum, taking readings of the telescope setting for each of the lines that can be recognized.

6. Turn off the discharge tube power supply and replace the helium discharge tube with the mercury tube. Turn the supply on and repeat Procedures 4 and 5 for the mercury spectrum.

7. Repeat Procedure 6 using the hydrogen tube. Four lines should be observed and telescope settings obtained for them. Record the longest wavelength (red) line as H_α, the next as H_β, and so on.

8. Turn off the power supply, remove the discharge tube, and set up a sodium light source in front of the spectroscope slit. Turn on the source and wait for it to warm up so that a brilliant yellow light is produced. Look through the telescope and adjust it so that the yellow sodium doublet falls exactly on the intersection of the cross hairs. Record the telescope setting.

9. Examine the spectrum of the light from a tungsten lamp. Record the setting of the telescope for the upper and lower limits of the visible spectrum.

DATA

Spectrum	Color	Known Wavelength	Setting of Telescope	Wavelength from Calibration Curve
Helium				

Mercury				
Sodium				
Visible				

● Hydrogen Spectrum

Line	Color	Setting of Telescope	Wavelength from Calibration Curve	Reciprocal of Wavelength	$1/n^2$
H_α; $n = 3$					
H_β; $n = 4$					
H_γ; $n = 5$					
H_δ; $n = 6$					

Measured value of Rydberg constant from slope   _____

Measured value of Rydberg constant from intercept   _____

Balmer series limit   _____

CALCULATIONS

1. Draw a graph with wavelengths in angstrom units as abscissas and the settings of the telescope as ordinates for the lines of the helium spectrum from the data of Procedures 4 and 5. The smooth curve so obtained is the calibration curve of the instrument. This curve is obtained by plotting the scale readings for the known lines of an element against their corresponding (known) wavelengths. From this curve the wavelengths of the lines observed in other spectra may be determined.

2. From the calibration curve, read off the wavelengths of the mercury lines by using the readings of the telescope obtained in Procedure 6; tabulate the results. How do these values compare with the known values?

3. Repeat Calculation 2 for the hydrogen readings obtained in Procedure 7. No table of known wavelengths is given for the hydrogen lines in this experiment. However, the hydrogen spectrum observed using a glass discharge tube and prism is part of the so-called *Balmer series* in hydrogen. The wavelengths of lines in this series are given by the Balmer formula

$$\frac{1}{\lambda} = R\left(\frac{1}{2^2} - \frac{1}{n^2}\right) = \frac{R}{4} - \frac{R}{n^2} \tag{157}$$

where λ is the wavelength in angstrom units, R is the famous Rydberg constant, and n is an integer having the value 3 for the longest wavelength (H_α) line in the series, 4 for the next longest (H_β), and so on. Calculate $1/\lambda$ for your four lines and plot the results as ordinates against the corresponding values of $1/n^2$ as abscissas. To obtain more accurate results, construct the graph using as large a scale as possible; that is, utilize the entire sheet of graph paper. Extend the line to $1/n^2 = 0$.

4. According to Equation 157 your graph in Calculation 3 should be a straight line of slope $-R$. Obtain a measured value of the Rydberg constant in this way.

5. Equation 157 also indicates that $1/\lambda$ tends to a maximum value $R/4$ and hence λ to a minimum value $4/R$ as n gets very large. This minimum value of λ is called the *series limit*. Find it from the $1/n^2 = 0$ intercept on the graph in Calculation 3. Also obtain a second measured value of the Rydberg constant from this intercept and compare it with that found in Calculation 4.

6. Measure the wavelength of the sodium doublet by using the reading of the telescope in Procedure 8 and reading off the wavelength from the calibration curve. Compare your results with the known value (use 5893 Å, the mean of 5890 Å and 5896 Å, which are the wavelengths of the doublet).

7. From the data of Procedure 9 and the calibration curve determine the range of the visible spectrum in angstrom units.

QUESTIONS

1. Compare the appearance of the bright-line spectrum with that of the continuous spectrum and explain the difference.

2. Compare your results for the range of the visible spectrum with the known value, which you should look up and record in the spaces provided in the data table.

3. Does the index of refraction of glass increase or decrease with increasing wavelength? Explain your answer.

4. Of what use is the calibration curve and why is it needed?

5. Explain how the Doppler effect may be adapted to light waves in determining the speed of stars toward the earth or away from the earth.

6. The Balmer formula, Equation 157, predicts spectrum lines in hydrogen corresponding to all integer values of n equal to or greater than 3. Why, then, did you see only four lines in Procedure 7?

7. (a) What are the frequencies in hertz of the limits of the visible spectrum whose wavelengths you found in Calculation 7? (b) How many times the lower limit frequency is the upper limit frequency? Is it as much as twice (an octave)?

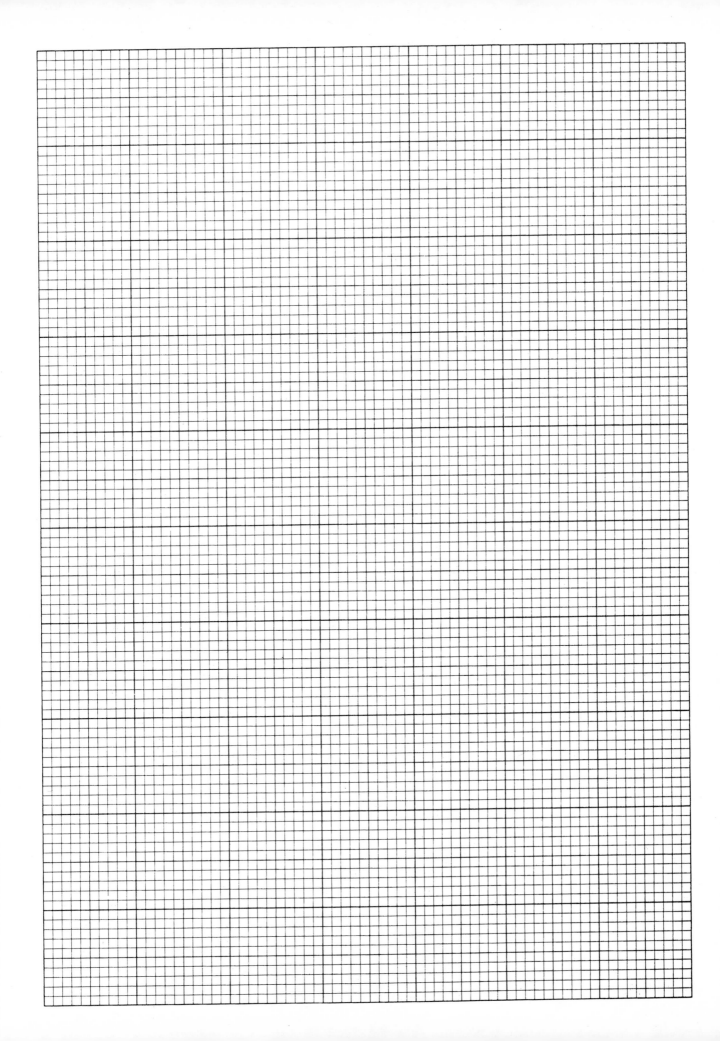

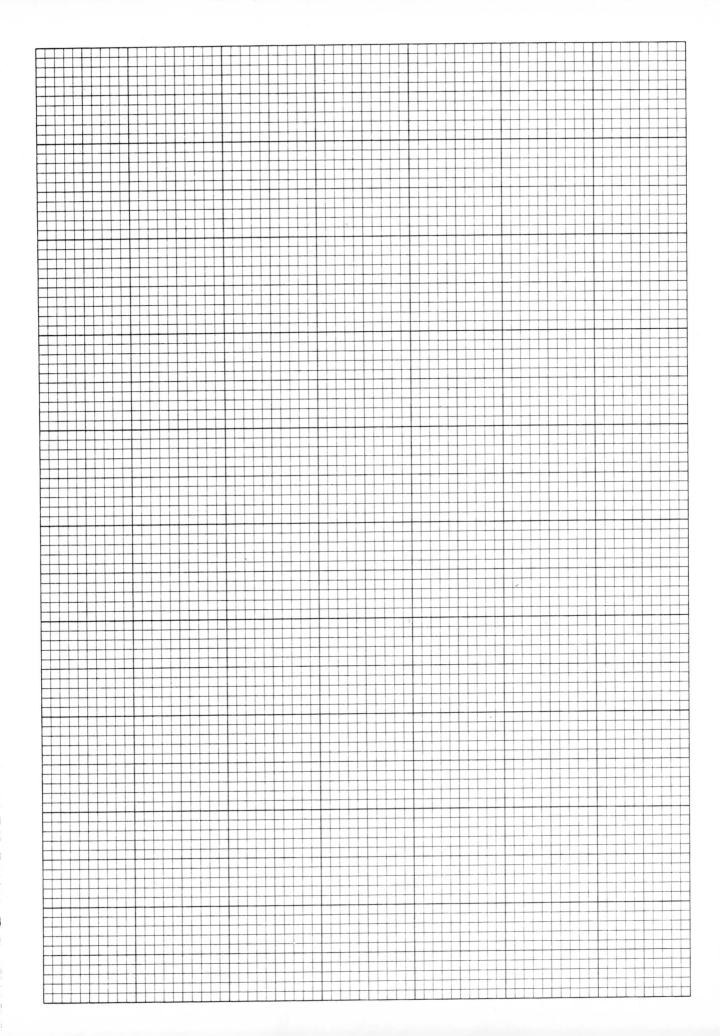

Interference of Light 43

One of the important methods of measuring the wavelength of light is to allow monochromatic light to be reflected from the two surfaces of a very thin film of varying thickness. The light reflected from the surfaces of the film will produce interference fringes, and from the dimensions of the film and the number of fringes, the wavelength of the light can be evaluated. The object of this experiment is to study the interference of light and to measure the wavelength of the yellow light from a sodium lamp.

THEORY

Interference is the effect produced when two or more wave trains are superimposed so as to either reinforce or cancel each other, the effect's magnitude depending on the relative phases of the two trains. With light, interference occurs when two monochromatic beams having identical frequencies are superimposed, darkness (destructive interference) being produced when the two beams are united so that the positive vibration peaks of one match the negative ones of the other (a 180° phase difference). Conversely, a bright spot (constructive interference) is seen when the two waves are matched peak to peak (zero phase difference). In most interference experiments the two beams are produced by splitting a single beam to ensure that the frequencies will be the same. The phase difference is obtained by making the two beams traverse paths of different length before being recombined. A path difference of half a wavelength produces a relative phase shift of 180°, hence a path difference of a whole wavelength or any integer number of wavelengths produces no phase shift at all. Therefore a path difference of any integer number of wavelengths (an even number of half wavelengths) gives rise to constructive interference, whereas a difference of an integer number of wavelengths plus a half wavelength (an odd number of half wavelengths) gives rise to destructive interference.

One method of producing interference is to allow monochromatic light to fall upon two pieces of plate glass placed at a very small angle to each other (see Fig. 119). Light will be reflected from each of the glass surfaces. The light reflected from the lower surface of the upper plate and the light reflected from the upper surface of the lower plate will interfere. The second reflected beam will travel a longer distance, namely, across the air film and back. Since the air film varies regularly in thickness, the interference pattern produced will consist of a series of parallel bright and dark lines which are called *interference fringes*. However, one point should be noted in connection with this method: reflection of the light at the upper surface of the lower glass plate (where the light strikes a denser medium) causes a 180° phase shift, which is equivalent to an extra half wavelength of path for this beam. Consequently, with this setup the conditions for constructive and destructive interference are reversed; an actual path difference of an even number of half wavelengths results in destructive interference while an odd number gives the constructive effect.

The wavelength of the yellow light from a sodium lamp may be determined by using interference fringes. At any place where the path difference is an odd number of half wavelengths, the two reflected rays will be in phase because of the phase reversal at the lower plate, and the spot will be bright. Where the path difference is an even number of half wavelengths the spot will be dark. Thus from one bright fringe to the next one, the difference in path increases by one whole wavelength of light. This means that the thickness of the air film increases by one half of a wavelength of light in this distance, since the light goes across the film and back. But in the entire length L of the air film (see Fig. 118), its thickness increases from zero to T, the thickness of the thin strip of steel that separates the glass plates at one end. Therefore, the rate at which the air film thickness increases as we move across the plates is T/L. If n is the number of bright fringes per centimeter, the distance between adjacent bright fringes must be $1/n$ centimeters. If we multiply the rate T/L by the spacing $1/n$ between bright fringes, we must get the necessary half wavelength increase in the air film thickness. Thus $(T/L)(1/n) = \lambda/2$ or

$$\lambda = \frac{2T}{nL} \qquad (158)$$

Note that if n is in reciprocal centimeters (cm^{-1}), L must be in centimeters, and λ will be in the same units of length as T. If T is in centimeters, λ will be in centimeters and must be multiplied by 10^7 for nanometers or 10^8 for angstrom units.

319

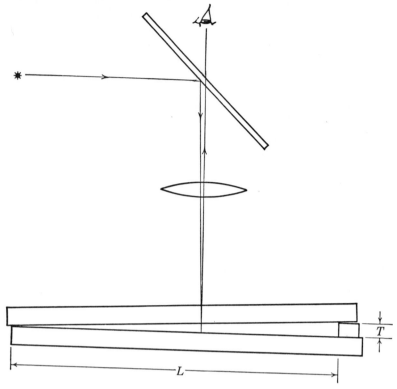

Figure 118 Apparatus for Viewing Interference Fringes

APPARATUS

1. Sodium light source
2. Two plane-parallel glass plates
3. Mirror, lens, and supporting stand
4. Thin strip of steel

5. Centimeter scale
6. Micrometer caliper
7. Sheet of black optical paper

PROCEDURE

1. Measure the thickness of the thin strip of steel with the micrometer caliper and record the reading.

2. Be sure that the glass plates are clean. Be careful not to get finger marks on the surfaces that are to be placed together. Lay the sheet of optical paper on the laboratory table and place one glass plate down flat on

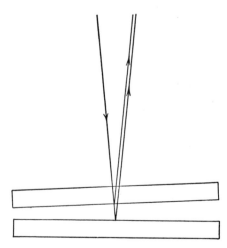

Figure 119 Interference from Two Plane Surfaces

it. Carefully lay the other glass plate down flat upon the first with the best surfaces in contact, so that the upper plate projects slightly at the thin end of the air film as shown in Fig. 118. Separate the plates at the other end by the thin steel strip, placing the edge of the strip parallel to the ends of the glass plates.

3. Measure the length L of the air film (the distance between the edge of the steel strip and the line along which the two plates rest on each other) and record this reading.

4. Place the supporting stand beside the glass plates so that the mirror, which is set at 45°, and the lens are positioned over the center of the wedge-shaped air film, as shown in Fig. 118. Set up the sodium light source so that it shines horizontally into the mirror. Turn the source on and wait for it to warm up so that a brilliant yellow light is produced. The image of the source, when viewed through the small hole in the mirror, should be parallel to the length of the air film. The lens is used to magnify the fringes; the hole in the mirror is placed at the focus of the lens. *Note:* A plain glass plate may be used instead of the mirror with the hole at the cost of a reduction in the observed light intensity. In this case only some of the light from the source is reflected down to the air wedge, the rest being lost by transmission through the plate, and only a part of the beam returning from the wedge is transmitted to the eye, the rest being reflected back to the source. The sodium light source has sufficient intensity, however, so that the amount of light reaching the eye is quite adequate for the required observations to be made.

5. Observe the interference fringes. They consist of a series of dark parallel lines across the image of the light source and are perpendicular to the length of the air film. Lay the centimeter scale flat upon the upper plate, so that the edge of the scale is perpendicular to the fringes. Looking through the lens, count the number of dark fringes in 2 centimeters of length, estimating to tenths of a fringe. Record this reading.

6. Repeat Procedure 5, making a second determination of the number of dark fringes in 2 centimeters of length, and record the reading.

DATA

Thickness of the thin strip of steel	_____	Average Number of fringes per centimeter of length	_____
Length of the air film	_____		_____
Number of fringes in 2 centimeters of length:		Wavelength of yellow light from sodium lamp	
First reading	_____		_____
Second reading	_____	Percent error	_____

CALCULATIONS

1. Compute the number of fringes per centimeter, using the average of the two readings taken.

2. Using Equation 158, calculate the wavelength of the yellow light from the sodium lamp. Express your result in angstrom units. ($1\text{Å} = 10^{-8}$ cm)

3. Compare your result with the known value, taking the average wavelength of the sodium doublet as 5893 Å, and compute the percent error of your determination.

QUESTIONS

1. (a) Explain how the method used in this experiment could be employed to test whether a surface is optically plane or not. (b) What would be the shape of the fringes if a spherical surface were placed in contact with a flat surface?

2. Explain why the fringes sometimes run diagonally across the plates.

3. Suppose that white light were used to produce the interference fringes in this experiment. What would be the appearance of the fringes produced?

4. Two plane glass plates touch at one edge and are separated at the opposite edge by a strip of tinfoil. The air wedge is examined with sodium light reflected normally from its two surfaces, and a total of 35 dark interference fringes are observed. Compute the thickness of the tinfoil.

5. How is this experiment evidence for a wave theory of light? Explain.

6. Instead of laying one glass plate on another with a thin strip separating them along one side as in Fig. 118, an air film may be produced by substituting a convex lens for the upper glass plate, this lens being simply set down on the lower glass plate. The thickness of the film thus formed increases with the radial distance from the contact point, but the increase is not linear. The resulting pattern is a series of concentric circles called *Newton's rings,* whose spacing decreases with their radius. Derive an expression for the difference in the radii of two adjacent interference rings as a function of these radii and the radius R of the spherical lens surface. *Note:* Make approximations based on the fact that the thickness of the air film is everywhere very small compared to the radius of curvature R. This is particularly true when a long focal length (large radius of curvature) lens is used, as is usually done.

7. Explain why interference patterns caused by light reflected from the top surface of the upper plate or from the bottom surface of the lower plate (see Fig. 119) are not observed.

The Diffraction Grating 44

The diffraction grating is one of the most important optical instruments. It is among the simplest devices for producing spectra so that the wavelengths contained in a given light beam may be measured and has the advantage of making calibration with a known spectrum unnecessary. The object of this experiment is to study the operation of a diffraction grating and, in particular, to see how it may be used to measure the wavelength of light from a helium–neon laser.

THEORY

The diffraction grating provides the simplest and most accurate method for measuring wavelengths of light. It consists of a very large number of fine, equally spaced parallel slits. There are two types of diffraction gratings: the reflecting type and the transmitting type. The lines of the reflection grating are ruled on a polished metal surface; the incident light is reflected from the unruled portions. The lines of the transmission grating are ruled on glass; the unruled portions of the glass act as slits. Gratings usually have about 10,000 to 20,000 lines per inch. The grating used in this experiment is a transmission grating replica.

The principles of diffraction and interference are applied to the measurement of wavelength in the diffraction grating. Let the broken line in Fig. 120 represent a magnified portion of a diffraction grating, and let a beam of parallel rays (a plane wave) of monochromatic light impinge on the grating from the left. By Huygens' principle, the light spreads out in every direction from the apertures of the grating, each of which acts as a separate new source of light. The envelope of secondary wavelets determines the position of the advancing wave. In the figure are seen the instantaneous positions of several successive wavelets after they have advanced beyond the grating. Lines drawn tangent to these wavelets connect points in phase, hence they represent the new wave fronts. One of these wave fronts is tangent to wavelets that have all advanced the same distance from the slits; this wave front is parallel to the original one. A converging lens placed in the path of these rays would form the central image. Another wave front is tangent to wavelets whose distances from adjacent slits differ by one wavelength. This wave front advances in direction 1 and forms the first-order spectrum. The next wave front is tangent to wavelets whose distances from adjacent slits differ by two wavelengths. This wave front advances in direction 2 and forms the second-order spectrum. Spectra of higher orders will be formed at correspondingly greater angles.

Looking at Fig. 120, we note that ray 1 has been drawn perpendicular to its wave front at the wave front's point of tangency with a Huygens wavelet that is coming

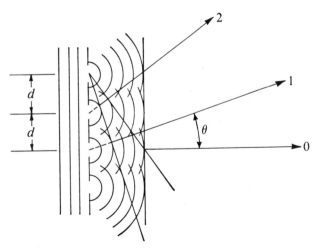

Figure 120 The Diffraction Grating

325

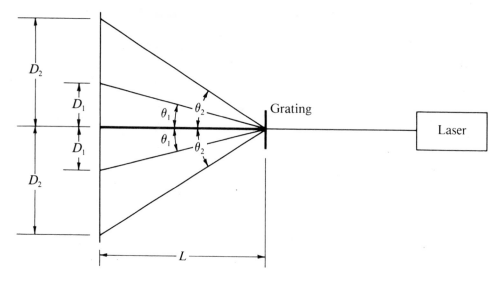

Figure 121   Experimental Arrangement for
the Diffraction Grating

from the third slit down from the top. A projection of this ray backward will pass through the center of the slit, because a line drawn perpendicular to a tangent at the point of tangency is a radius of the circle in question. In the right triangle of ray 1, the associated wave front, and the line representing the grating, we see that the angle between the wave front and the grating is the same as θ, the angle between rays 1 and 0. Consideration of this triangle reveals that $\sin \theta = 2\lambda/2d = \lambda/d$. Similarly, for ray 2, which for convenience has been drawn to the center of the second slit down, we note that $\sin \theta = 2\lambda/d$, where now θ is the angle between rays 2 and 0. In general we will find that $\sin \theta = n\lambda/d$, or

$$n\lambda = d \sin \theta \qquad (159)$$

where n is called the *order* of the spectrum. In this equation, λ is of course the wavelength of the light and will be in the same units as d, the spacing between the grating lines. The wavelength of visible light is usually expressed in angstrom units; one angstrom unit is 10^{-8} centimeter. However, the nanometer (10^{-9} meter, hence 10^{-7} centimeter or 10 angstrom units) is becoming increasingly popular in expressing optical wavelengths. The grating constant d is obtained by noting the number of lines per inch or per centimeter marked on the grating and taking the reciprocal to find the spacing between

lines. This spacing will then be in inches or centimeters, so that a conversion must be made to obtain λ in appropriate units.

In this experiment a helium–neon laser will be used to provide a narrow beam of highly monochromatic light ($\lambda = 6328$ Å). A plan view of the experimental arrangement is shown in Fig. 121. The laser beam strikes the grating normally and the direct ray proceeds in a straight line to the screen. The first-order ($n = 1$) diffracted rays are deflected through angles θ_1 on either side of the central ray and strike the screen at points spaced D_1 on either side of the central spot. Similarly, the second-order ($n = 2$) diffracted rays each make angle θ_2 with the central ray and strike the screen at points separated from the central spot by D_2. These angles are determined by measuring D_1, D_2, and L, the distance between the grating and the screen, and noting that $\tan \theta_1 = D_1/L$ and $\tan \theta_2 = D_2/L$.

Because the laser light contains only one wavelength, only one pair of spots will appear on the screen in each order. However, if the experiment is to run with light that is not monochromatic, there will be as many pairs of spots in each order as there are different wavelengths in the light from the source, the diffracting angle θ for each wavelength being given by Equation 159. We shall observe these results when white light from a tungsten lamp is substituted for the laser light.

APPARATUS

1. Low-power helium–neon laser
2. Diffraction grating
3. Optical bench
4. Long-focus convergent lens (about 25 cm)
5. Ground-glass screen
6. Lens holders and optical bench clamps
7. Meter stick
8. Masking tape
9. Tungsten lamp
10. Stand and clamps for mounting the light source

PROCEDURE

1. Set up the optical bench and mount the ground-glass screen at one end. Mount the grating about 30 centimeters in front of it using one of the lens holders and optical bench clamps. The laser is then mounted so that the beam is directed normally at the grating, as shown in Fig. 121.

2. Turn the laser on and observe the central spot on the screen. CAUTION: The low-power laser used in this experiment cannot hurt you even if the beam is allowed to fall on your skin for a prolonged period. However, you must *not look into it or allow anyone else to do so*. Keep the laser beam down below eye level and do not let it shine in anyone's face including your own.

3. With the usual size ground-glass screen and a grating of about 15,000 lines per inch placed 30 cm away, the first-order spots will appear on the screen but the second-order ones will miss it on either side. To display both orders and facilitate measurement of the separations D, tape the meter stick to the screen so that the central spot falls approximately at the 50-centimeter mark and the diffracted spots are displayed right on the stick. Adjust the orientation of the screen so that it and the meter stick are perpendicular to the central ray. Do this by rotating the screen in its mount until the two second-order spots are equidistant from the central spot.

4. Read and record the positions of the two first-order spots on the meter stick.

5. Similarly, read and record the positions of the two second-order spots. These spots will be fainter than the first-order ones; you may need to darken the room in order to see them clearly.

6. Measure and record the distance L from the grating to the surface of the meter stick. If the length graduations on the optical bench are used for this purpose, allowance must be made for the thickness of the meter stick and one-half the thickness of the ground-glass screen.

7. Shut off the laser, replace it with the tungsten lamp, and mount the converging lens on the optical bench between the lamp and the grating. Align the lamp and the lens so that the light passing through the lens falls on the middle of the grating. The lens should be moved back and forth along the optical bench until the central spot observed on the screen is a focused image of the lamp filament. The lamp must be mounted at a distance greater than the lens's focal length behind the grating to make this adjustment possible.

8. Examine the continuous spectrum displayed on the meter stick in first order, noting the relative position of the different colors. The second-order spectrum will probably be too faint to be seen. Record the positions of the inner ends of the first-order spectra observed on either side of the central white spot.

9. Record the positions of the outer ends of the displayed spectra.

DATA

Number of lines per inch on the _____   Distance L from grating to meter _____
  grating   stick in centimeters
Grating constant d in centimeters _____

| Source | Meter-stick Reading | | Calculated Values | | | | | | |
	Right	Left	Difference	D	$\tan \theta$	θ	$\sin \theta$	λ	%Error
Laser (First Order)									
Laser (Second Order)									

Source	Meter-stick Reading		Calculated Values						
	Right	Left	Difference	D	$\tan \theta$	θ	$\sin \theta$	λ	%Error
Lamp (Violet End)									
Lamp (Red End)									

Range in wavelength of visible spectrum _____

CALCULATIONS

1. Compute the grating constant (that is, the distance between the lines on the grating) in centimeters from the number of lines per inch stated on the grating.

2. Find D_1, the displacement of the first-order spot from the central spot, by computing the difference between the two spot positions recorded in Procedure 4 and dividing by 2.

3. Calculate θ_1, the deviation of the first-order diffracted laser beam, from values of D_1 and L.

4. Calculate the laser light wavelength by using Equation 159 with $n = 1$, $\theta = \theta_1$ (from Calculation 3), and the grating constant d found in Calculation 1. Note that with d in centimeters, λ will come out in centimeters and must be multiplied by 10^8 to give angstrom units.

5. Find the percent deviation of your computed value of the laser light wavelength from the known value (6328 Å).

6. Repeat Calculations 2–5 for the second-order diffracted laser light, using the data of Procedure 5.

7. Repeat Calculations 2–4 for the violet end of the continuous spectrum obtained with the tungsten lamp, using the data of Procedures 7 and 8.

8. Repeat Calculations 2–4 for the red end of the continuous spectrum, using the data of Procedure 9.

9. Calculate and record the range in wavelengths of the visible spectrum as measured in this experiment.

QUESTIONS

1. (a) Discuss the difference between the continuous spectrum obtained with a grating and one obtained with a prism. (b) Which color is deviated most in each case? Why?

2. Why does a grating produce several spectra, while a prism produces only one?

3. Why are the higher-order spectra more accurate than the first-order spectrum in determining wavelengths of light?

4. A monochromatic beam of light impinges normally on a grating and forms an image of the slit 10 cm from the central image on a screen. The screen is 50 cm from the grating and is set parallel to it. The grating has 10,000 lines per inch. What is the wavelength of the light?

5. A grating having 15,000 lines per inch is used to produce a spectrum of the light from a mercury arc shining through a slit. What is the angle between the first-order green line and the second-order green line? See Experiment 42 for the wavelength of the mercury green line.

6. Consider two rays coming from adjacent slits in a grating of grating constant d and intersecting at the point where they strike a distant screen. Use the rule for constructive interference (see Experiment 43) to derive Equation 159. *Note:* Because the distance from the grating to the screen is very much larger than the separation between the slits, the two rays are almost parallel.

7. What would be the effect of the plane of the grating not being perpendicular to the beam of light coming from the laser? If you can, show how Equation 159 should be modified to take account of an angle ϕ between this beam and the normal to the grating plane.

8. Is it possible for the red end of the visible spectrum viewed in first order to overlap the violet end viewed in second order? Assume that the visible spectrum runs from 7000 to 4000 Å.

9. (a) Why in the present experiment did you not observe a third-order diffracted beam of laser light? (b) What is the highest order in which you would expect to be able to view the mercury green line (see Experiment 42 for λ) when using a grating having 15,000 lines per inch?

10. What is the effect on the observed spectrum of using a larger grating, *i.e.*, one having the same grating constant but a greater total number of lines? Assume an incident beam whose cross section covers the grating's entire active surface.

Polarized Light **45**

Experiments on interference and diffraction show that light is composed of waves, but they do not offer any evidence as to whether these waves are transverse or longitudinal. However, experiments on polarization show that light waves must be transverse waves. The purpose of this experiment is to study some of the fundamental phenomena of polarized light.

THEORY

Light waves consist of transverse electromagnetic vibrations that travel with finite velocity. The electric vibrations and the magnetic vibrations are perpendicular to the direction of propagation of the light, and also perpendicular to each other. In ordinary light, the electric vibrations occur in all directions in a plane perpendicular to the direction of propagation as shown in Fig. 122(a). This light is unpolarized. It is possible to produce plane-polarized light, that is, light in which the electric vibrations all occur in some one direction perpendicular to the ray of light. Vertical plane polarization is illustrated in Fig. 122(b) and horizontal plane polarization in Fig. 122(c). Notice that in these cases the direction of the electric vibrations (the electric vector **E**) and the direction of propagation (the light ray) determine a plane.

Because the electric vibrations are restricted to this plane, it is called the *plane of polarization*.

Light may be plane polarized by reflection from a nonconducting reflector or by passing through a doubly refracting crystal. When light is reflected from the surface of a transparent dielectric medium such as glass or water, it is polarized to some extent if the angle of incidence is greater than $0°$ but less than $90°$. The *reflected* light is *completely* polarized if the ray reflected from the medium is at right angles to the refracted ray proceeding into the medium. Calculation of the angle of incidence i_p for which this condition is satisfied may be carried out by reference to Fig. 123. Clearly $r_1 + r_2 + 90° = 180°$ or $r_1 + r_2 = 90°$. But by the law of reflection $r_1 = i_p$, and by the law of refraction (Snell's law, Equation 150) sin

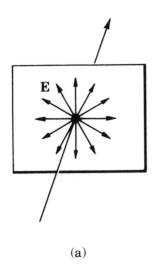

(a)

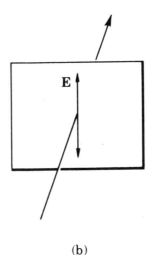

(b)

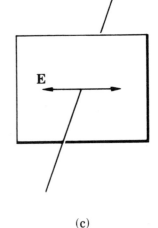

(c)

Figure 122   Polarization of Light

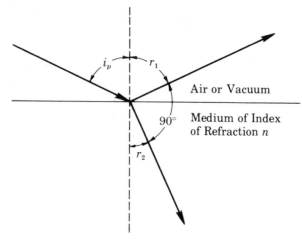

Figure 123   Brewster's Angle

$i_p = n \sin r_2$. Then, since $r_2 = 90° - r_1 = 90° - i_p$, $\sin i_p = n \sin (90° - i_p) = n \cos i_p$ and there results

$$n = \frac{\sin i_p}{\cos i_p} = \tan i_p$$

This is *Brewster's law*, and the angle of incidence i_p whose tangent is equal to the index of refraction of the medium and for which the reflected beam is completely polarized is called *Brewster's angle*. For a piece of flat, unsilvered glass it is about 57°. Note that the direction of polarization of the reflected beam is parallel to the reflecting surface (perpendicular to the plane of the paper in Fig. 123). Note also that although this beam is completely polarized, the refracted beam in the dielectric medium is not. This situation may be understood by considering that the incident (unpolarized) light is made up of two components, one polarized perpendicular to the plane of the paper in Fig. 123 and the other lying in that plane. The former is largely reflected, although a part proceeds into the new medium, but at Brewster's angle *all* of the component polarized in the plane of the paper is transmitted into the medium. Thus the transmitted beam is "enriched" in this latter component, while the reflected beam contains none of it.

In a doubly refracting crystal such as calcite, light propagates at two different velocities depending on whether it is polarized along or perpendicular to the so-called *optic axis*. The electric vibrations of a light ray propagating along the optic axis are perpendicular to this axis even if the light is unpolarized, because light waves are transverse. Such a ray travels with a speed $v_\perp$ regardless of its polarization. However, if a ray is directed perpendicular to the optic axis in a doubly refracting crystal, we must consider two components. One is polarized perpendicular to the optic axis and travels with the same speed $v_\perp$ as does a wave traveling along the optic axis and thus is called the *ordinary ray*. The other component, polarized along the optic axis, travels with a different speed $v_\parallel$ and is called the *extraordinary ray*. The crystal

then exhibits two indices of refraction, $n_\perp = c/v_\perp$ and $n_\parallel = c/v_\parallel$. If a ray of unpolarized light travels through a doubly refracting crystal in a direction making some angle (other than 0° or 90°) with the optic axis there will still be an ordinary ray polarized perpendicular to this axis that obeys Snell's law in the usual manner. In this case, however, the extraordinary ray's electric vibrations are neither parallel nor perpendicular to the optic axis and thus involve both indices of refraction. Therefore, this ray does not obey the simple form of Snell's law and is consequently bent away from the ordinary ray. Hence a doubly refracting (also called a birefringent) crystal has the property of dividing a beam of unpolarized light into two beams plane polarized at right angles to each other.

Some crystals have the property of absorbing a light beam polarized along the optic axis and transmitting one polarized at right angles to this axis. Unpolarized light incident perpendicular to the optic axis can be plane-polarized in this manner. This property is called *dichroism*. Tourmaline is the most common example of a dichroic crystal.

Natural crystals are usually too small to be used in an experiment where a wide beam of polarized light is needed; for this purpose, Polaroid sheets are used. In a Polaroid, very minute crystals are distributed uniformly and densely in a cellulose film mounted between two glass plates. The individual crystals are needle-shaped and submicroscopic in size and are turned so that their axes are parallel by stretching the film before it is mounted on glass. Like tourmaline, Polaroids are dichroic and produce polarized light in the same manner.

A Polaroid has an optic axis in the plane of the sheet along which the electric vibrations of a normally incident light ray must lie in order for the ray to get through without absorption. If there is an angle θ between the optic axis and the direction of the electric vibrations, then the electric vector **E** in this direction may be broken up into a component $E_\perp = E \sin \theta$, which is perpendicular to the optic axis and gets absorbed by the Polaroid, and a component $E_\parallel = E \cos \theta$, which is parallel to the optic axis and is therefore transmitted with little or no absorption. Since the intensity of a light beam is proportional to the square of the amplitude of the electric vibrations, the ratio of the intensity of the polarized light transmitted by the Polaroid to the incident intensity is

$$\left(\frac{E_\parallel}{E}\right)^2 = \cos^2 \theta$$

This result is called the law of Malus after its discoverer, E. L. Malus (1775–1812), and states that if a plane-polarized light is normally incident on a Polaroid or other dichroic material, the transmitted intensity is proportional to the square of the cosine of the angle between the optic axis and the plane of polarization. Note that the intensity becomes zero when this angle is 90°.

APPARATUS

1. Two Polaroid disks, each mounted in a holder in which it can rotate
2. Optical bench with clamps and holders
3. Light source with lens to produce a parallel beam of light
4. Desk lamp
5. Photronic cell connected to a microammeter
6. Calcite crystal
7. Pile of thin glass plates
8. Black glass mirror
9. Piece of cellophane
10. Thin piece of mica
11. Unannealed glass plate
12. Annealed glass plate
13. Piece of celluloid
14. Iris diaphragm

PROCEDURE

1. Place the light souce at one end of the optical bench. Place the photronic cell at the other end in such a position as to receive the light. Place one of the Polaroid disks on the optical bench in front of the photronic cell so that the light has to pass through it to get to the cell. This Polaroid is the analyzer. Rotate the analyzer and notice the reading of the microammeter. If the microammeter goes off scale, decrease the size of the beam of light by placing a diaphragm in front of the lens, regulating the size of the beam so that the microammeter reads nearly full scale. The light coming directly from the lamp is unpolarized. The intensity of the light transmitted by the analyzer should be nearly independent of the angle that its axis makes with the vertical.

2. Place the second Polaroid disk on the optical bench in front of the lamp and about 4 inches away from it, in such a position that all the light passes through it. This is the polarizer. Place it with its axis in the vertical direction.

3. Now rotate the analyzer and take readings of the microammeter for different angular settings. Begin with the axis vertical and take readings of the microammeter for angles 0°, 10°, 20°, 30°, etc., up to 180°. You will observe that the intensity of the light transmitted by the analyzer depends upon the angle between the axis of the analyzer and that of the polarizer.

4. Remove the photronic cell. Allow the light to pass through both Polaroids, leaving a space of about 6 inches between them. Set the two Polaroids for extinction, that is, so that the minimum amount of light passes through them. Now place a piece of cellophane in the space between the Polaroids, look at the cellophane through the analyzer, and observe the field of view. Fold the cellophane into several thicknesses and observe the effect on the color produced.

5. Place a thin piece of mica in the space between the Polaroids, look at the mica through the analyzer, and observe the field of view.

6. Repeat Procedure 5 using the unannealed glass plate.

7. Repeat Procedure 5 using the annealed glass plate.

8. Repeat Procedure 5 using a piece of celluloid. Twist the celluloid slightly out of shape and observe the regions where strains are produced.

9. Remove the polarizer from the optical bench. Place the black glass mirror in a holder on the optical bench and allow the unpolarized light from the desk lamp to fall on it. Rotate the mirror holder so as to make the angle of incidence about 55°. Observe the light reflected from the mirror by looking at it through a Polaroid. By rotating the Polaroid, notice whether the reflected light is polarized.

10. Repeat Procedure 9 using a pile of thin glass plates as the mirror. Determine whether the reflected light is polarized and also whether the transmitted light is polarized.

11. Make a dot on a piece of paper. Place a calcite crystal over the dot and look through it. You will observe two images. Rotate the crystal and notice what happens to the two dots. Now look at the dots through the calcite and a Polaroid. Rotate the Polaroid and notice the result. The two dots represent two beams of light coming from the calcite crystal. Determine whether these two beams are polarized and find the relative direction of vibration of the two beams.

12. If the sun is shining, look at the blue light from the sky through the analyzer. Notice particularly the light coming from a direction roughly perpendicular to the sun's direction. Determine whether this light is polarized.

DATA

Analyzer Angle θ	$\cos^2\theta$	Microammeter Reading	Analyzer Angle θ	$\cos^2\theta$	Microammeter Reading
0°			100°		
10°			110°		
20°			120°		
30°			130°		
40°			140°		
50°			150°		
60°			160°		
70°			170°		
80°			180°		
90°					

CALCULATIONS

1. The percentage polarization is given by the expression

$$P = 100 \frac{I_{max} - I_{min}}{I_{max} + I_{min}}$$

where P is the percentage polarization, I_{max} is the largest value of the microammeter reading observed, and I_{min} is the smallest value observed. Calculate the percentage polarization from your readings in Procedure 3.

2. Calculate and record the value of $\cos^2 \theta$ for each of the angles between the polarizer and analyzer optic axes used in Procedure 3.

3. Plot a graph using the microammeter readings obtained in Procedure 3 as ordinates and the values of $\cos^2 \theta$ as abscissas. Note that $\cos^2 \theta$ will be the same for $\theta = 80°$ and 100°, for $\theta = 70°$ and 110°, etc., and you should therefore get the same microammeter readings for the two angles in each of these pairs. Differences in these readings represent errors in the experiment. Plot all points and draw the best straight line through them.

QUESTIONS

1. Explain the difference between transverse waves and longitudinal waves.

2. What is meant by a plane-polarized beam of light?

3. Mention some of the uses of polarized light.

4. (a) Describe the construction of a nicol prism. (b) Explain how it produces a polarized beam of light.

5. Two Polaroids are set for extinction, that is, with their optic axes at right angles, and a third Polaroid is then placed between them. (a) Explain how the third Polaroid can be adjusted so that light can pass through the set of three even though no light can get through the original pair. (b) What setting of the third (middle) Polaroid results in maximum transmitted intensity?

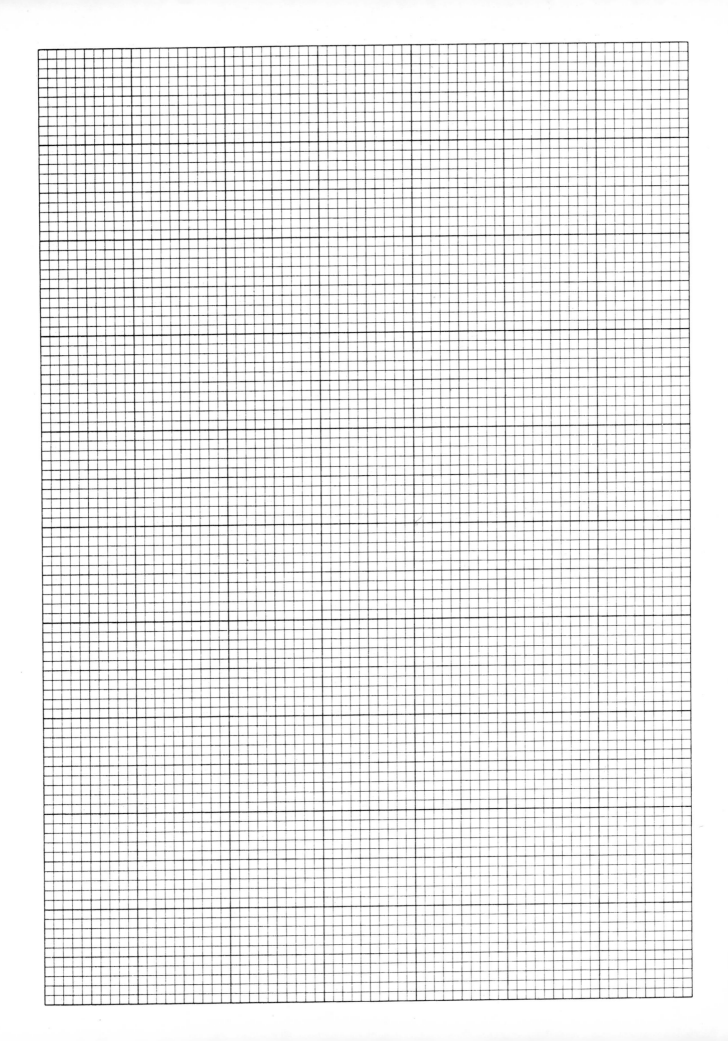

Radioactivity and 46
the Geiger Counter

The discovery of radioactivity has revolutionized our concept of the nature of the physical world and has given us the means by which we can study the innermost structure of an atom, that of the nucleus itself. Radioactive disintegration involves the actual transmutation of elements and the formation of new isotopes. This experiment contains a discussion of the nature of radioactivity and of the radiations from radioactive substances. One of the most important relations involved is the inverse square law of intensity. One purpose of this experiment is to show that the intensity of beta and gamma radiation varies inversely with the square of the distance from the source. To detect the radiation from radioactive sources, several methods are used. Some of these will be discussed in this experiment, a further purpose of which is to introduce the student to one of the most important of them — the Geiger-Müller counter — and to study its operating characteristics.

THEORY

In 1896 the French physicist Henri Becquerel discovered that minerals containing uranium gave out a radiation that resembled X rays, since it could penetrate an envelope of black paper and affect a photographic plate. These rays could also produce ionization in air and were thus able to discharge electrified bodies. They could easily be detected with a sensitive electroscope and their intensity measured. This property of uranium was later found to be possessed by a number of the heavy elements, and the name radioactivity was applied to it. It was soon found that radioactivity is an atomic phenomenon, for the activity of any given compound of uranium is proportional to the amount of uranium in the substance and does not depend on the physical or chemical condition of the substance.

Soon more powerful sources of radioactivity were discovered. Madame Marie Curie and her husband, Pierre, began to suspect the existence of other radioactive elements in the uranium ores. After two years of hard labor, during which nearly a ton of uranium ore was worked over chemically, they discovered two new chemical elements, polonium and radium. Both of these elements were found to be thousands of times more radioactive than uranium.

The nature of the radiations from radioactive substances was investigated by Ernest Rutherford (1871–1937) and his collaborators. They found that the radiation consisted of three different kinds of rays. These were called α (alpha), β (beta), and γ (gamma) rays. They differ in their power of penetrating matter and of producing ionization and also in their behavior under the influence of electric or magnetic fields.

It is easy to show that these rays are different in nature. Suppose a piece of radioactive material is placed at the bottom of a small hole drilled in a lead block, as shown in Fig. 124. A photographic plate is placed at a convenient distance above the lead block, and the whole apparatus is highly evacuated. This arrangement produces a narrow beam of rays emerging from the top of the lead block. A strong magnetic field is then applied at right angles to the plane of the figure. When the photographic plate is developed, three distinct spots are found. One of these was produced by the alpha rays, which were deflected slightly to one side; from the known direction of the magnetic field, we know that the alpha rays consist of positively charged particles. The second spot was produced by the beta rays, which were deflected more strongly in the opposite direction; this shows that they consist of negatively charged particles. The third spot was produced by the gamma rays, which were not

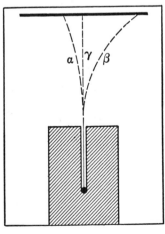

Figure 124   Separation of Alpha Rays, Beta Rays, and Gamma Rays by a Magnetic Field

deflected at all; this shows that they do not carry any electrical charges.

Rutherford showed in 1907 that *alpha particles* are the same as the nuclei of helium atoms. Each particle has a mass about four times that of a hydrogen atom and each carries a positive charge equivalent to two electron charges. Alpha particles are emitted with a speed of about one twentieth the velocity of light. The *beta particles* have the same mass and the same charge as electrons; they are in fact electrons emitted from the nuclei of radioactive atoms with speeds that nearly approach the velocity of light. The *gamma rays* consist of electromagnetic waves of extremely short wavelength, about $\frac{1}{100}$ that of X rays, and are similar to X rays in their properties.

It was found that no physical or chemical process, such as raising the temperature or producing a new chemical compound with some other radioactive substance, could change the radioactivity of a given sample. As a result, it was concluded that radioactivity is a nuclear process and that the emission of a charged particle from the nucleus of an atom results in leaving behind a different atom, occupying a different place in the periodic table. Radioactivity, then, involves the transmutation of elements.

When an alpha particle is given off, the original atom loses four units of mass and two units of charge and thus becomes a new atom of lower atomic weight and lower atomic number. When a beta particle is given off, the new isotope produced will have the same mass number as the parent atom, since the mass of an electron is negligible in comparison with the mass of the nucleus, but the atomic number will be raised by one, since the change produced is equivalent to increasing the net positive charge on the nucleus. In the emission of gamma rays no new elements are produced, since both the atomic number and the atomic weight of the element remain the same. The gamma rays are produced during the transitions from higher to lower nuclear energy states of the same isotope. The emission of alpha particles may be explained by assuming that some of the nuclear protons and neutrons may exist together as alpha particles within the nucleus. To explain the emission of beta particles is not so simple. According to the present belief, electrons do not form part of the nuclear structure. However, when a beta particle is emitted, a neutron within the nucleus becomes a proton and emits an electron.

The occurrence of radioactive disintegration is statistical in nature. For any individual atom of a radioactive substance there is a definite probability that it will disintegrate within a certain time interval. This statement means that when a very large number of atoms are involved, the emission obeys the laws of probability. The rate of disintegration is proportional to the number of atoms present. Accordingly, when the mathematical relation is set up and solved, the radioactive decay is found by using the expression

$$N = N_0 e^{-\lambda t} \tag{160}$$

where N is the number of atoms present at any time t, N_0 is the number present at $t = 0$, e is the base of natural logarithms, and λ, often called the decay constant, is the probability per unit time that a particular atom will disintegrate.

A convenient way of representing the rate of radioactive decay is by the *half-life* of a radioactive element, which is defined as the time required for half of a given quantity of the element to disintegrate into a new element. The rate at which a given quantity of a radioactive element decays is found by measuring the intensity of radiation given off by the sample over a period of time and by plotting a graph like the one shown in Fig. 125. This figure shows the decay curve for the radioactive element polonium. Thus, for polonium the intensity drops to half of its original value in 140 days; in another 140 days the intensity again drops to half the value it had at the beginning of this time interval, and so on. Since it would take an infinite time for the intensity to drop to zero, the rate of decay is usually specified by the half-life of the atom, and this is one of the important constants to be measured. The same law of decay applies to all the other radioactive elements, except that each one has its own half-life and its own decay constant. The half-life may be extremely short, such as less than one millionth of a second for thallium-208 (Te^{208}, sometimes called thorium C'), or it may be quite long, such as 1620 years for radium itself.

The intensity of the rays emitted by a radioactive substance decreases as the distance from the source is increased. Because the rays spread out in all directions, their intensity follows *the inverse square law of radiation*. This law states that the radiation received on a small surface held perpendicular to the rays varies directly with the source's intensity and inversely as the square of the surface's distance from the source. This law applies strictly for a point source or, in practice, when the distance is very great compared to the source size.

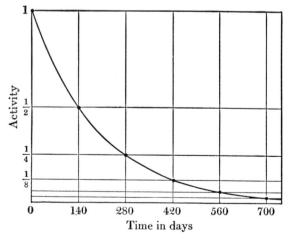

Figure 125   Decay Curve for the Radioactive Element Polonium

Experimentally, the radiations given off by radioactive substances may be detected and measured in several ways. These include the use of photographic plates, the ionization chamber, the Wilson cloud chamber, and the Geiger-Müller counter.

One method of measurement is to use a *photographic plate*. The rays that penetrate the sensitive emulsion produce a latent image that can be developed. The density of the film is a measure of the intensity of radiation. The *Wilson cloud chamber* is another instrument used to study penetrating radiation. The cloud chamber works on the principle that when air saturated with water vapor is expanded suddenly, the cooling effect causes the formation of a cloud of tiny water drops. Drops of water form around dust particles or any ionized particles present in the chamber. When a ray of penetrating radiation passes through the chamber, it produces ions along its path. The air is then suddenly expanded and liquid droplets are formed on the ions. These water drops appear momentarily as a white line that may be seen or photographed.

Another method of measurement is to use an *ionization chamber*. When a gas is exposed to radioactive radiations, it becomes ionized, and the amount of ionization produced is a measure of the intensity of the rays. The ionization chamber consists of a closed metal box provided with a thin aluminum or celluloid window and containing an insulated metal electrode maintained at a fairly high potential (either negative or positive) with respect to the box itself. If some penetrating radiation enters the chamber, it will ionize the gas therein, producing a large number of electron–ion pairs. Under the influence of the applied voltage, the positive ions go one way and electrons the other, so that an electric current flows through the chamber. The amount of ionization and hence the magnitude of the current is proportional to the intensity of the radiation, which can thus be measured by measuring the current.

The Geiger-Müller counter is one of the most important instruments of modern physics, giving us the most convenient method of measuring penetrating radiation. A Geiger counter consists of a metal tube fitted inside a thin-walled glass cylinder, with a fine tungsten wire stretched along the axis of the tube. The arrangement is illustrated in Fig. 126. The cylinder contains a gas such as air or argon at a pressure of about 10 centimeters of mercury. A difference of potential of about 1000 volts is applied between the wire and the metal cylinder, the positive voltage being applied to the wire and the negative to the cylinder. This difference of potential is slightly less than that necessary to produce a discharge. When any ionizing radiation, such as a high-speed particle from a radioactive source, enters the tube, ions will be produced by the freeing of electrons from the gas molecules. These ions and electrons are accelerated by the electric field and produce more ions by collision. The process repeats itself and the ionization current is rapidly

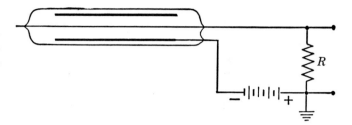

Figure 126 The Geiger Counter (Schematic Diagram)

built up. However, the current drops to zero just as rapidly. As the current flows through the external resistance R, it produces a potential drop across R and this causes the net voltage across the tube to fall below the extinction voltage. The current then ceases to flow. But this sudden pulse of current, which produces a sudden pulse of voltage across R, can be amplified electronically and made to operate a mechanical counter or to produce a click on a loudspeaker. In this manner the passage of a single particle through the counter can be registered.

The electrical pulses from the Geiger tube are usually fed to a counting circuit. In some instruments the individual pulses are actually counted. In other instruments a count-rate meter is used that integrates the pulses and gives a reading in terms of counts per minute (CPM). The count-rate meter operates as follows: the current pulses from the Geiger tube are used to charge a capacitor C that is simultaneously being discharged by a resistor R connected across it as shown in Fig. 62 (Experiment 32). An equilibrium is reached when C is being charged by the incoming pulses at the same rate that it is being discharged through R. Each current pulse lasts for some fixed time and thus delivers charge q. If the count rate is r pulses per second, then the charging current is rq amperes and is equal to the discharging current V_c/R where V_c is the voltage across the capacitor. Hence $rq = V_c/R$; $r = V_c/qR$; and the count rate is proportional to V_c, which can be measured by an appropriate voltmeter calibrated in counts per minute. The purpose of the capacitor is to smooth out the current pulses delivered by the Geiger tube so that a constant average current flows through resistor R and results in a constant voltage V_c across R and C. This voltage thus represents charge flow per second and is proportional to the number of pulses per second as already noted. The capacitance value must be large enough so that V_c is constant when the pulse rate r is constant but not so large that V_c cannot follow normal pulse rate changes. This condition is satisfied by making the time constant RC about ten times the average interval between pulses (the reciprocal of the average pulse rate). In addition, each current pulse is fed through a power amplifier to a neon indicator lamp and loudspeaker. A Geiger counter of this general type is shown in Fig. 127.

Different types of Geiger counter tubes are used for the detection of different kinds of particles. When the

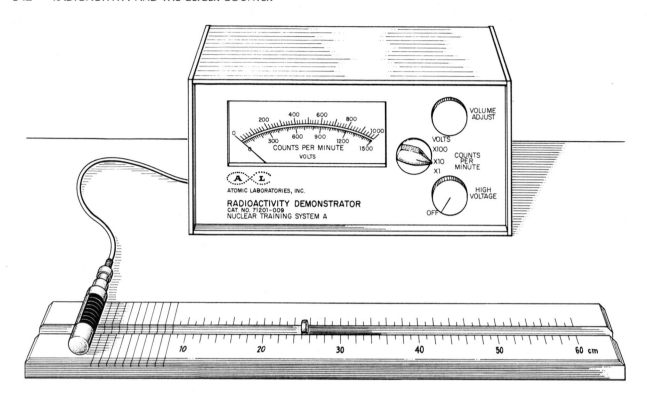

Figure 127   A Typical Geiger Counter Tube Count-Rate Meter

walls of the tube are made comparatively thick, the tube is used for the more penetrating radiations, such as X rays or gamma rays. If the tube is to be used for less penetrating radiations, such as alpha or beta particles, then very thin aluminum windows or glass bubble windows are used to permit the particles to enter the tube. One type of Geiger counter uses a thin-walled stainless steel tube for the detection of beta and gamma rays.

The proper operation of the tube as a counter is dependent upon the voltage applied to it. If the voltage is too low, the passage of radiation through the tube will not cause a current pulse. If the voltage is too high, on the other hand, there will be a discharge within the tube even without any radiation. Hence it is very important to determine the characteristic curve of the Geiger tube which is to be used in a set of measurements. This curve shows the effect of the voltage applied to the tube on its operation as a counter and clearly shows the voltage needed for proper results. Each tube has its own characteristic curve and its own best operating voltage. The curve also provides an understanding of how the Geiger counter operates.

The *characteristic curve* of a Geiger tube is obtained by measuring its counting rate when placed near a radioactive source of moderate and constant intensity.

The distance between the tube and the source is kept fixed. The counting rate is then determined for each of a series of increasing applied voltages. At low voltages no counts are produced. As the voltage is increased, a particular voltage is reached at which the tube just begins to produce counts, at a slow rate. This is the *threshold voltage* or the starting voltage. As the applied voltage is increased, the counting rate also increases, until a region is reached where the counting rate is practically constant for an increase in voltage of 100 to 200 volts. This region is known as the *plateau,* or the flat portion of the curve. As the applied voltage is increased still further, the counting rate increases rapidly, because of spurious counts. The *operating voltage* for the Geiger tube is usually taken at approximately the center of the plateau region.

When the intensity of radiation produced by a beta-gamma source is to be measured by means of a Geiger counter, the counter tube is placed at a definite distance from the source and the count rate observed. For a study of the variations in intensity with distance, the source is first placed a short distance from the tube, and this distance is then increased in regular steps. The corresponding intensity of radiation reaching the tube at each distance is recorded.

APPARATUS

1. Geiger counter tube in holder (probe)
2. Count-rate meter
3. Calibrated mounting board
4. Source of beta-gamma radiation
5. Stop watch or stop clock

PROCEDURE

1. Make sure that the count-rate meter is turned off by checking that its high-voltage control is turned fully counterclockwise so that the switch has clicked off. The instrument's line cord may then be plugged into a 115-volt, 60-hertz A.C. outlet.

2. Place the probe containing the Geiger tube in the holder on the mounting board. Leave the beta shield open and have the probe openings face along the board towards where the radiation source will be placed. The cable from the probe should be plugged into the appropriate jack on the count-rate meter.

3. Turn the meter selector switch to the "Volts" position so that the meter will indicate the voltage applied to the Geiger tube.

4. Turn the high-voltage control clockwise until the switch clicks on to turn on the instrument. Continue rotating this control clockwise while watching the meter and set the high voltage at the proper initial value. This will be given you for your particular counter tube by the instructor.

5. Turn the meter selector switch to one of the counting positions to get the count-rate meter into operating condition. The proper position will be one that sets the meter for a full-scale reading of 5000 to 10,000 cpm. For the Central Scientific Company's No. 71201, this will be the ×10 position. The instrument is now ready to indicate the rate at which the Geiger tube is counting in counts per minute.

6. Place the source of beta-gamma radiation on the mounting board and about 20 centimeters from the Geiger tube. Hold the beta-gamma source perpendicular to the table and facing the probe openings. Move the source slowly toward the tube while watching the meter reading until the meter reads about one-half full scale. Keep the positions of the source and probe fixed for Procedures 7–10.

7. Turn the meter selector switch to the "Volts" position. Turn the high voltage control to the minimum voltage. Then increase the voltage on the Geiger tube slowly until the first indication of counting appears. This is shown by a flashing of the neon bulb and by clicks from the loudspeaker. The value of the voltage at which these pulses first appear is called the Geiger tube threshold voltage. Record this voltage.

8. Increase the voltage by 20 volts and record it. Now turn the meter selector switch to the proper counting position. Observe the meter reading for several seconds and record the reading in counts per minute.

9. Turn the meter selector switch back to the "Volts" position. Increase the voltage by another 20 volts and record it. Return the switch to the counting position. Observe and record the meter reading corresponding to this voltage.

10. Repeat Procedure 9 several times, increasing the voltage by 20 volts each time and recording the meter reading corresponding to each voltage. Notice that after the first big increase in the count rate, the readings remain very nearly the same over a fairly wide range of voltage. This is the plateau region of the Geiger tube. Then, with a further increase of voltage, there may be a considerable increase in the count rate. Do *not* increase the voltage any further; lower it quickly to avoid damaging the tube and stop the taking of readings.

11. From the data of Procedures 7–10, choose a voltage in the middle of the plateau region. Record this as the operating voltage for your Geiger tube. Set the count rate meter for this voltage.

12. Hold the beta-gamma source perpendicular to the table and facing the probe openings. Move the source slowly until it is 5 centimeters from the center of the Geiger tube. Do this very carefully, and watch the meter reading to make sure that the needle does not go off scale. If the reading gets too close to full-scale deflection, use the shortest suitable distance or change scales. Observe the meter reading for this distance for a period of about 30 seconds, taking the average of the maximum and minimum values as the reading. Record this reading, which will be the counting rate of the Geiger tube at this distance.

13. Move the source until it is 6 centimeters from the center of the Geiger tube. Observe the meter reading at this distance for a period of about 30 seconds and record the average of the maximum and minimum values. This value will be the counting rate for this distance.

14. Repeat Procedure 13 for distances of 7, 8, 9, 10, 12, 14, 16, 19, 22, and 25 centimeters and record the counting rate for each distance.

15. Measure the background radiation. To do this, remove the radioactive source from the vicinity of the Geiger tube. It should be placed at least 10 feet away so that the intensity of radiation produced by it at the Geiger tube will be negligible. It will be noticed that the tube still continues to count at a slow rate. The number of pulses produced may be counted by listening to the loudspeaker or by watching the neon flasher lamp on the panel. These pulses are produced by the background radiation due mostly to cosmic rays. Count the number of pulses which occur in a one-minute interval and record this value. Do this four separate times.

DATA

Type of Geiger counter used _____   Plateau region _____ to _____

Operating voltage (recommended) _____   Operating voltage (measured) _____

Threshold voltage _____

Geiger Tube Voltage	Counting Rate in Counts Per Minute	Geiger Tube Voltage	Counting Rate in Counts Per Minute

Cosmic ray background:

Readings   1. _____ 2. _____ 3. _____ 4. _____

Average value   _____

Distance in Centimeters	Intensity of Radiation in Counts per Minute	Distance in Centimeters	Intensity of Radiation in Counts per Minute
5		12	
6		14	
7		16	
8		19	
9		22	
10		25	

Value of slope from the graph _____   Percent deviation from the theoretical value _____

CALCULATIONS

1. Plot a graph showing the relation between the counting rate and the voltage applied to the Geiger tube. Plot the voltage along the x axis and the corresponding values of the counting rate along the y axis. Choose the scales on the coordinate axes in such a way that the curve will extend over almost all of a full sheet of graph paper. Indicate the position of each point on the graph by a small pencil dot surrounded by a small circle. Then draw a *smooth* curve through the plotted points.

2. Observe the plateau region of your curve; that is, notice the region where the counting rate remains nearly the same for a wide range of voltage. This is the region where the curve is nearly horizontal. Record the width of this plateau region; that is, record the value of the voltage at the beginning and at the end of the flat portion of the curve.

3. Mark the operating voltage selected in Procedure 11 on your curve. Confirm that it is located at or near the middle of the plateau region.

4. Using the data of Procedures 12–14, plot the values of the distance of the source of radiation from the Geiger counter against the corresponding values of the radiation intensity in counts per minute on a sheet of 2-cycle full logarithmic graph paper. Read the instructions given in the Introduction for plotting such a graph. Indicate the position of each point on the graph by a small pencil dot surrounded by a small circle. Then draw the best fitting straight line through the plotted points. This line does not necessarily have to pass through all the points, but it should have, on the average, as many points on one side of it as on the other.

5. Theoretically, if the radiation intensity is inversely proportional to the square of the distance between the source and the Geiger tube, then the logarithm of the intensity should be a linear function of the logarithm of the distance with slope -2. Find the slope of the graph drawn in Calculation 4 and compare it with this theoretical value by computing the percent deviation.

6. From the data of Procedure 15, calculate the number of counts per minute produced by the background radiation by finding the average of the four readings taken.

QUESTIONS

1. How does the operating voltage that you obtained from your measurements compare with that recommended by the manufacturer?

2. Suppose that the voltage applied to the Geiger tube fluctuated by ± 20 volts about the value of the operating voltage that you determined. What percent fluctuation in the counting rate would this produce?

3. Strictly speaking, the number of counts per minute produced by the background radiation should have been subtracted from the count rates recorded in Procedures 12–14 in order to obtain the count rates due to the source alone. How big an error was introduced by neglecting to do this?

4. A source of gamma radiation is of such intensity as to produce 30,000 counts per minute on a Geiger counter placed 5 cm from the source. At what distance would the source produce 30 counts per minute?

5. Find the relation between the half-life of a radioactive isotope and its decay constant λ.

6. An example of radioactive decay is given by the radioactive gas radon, a product of radium. Radon loses its activity quite rapidly, its half-life being 3.82 days. A practical use is made of this fact in that thin-walled glass or metal tubes filled with radon are used for therapeutic work. These tubes are enclosed in hollow steel needles that may be inserted into a diseased tissue, which is thus exposed to the radiation given off by the radon. The short life of the radon is a safeguard against overexposure. Calculate the activity of the radon in these tubes after 19.1 days as compared with the initial activity.

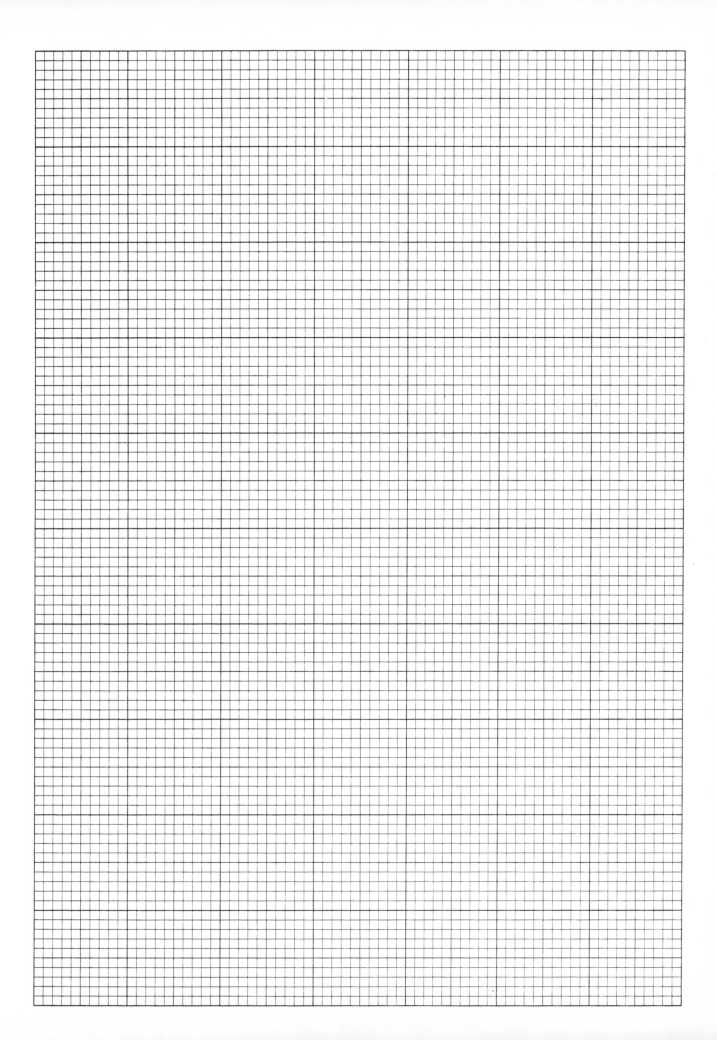

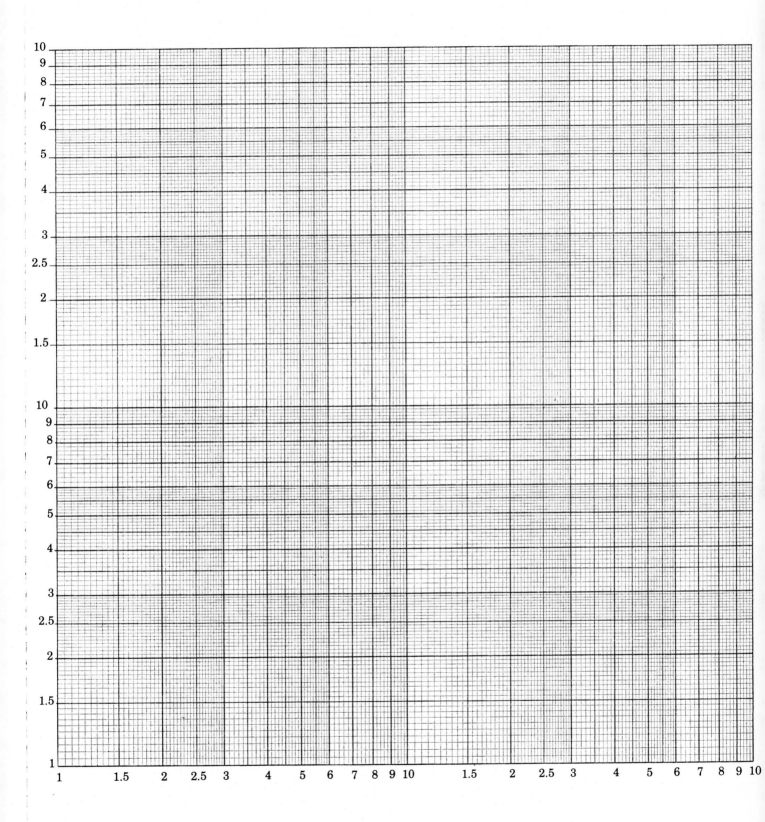

The Absorption of 47 Gamma Rays

Radioactive substances emit one or more radiations known as alpha rays, beta rays, or gamma rays. These rays differ widely in their penetrating power. Thus, a convenient way of distinguishing them is by measuring their intensity after passing through different thicknesses of material; that is, by measuring their absorption in matter. The absorption of beta rays is used in industrial applications for the automatic control of the thickness of metal or plastic films. Absorption measurements of gamma rays are used to determine the energy of the rays. It is the purpose of this experiment to study the absorption of gamma rays by aluminum plates and to determine the linear absorption coefficient and the mass absorption coefficient of aluminum for the gamma rays used.

THEORY

The radiation from radioactive materials, such as uranium, thorium, or radium, is found to contain three different kinds of rays known as alpha rays, beta rays, and gamma rays. These rays can be distinguished from each other by their penetrating power, by their relative ionizing power, and by their deflection in a magnetic or electric field. Under the influence of a transverse electric or magnetic field, the alpha rays are bent slightly in one direction, the beta rays are bent more strongly in the opposite direction, and the gamma rays are not affected at all. The alpha rays are completely absorbed by a few centimeters of air, or by an aluminum foil 0.006 centimeter thick, or by a sheet of ordinary writing paper. The beta rays are about 100 times more penetrating than the alpha rays; they are absorbed by a sheet of aluminum about 3 millimeters thick. The gamma rays are the most penetrating of all; they can be detected after passing through a block of iron 30 centimeters thick. The ionizing power of the three kinds of rays is roughly inversely proportional to their penetrating power.

The *alpha rays* consist of a stream of positively charged particles, each having a mass about four times that of a hydrogen atom and each carrying a positive charge equivalent to two electron charges. In fact, an alpha particle is identical with the nucleus of a helium atom. Alpha particles are emitted with a velocity of about one twentieth the velocity of light. *Beta rays* consist of a stream of negatively charged particles traveling at very high speed. Actually, they are nothing but high-speed electrons emitted by radioactive substances with velocities varying from about 0.2 to 0.9995 the velocity of light. *Gamma rays* are electromagnetic radiations of the same nature as X rays, but of shorter wavelength and higher penetrating power. The wavelengths of gamma rays have been measured by crystal diffraction methods and more indirectly from measurements of the velocities of ejected photoelectrons. Their wavelength varies from about 0.005 to 1.40 Å. All radioactive elements emit either alpha or beta rays, and either of these rays may be accompanied by gamma radiation.

The measurement of the penetrating power of gamma rays is carried on in the following manner: the intensity of radiation produced by a source of gamma rays is measured by means of a Geiger counter, without any absorber in the path. Then layers of various thicknesses of the absorber are placed between the source and the Geiger tube. The intensity of radiation for each thickness of the absorber is then recorded.

The intensity of a beam of gamma rays after passing through a certain thickness of the absorber is given by the relation

$$I = I_0 e^{-\mu d} \qquad (161)$$

where I_0 is the original intensity, I is the intensity after passing through the absorber, d is the thickness of the absorber in centimeters, e is the base of natural logarithms, and μ is the *linear absorption coefficient*, which is a constant characteristic of the absorbing material and of the wavelength of the gamma rays. The law of absorption is illustrated in Fig. 128.

From the above equation, by taking the natural logarithms of both sides, we get the relation

$$\ln I = \ln I_0 - \mu d \qquad (162)$$

This is the equation of a straight line. If the values of the natural logarithm of the intensity, $\ln I$, are plotted along the y axis, and the values of the thickness of the absorber are plotted along the x axis, then the intercept on the y axis is the natural logarithm of the original intensity, and the slope of the straight line is the linear absorption coefficient μ. The negative sign for the slope merely shows that the line slopes downward from left to right.

347

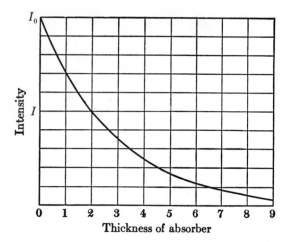

Figure 128 Absorption of Gamma Rays

As described in the Introduction, Equation 162 may also be written

$$\ln \frac{I}{I_0} = -\mu d \qquad (163)$$

and the ratio I/I_0 plotted against the absorber thickness d on semilogarithmic paper, which automatically does the job of taking the logarithm. A straight line is then obtained without the need for looking up the logarithm of each measured value of I/I_0.

The absorption of gamma rays by a certain element is simply proportional to its density in the absorbing body, irrespective of chemical combination or physical state of the element. It is convenient, then, to introduce another absorption coefficient, called the *mass absorption coefficient*. This is defined by the relation

$$\mu_m = \frac{\mu}{\rho} \qquad (164)$$

where μ_m is the mass absorption coefficient, μ is the linear absorption coefficient, and ρ is the density of the absorber. The use of the mass absorption coefficient is convenient because its value is the same, for a given element, irrespective of physical state or chemical combination of the element. However, it does depend on the wavelength of the gamma rays.

A convenient way of representing the penetrating power of a radiation is to use the half-value thickness. This is defined as the thickness of the absorber necessary to reduce the intensity of radiation by a factor of two. The half-value thickness is easily read off directly from a plot of Equation 163. It will be the value of d for which the intensity of radiation is one half of the original value, that is, for which $I/I_0 = 0.5$.

When a sensitive instrument is used to measure radiation, it is found that it will show a small reading even when no radioactive sources are in the vicinity. This small reading, known as the background, is due to cosmic rays coming from outer space and to minute traces of radioactive materials present in the laboratory. To obtain an accurate measurement of the intensity of radiation produced by a sample, a measurement is first made with the sample in position, then with the sample removed. The intensity of radiation due to the sample is obtained by subtracting the second reading, which is the background measurement, from the first.

APPARATUS

1. Geiger counter tube in holder (probe)
2. Count-rate meter
3. Source of beta-gamma radiation
4. Calibrated mounting board
5. A set of 12 aluminum plates, each about 1 millimeter thick and about 10 centimeters square
6. Vernier caliper
7. Stop watch or stop clock

PROCEDURE

1. Measure the thickness of a stack of 10 aluminum plates with a vernier caliper and record the value.

2. Make sure that the count-rate meter is turned off by checking that its high-voltage control is turned fully counterclockwise so that the switch has clicked off. The instrument's line cord may then be plugged into a 115-volt, 60-hertz A.C. outlet.

3. Place the probe containing the Geiger tube in the holder on the mounting board. Close the beta shield. The cable from the probe should be plugged into the appropriate jack on the count-rate meter.

4. Turn the meter selector switch to the "Volts" position so that the meter will indicate the voltage applied to the Geiger tube.

5. Turn the high-voltage control clockwise until the switch clicks on to turn on the instrument. Continue rotating this control clockwise until the proper Geiger tube operating voltage is obtained as read on the meter. If you have done Experiment 46 and are now using the same Geiger counter, the proper operating voltage is the one determined in that experiment. Otherwise the instructor will recommend an operating voltage for your counter.

6. Turn the meter selector switch to one of the counting positions to get the count-rate meter into operating condition. A full-scale range of about 5000 counts per minute should be chosen. For the Central Scientific Company's No. 71201 count-rate meter, either the ×1 (1000 counts per minute) or the ×10 (10,000 counts per minute) scales may be used, depending on the activity of the source. The instrument is now ready to indicate the intensity of radiation received by the Geiger tube.

7. Place the source of beta-gamma radiation on the mounting board about 20 centimeters from the Geiger tube. Hold the source perpendicular to the board and facing the tube. Move the source slowly toward the tube while watching the meter reading until the meter reads about three-quarters full scale. Keep the positions of the source and probe fixed for the remainder of the experiment. Observe the meter reading for a period of about 30 seconds, taking the average of the maximum and minimum values as the reading. Record this reading in counts per minute. This value will be the counting rate with zero as the number of plates in the absorber.

8. Place an aluminum plate between the Geiger tube and the beta-gamma source. The plate should be placed perpendicular to the board in such a way as to shield the Geiger tube from the beta-gamma source. Again, observe the meter reading for a period of about 30 seconds and record the average of the maximum and minimum values. This will be the counting rate with one plate as the absorber.

9. Repeat Procedure 8 using successive sheets of absorbing material, that is, increase the number of plates by one each time, until all of the 12 plates are used. Record the counting rate for each set of plates used.

10. Measure the background radiation. To do this, remove the radioactive source and the aluminum plates from the vicinity of the Geiger tube. The source should be placed at least 10 feet away so that the intensity of radiation produced by it at the Geiger tube will be negligible. The tube will still continue to count at a slow rate. The number of pulses produced may be counted by listening to the loudspeaker or by watching the neon flasher lamp on the panel. These pulses are produced by the background radiation due mostly to cosmic rays. Count the number of pulses which occur in a one-minute interval and record this value. Do this four separate times.

DATA

Type of Geiger counter used _____ Operating voltage _____

Cosmic ray background:
  Readings 1. _____ 2. _____ 3. _____ 4. _____

  Average value _____

Thickness of 10 aluminum plates _____ Linear absorption coefficient _____

Thickness of one plate _____ Mass absorption coefficient _____

Half-value thickness _____ Density of absorber _____

Number of Plates	Intensity of Radiation in Counts per Minute		Fraction I/I_0 of Intensity with no Plates
	Observed	Corrected	
0			1.000
1			
2			
3			

Number of Plates	Intensity of Radiation in Counts per Minute		Fraction I/I_0 of Intensity with no Plates
	Observed	Corrected	
4			
5			
6			
7			
8			
9			
10			
11			
12			

CALCULATIONS

1. From your readings of the cosmic ray background, calculate the average value of the background.

2. From each observed value of the intensity of radiation, subtract the average value of the cosmic ray background obtained in Calculation 1 to get the corrected values of the intensity of radiation.

3. Find the fractional intensity observed with each number of absorbing plates by dividing the intensity observed with the plates in position by the intensity observed with no plates. Use the corrected values of intensity. Record your results.

4. Plot an absorption curve for gamma rays on semilogarithmic graph paper. Read the instructions given in the Introduction for plotting a graph of this kind. Plot the absorber thickness (the number of plates) along the linear axis and the fractional intensity along the logarithmic axis. Indicate the position of each point on the graph by a small pencil dot surrounded by a small circle. Then draw the best fitting straight line through the plotted points. *Note:* If the gamma rays used are made up of different wavelengths, the absorption curve will not be a straight line because each wavelength will have its own characteristic value of μ. In this case, the best fitting straight line will give an average value of μ.

5. Determine the half-value thickness of aluminum for the gamma rays used. Read the number of plates corresponding to $I/I_0 = 0.5$ from the absorption curve plotted in Calculation 4. Then multiply this number (not necessarily an integer) by the thickness of one plate to obtain the half-value thickness in centimeters. Express your result in centimeters of aluminum.

6. From the slope of your absorption curve, determine the linear absorption coefficient of aluminum for the gamma rays used.

7. Use Equation 164 and your value of the linear absorption coefficient to calculate the mass absorption coefficient of aluminum for the gamma rays used.

QUESTIONS

1. Use Equation 163 to calculate the half-value thickness of aluminum for the gamma rays used in your experiment. Use the value of the linear absorption coefficient obtained in Calculation 6. To obtain the natural logarithm of a number, multiply the logarithm to the base 10 by the factor 2.303. How does this half-value thickness compare with the value you obtained in Calculation 5?

2. A beam of gamma rays of energy 1.5 Mev (million electron volts) is passed through an iron plate 10 centimeters thick. If the linear absorption coefficient of iron for these rays is 0.40 per cm, find what fraction of the beam is transmitted by the iron plate. The value of e, the base of natural logarithms, is approximately 2.718.

3. Bismuth-214 (Bi^{214}, sometimes called radium C) sends out a group of beta rays whose linear absorption coefficient in aluminum is 13.2 per cm. A beam of these beta rays is passed through an aluminum plate 3.03 millimeters thick. Find what fraction of the beam is absorbed by the aluminum plate. For the absorption of beta rays use the same relation as was used for the absorption of gamma rays.

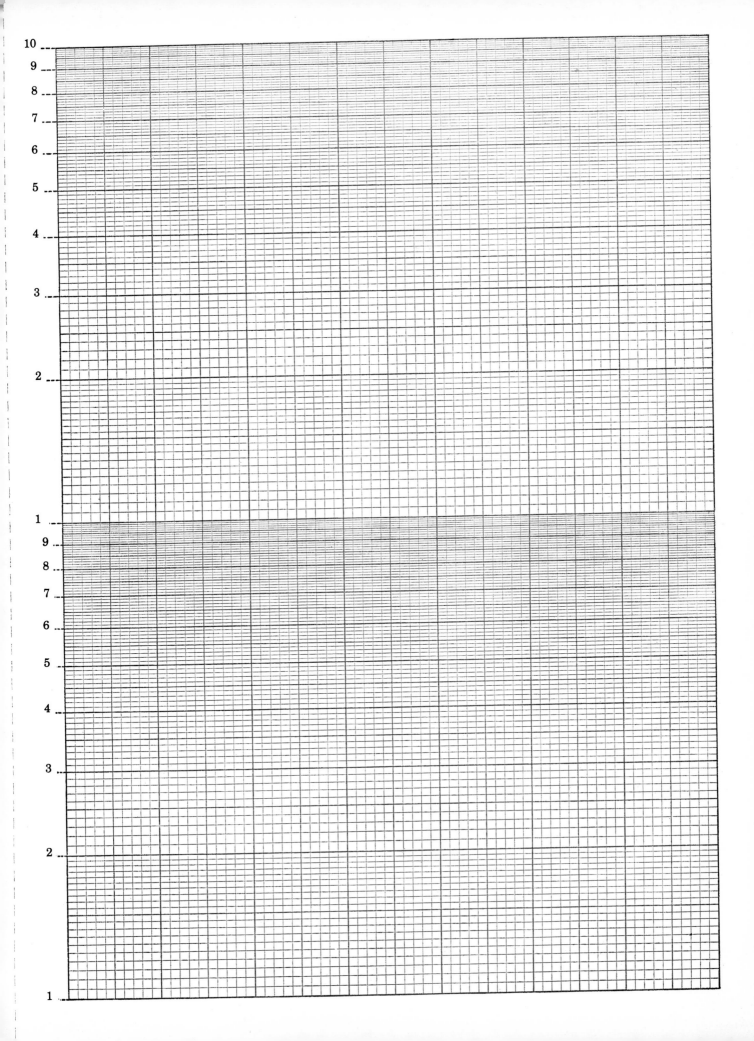

Half-Life of a **48** Radioactive Isotope

In this modern age, everyone is affected by the utilization of radioactive materials, hence some understanding of the nature of radioisotopes and their potential uses is very essential. Radioactive isotopes are widely used in a variety of practical applications in industry, medicine, agriculture, and in the physical and biological sciences. Some of the techniques that can be employed enable us to perform experiments that seem almost uncanny and permit the detection of amounts of materials far smaller than is possible by ordinary chemical means. One of the most important properties of radioisotopes is their rate of radioactive decay, which is measured in terms of their half-life. It is the object of this experiment to determine the half-life of a radioactive isotope from a series of activity measurements taken over a suitable period of time, and to calculate this isotope's decay constant.

THEORY

The nucleus of an atom is made up of a certain number of protons and neutrons. The number of protons in the nucleus is equal to the atomic number of the element, and all atoms of the same element have the same atomic number. But the atomic weight of an element is equal to the sum of the protons and neutrons in the nucleus of the element. Atoms of the same atomic number but different atomic weights are called *isotopes*. Certain combinations of protons and neutrons produce *stable* isotopes, or isotopes which do not change into other elements. Other combinations of protons and neutrons produce unstable, or *radioactive,* isotopes. The radioactive isotopes reach a stable state by a process of decay which involves the emission of an alpha or a beta particle, usually accompanied by gamma radiation. This spontaneous emission of radiation by the nucleus is called radioactivity. Hence a *radioactive isotope* is an isotope that will spontaneously disintegrate into a new element by the emission of radiation. Some isotopes of the same element are stable, such as carbon-12 and -13, and some are radioactive, such as carbon-11 and -14. Radioactive atoms have the same number of protons as the stable atoms of the same element but a different number of neutrons in the nucleus. They behave chemically and biologically the same as the stable atoms of the same element. This is what makes the tracer technique possible with radioisotopes. Since the mass of a radioactive isotope is slightly different from that of a stable isotope of the same element, their separation involves very special techniques. For example, radioisotopes may be isolated by means of a mass spectrometer, or an ultra high speed centrifuge, or by gaseous diffusion techniques.

Some radioactive isotopes occur in nature, but most of those that are now being used have to be produced artificially. Those which occur in nature are isotopes of the heavy elements which are naturally radioactive, or they are products of the disintegration of these elements. But since the discovery of artificial radioactivity, hundreds of new radioisotopes have been produced. In fact, radioisotopes of all the known elements have been made, including several new elements heavier than uranium and thus known as the *transuranic elements*. Radioisotopes can be produced artificially in one of two ways. One method is by bombarding stable elements with high energy atomic particles such as electrons, protons, neutrons, deuterons, or alpha particles. The bombardment results in adding one or more particles to the nucleus of the element, or in taking one or more particles away. In either case, the result is a change in the mass number or in the atomic number of the element. Thus a new isotope is formed, either of the same, or of a different element, and it is very likely to be a radioactive isotope. The other method is by breaking elements into pieces in a process known as *nuclear fission*. In the first process, if charged particles are used to bombard the elements, a cyclotron or some other type of high voltage accelerator has to be employed. If neutrons are used, an atomic reactor is employed. Similarly, in producing radioisotopes by nuclear fission, an atomic reactor has to be used. The fission fragments are elements near the middle of the periodic table, and they are likely to have unstable combinations of protons and neutrons; hence they themselves are very likely to be radioactive.

As the number of manufactured radioactive isotopes has increased, very many uses have been found for them. They are used in *industry* to measure radiation intensity, to study diffusion and mixing operations, in thickness gauges for paper and plastics, and in toxicology studies to

353

check the poisonous effects of many products. Radio-isotopes are used in *agriculture* to study metabolism, nutrition, soil fertility, and the action of insecticides. They are used in *clinical research* to study vitamins, hormones, the action of drugs such as antibiotics and anesthetics, and the action of viruses, bacteria, and toxicants. They are used in *medicine* in diagnosis, such as in tumor detection, and in the treatment of thyroid and heart disorders, cancer, and leukemia. In *physics,* radioisotopes are employed in studying the nature of radiations and of radioactivity, and in *physical chemistry* in studying reaction rates, molecular structure, and diffusion rates in solids and in solutions. In *biological chemistry* they are useful in research on genetics, metabolic transformations, and the mechanism of enzyme reactions.

In *analytical chemistry* different radioisotopes can be detected from their half-lives and from the different radiations they emit. In quantitative analysis, minute traces of some elements present in compounds or mixtures can be detected by these methods. In some instances quantities smaller than a billionth of a gram can be detected. In *inorganic chemistry* radioisotopes are used as *tracers* to detect impurities. In this way the degree of separation of contaminants from the desired substance can be measured. The *tracer* technique has been one of the most important developments in the studies made with radioisotopes. By means of this technique, the progress of tagged atoms (radioactive atoms) through a system can be traced. To follow the movements of individual atoms or molecules in a system, one must be able to identify them. If the system is nonhomogeneous, the changes in composition can be measured. If the system is homogeneous, the motion of individual atoms or molecules can be followed by using radioactive isotopes and detecting the radiation given off by means of a Geiger counter. For example, by using radioactive sodium it is possible to trace the motion of certain organic and inorganic chemicals passing through the human body, or through plants. The rate at which the blood stream circulates through the body is measured by feeding a person a salt solution containing radioactive sodium. As the solution enters the blood stream, it is distributed to all parts of the body. The presence of the radioactive sodium in the body can be demonstrated by detecting the radiation emitted by the sodium with a Geiger counter. The circulation of the blood through different parts of the body can thus be followed and the rate of circulation measured. A similar technique is used in the study of blood cells, viruses, bacteria, and in many other practical applications which have already been mentioned.

Even though the radioactivity of some materials, such as radium, appears to be quite constant, it has been found that all radioactivity becomes weaker as time goes on. This *radioactive decay,* or loss of activity, is quite apparent in some substances, such as polonium, for example. The rate at which a given quantity of a radioac-

tive element decays is found by measuring the intensity of radiation given off by the sample at regular intervals over a period of time, and plotting a graph like the one shown in Fig. 125. This figure shows the decay curve for the radioactive element polonium. It will be noticed that the intensity of radiation drops to half of its original value in 140 days; in another 140 days the intensity again drops to half the value it had at the beginning of this time interval; and so on. Since it would take an infinite time for the intensity to drop to zero, the rate of decay is specified in terms of the time required for the activity of a given sample to drop to half its original value. This is the half-life of the radioactive isotope and is a very convenient way of expressing its decay rate. The same law of decay applies to all the other radioactive isotopes, except that each one has its own characteristic half-life.

The half-life of a radioactive isotope is the time required for the disintegration of half of the atoms in a sample of the isotope. This is also the time required for a given quantity of radioactive material to lose half of its activity. Each isotope has a characteristic half-life that can be determined experimentally. Some isotopes have a very long half-life and some have an extremely short one. For example, the half-life of uranium-238 is 4.50 billion years, that of phosphorus-32 is 14.3 days, while that of thallium-208 is less than one millionth of a second.

The *rate* at which a radioactive material decays is an immutable physical property of each radioisotope and is characteristic of that isotope. The occurrence of radioactive disintegration is statistical in nature. For any individual atom of a given radioisotope, there is a definite probability that it will disintegrate within a certain time interval. This probability does not depend on the past history of the atom nor on the surrounding atoms. Under these conditions, the laws of statistics apply, and when we are dealing with a very large number of atoms, there is a definite probability that a specific proportion of disintegrations will occur in a given time interval.

We define the *decay constant* λ as the probability that any individual atom of a particular radioisotope will decay per unit time. Thus $\lambda \, dt$ is the probability that an atom will decay in the short time interval dt. If N is the number of such atoms present at time t, then on the average there will be $N\lambda \, dt$ disintegrations during the interval dt. This means a decrease dN in the number N in this interval, so that

$$dN = -N\lambda \, dt \qquad (165)$$

where the negative sign represents the fact that the change dN is a *decrease*.

Equation 165 may be put in the form

$$\frac{dN}{N} = -\lambda \, dt$$

and then integrated as follows:

$$\int_{N_0}^{N} \frac{dN}{N} = -\int_{0}^{t} \lambda \, dt$$

$$\ln N - \ln N_0 = -\lambda t$$

$$\ln \frac{N}{N_0} = -\lambda t \qquad (166)$$

$$\frac{N}{N_0} = e^{-\lambda t} \qquad (167)$$

Note that if Equation 167 is multiplied through by N_0, Equation 160 (already mentioned in Experiment 46) is obtained. This is the expression for radioactive decay. In these equations N is the number of radioactive atoms present at any time t, N_0 is the number present at some initial instant taken as $t = 0$, e is the base of natural logarithms, and λ is the decay constant for the particular radioisotope. Notice that with t in seconds, the dimensions of λ must be seconds $^{-1}$.

The determination of the half-life of a radioactive isotope is very important. In performing experiments with short-lived isotopes, the time that it takes to complete the experiment may be long enough to decrease appreciably the total number of radioactive atoms present and, at the same time, the intensity of radiation produced. This loss of activity has to be taken into account in making accurate measurements. Similarly, in making a chemical analysis of radioisotopes, a determination of the half-life of the radioactive sample will help to identify the isotope present, since each isotope has its own characteristic half-life.

The relation between the half-life of a radioactive isotope and its decay constant may be found by setting the ratio N/N_0 in Equation 166 equal to $\frac{1}{2}$ and solving for t. Since $\ln 0.5 = -0.6931$, $t_{\frac{1}{2}} = 0.6931/\lambda$. For isotopes with very long half-lives, such as thousands of years, λ is easily measured directly by counting the number of dis-

integrations in some short time (such as a minute) and using Equation 165, in which $-dN$ is the number of disintegrations counted, dt is the (short) counting interval, and N is the number of atoms in the sample, obtained by weighing it and dividing by the atomic weight in appropriate units. Note that dt must be in seconds in order to get λ in seconds^{-1}. For isotopes with short half-lives this procedure cannot be used because N changes by an appreciable amount during the counting period (dt is not very short compared to the half-life). In this case the activity of a given sample is measured at regular intervals over a period of hours, days, or weeks, that is, over several half-lives of the given isotope. The activity, in counts per minute, is then plotted as a function of the time, and a smooth curve is drawn through the plotted points. The half-life, being the time for the activity (i.e., the count rate) to drop to one-half its initial value, can then be read off from the curve directly. Alternatively, the logarithms of the activity measurements may be plotted as a function of the time, and the best fitting straight line then drawn through the plotted points. Because (as Equation 165 shows) the activity as represented by the count rate dN/dt is proportional to N, this procedure is tantamount to plotting Equation 166 with ln N/N_0 and t as variables. Hence the slope of the straight line is λ, from which the half-life can be calculated as already described. The need to calculate the logarithms of each count rate reading is avoided by using semilogarithmic graph paper as described in the Introduction, and this is what is usually done.

In this experiment, the half-life of radioactive iodine, I^{131}, will be determined by taking measurements of the activity of a sample at regular intervals over a period of several weeks. The net counting rate will be plotted as a function of time on semilogarithmic paper. From the resulting straight line, both the decay constant and the half-life will then be determined.

APPARATUS

1. Geiger counter tube in holder (probe)
2. Count-rate meter
3. Source of beta–gamma radiation
4. Short-lived radioisotope such as radioactive iodine (I^{131})
5. A holder such as a calibrated mounting board or a sample mount
6. Stop watch or stop clock

PROCEDURE

1. Make sure that the count-rate meter is turned off by checking that its high-voltage control is turned fully counterclockwise so that the switch has clicked off. The instrument's line cord may then be plugged into a 115-volt, 60-hertz A.C. outlet.

2. Place the probe containing the Geiger tube in the holder on the mounting board. Leave the beta shield open and have the probe openings face along the length of the board. The cable from the probe should be plugged into the appropriate jack on the count-rate meter.

3. Turn the meter selector switch to the "Volts" position so that the meter will indicate the voltage applied to the Geiger tube.

4. Turn the high-voltage control clockwise until the switch clicks on to turn on the instrument. Continue rotating this control clockwise until the proper Geiger tube operating voltage is obtained as read on the meter. If you have done Experiment 46 and are now using the same Geiger counter, the proper operating voltage is the one determined in that experiment. Otherwise the instructor will recommend an operating voltage for your counter.

5. Turn the meter selector switch to one of the counting positions to get the count-rate meter into operating condition. A full-scale range of about 5000 counts per minute should be chosen. For the Central Scientific Company's No. 71201 count-rate meter, either the × 1 (1000 counts per minute) or the × 10 (10,000 counts per minute) scales may be used, depending on the activity of the source. The instrument is now ready to indicate the intensity of radiation received by the Geiger tube.

6. Place the source of beta-gamma radiation, the long-lived reference source, on the mounting board about 20 cm from the Geiger tube. Hold the source perpendicular to the board and facing the Geiger tube. Move the source slowly toward the Geiger tube, while you are watching the meter reading, until the meter reads about three-quarters full scale. Record the exact position of the source, so that it may be placed in the same spot later. Keep the position of the source fixed while taking the reading. Observe the meter reading for a period of about 30 seconds, taking the average of the maximum and minimum values as the reading. Record this reading in counts per minute. This will be the counting rate for the long-lived reference source.

7. Remove the long-lived radioactive source from the vicinity of the Geiger tube. It should be placed at least 10 feet away so that the intensity of radiation produced by it at the Geiger tube will be negligible. Now place the short-lived source of beta-gamma radiation, the radioactive iodine, on the mounting board about 20 cm from the Geiger tube. Move the source slowly toward the Geiger tube, while you are watching the meter reading, until the meter reads about one-half full scale. Record the exact position of the source, so that it may be placed in the same spot later, keeping the position of the source fixed while taking the reading. Observe the meter for a period of about 30 seconds and record the average of the maximum and minimum readings in counts per minute. This will be the counting rate for the radioactive iodine. *Note:* If the source is in the form of a vial, it may be used in this form, provided a sufficiently large reading can be obtained on the Geiger counter and provided the source can be replaced in exactly the same position later. If the source is in a planchet, more accurate results can be obtained. But the planchet should have been already prepared by the instructor. The planchet may be kept in a small beaker, which in turn is kept in a wide-mouth glass container with a metal cap. The planchet may be handled with forceps. Provision should be made for locating the planchet accurately with respect to the Geiger tube by means of a suitable sample mount.

8. Measure the background radiation. Remove the radioactive iodine from the vicinity of the Geiger tube. It should be placed at least 10 feet away so that the intensity of radiation produced by it at the Geiger tube will be negligible. Note that the tube continues to count at a slow rate. The number of pulses produced may be counted by listening to the loudspeaker or by watching the neon glow lamp on the panel. These pulses are produced by background radiation due mostly to cosmic rays. Count the number of pulses that occur in a one-minute interval and record this value. Do this four separate times. Record the day and time of these observations.

9. Repeat Procedures 1–8 after two or three days. For Procedure 6, place the long-lived reference source in exactly the same position as before while taking the reading. Observe the meter for a period of about 30 seconds and record the average of the maximum and minimum readings in counts per minute as before. For Procedure 7, place the radioactive iodine in exactly the same position as for the first reading and measure the activity in an identical manner. Repeat Procedure 8 exactly as before. Record the day and time of these observations.

10. Repeat Procedure 9 after one week, counting from the time of the original observations. Record the day and time of the observations.

11. Repeat Procedure 9 after two weeks. Record the day and time of the observations.

12. Repeat Procedure 9 after three weeks. Record the day and time of the observations.

DATA

Type of Geiger counter used   _____   Decay constant of radioactive iodine   _____

Operating voltage   _____

Position of reference source   _____   Half-life of radioactive iodine from decay constant   _____

Position of radioactive iodine _____   Half-life of radioactive iodine, known
                                             value _____
Half-life of radioactive iodine from
decay curve _____   Percent error _____

Time of Observations			Intensity of Radiation from Reference Source in Counts per Minute		
Date	Time	Elapsed Time in Days	Counting Rate	Net Rate	Correction Factor

Time of Observations			Intensity of Radiation from Radioactive Iodine in Counts per Minute		
Date	Time	Elapsed Time in Days	Counting Rate	Net Rate	Corrected Values	Fraction I/I_0 of Initial Intensity

Time of Observations		Background Radiation				
Date	Time	Readings in Counts per Minute				Average

CALCULATIONS

1. From your readings of the cosmic ray background calculate the average value of the background radiation for each set of measurements. Use the nearest integral value for the average value.

2. Calculate the net counting rate for the long-lived reference source, at the time of each measurement, by subtracting the average value of the background reading at that time from the corresponding counting rate of the reference source.

3. Calculate the net counting rate for the radioactive iodine, at the time of each measurement, by subtracting the average value of the background reading at that time from the corresponding counting rate of the radioactive iodine.

4. If the net counting rate of the reference source is not the same at the time of all the measurements, then the values of the net counting rate of the radioactive iodine have to be corrected by multiplying each one by the proper correction factor. Calculate the ratio of the net counting rate of the reference source at the time of each set of measurements to the net counting rate of the reference source at the time of the first set of measurements. This will be the correction factor. Correct the net counting rate of the radioactive iodine by multiplying each value of the net counting rate by the value of the correction factor for that time of measurement. This will be the corrected value of the net counting rate for the radioactive iodine.

5. For each set of observations, calculate the time elapsed since the first set was taken. Calculate the time in days, express it in decimals, rounding off the figure to the nearest hundredth of a day, and record the values.

6. Find the fractional intensity observed each time the radioactive iodine was counted by dividing each corrected net count rate by the initial net corrected count rate observed on the first day. Clearly the fractional intensity for this first measurement must be 1.000. Record all your results.

7. Plot a decay curve for radioactive iodine on semilogarithmic graph paper. Read the instructions given in the Introduction for plotting a graph of this kind. Plot the time in days along the linear axis and the corresponding values of the fractional intensity along the logarithmic axis. Indicate the position of each point on the graph by a small pencil dot surrounded by a small circle. Then draw the best fitting straight line through the plotted points.

8. Determine the half-life of radioactive iodine. Read this time directly from the decay curve plotted in Calculation 7 by noting the point on the curve where the fractional intensity I/I_0 is equal to 0.5. This is the half-life of radioactive iodine as determined from the decay curve.

9. From the slope of your decay curve, find the decay constant λ for radioactive iodine.

10. Using the relation between $t_{\frac{1}{2}}$ and λ, calculate the half-life of radioactive iodine from your result in Calculation 9.

11. Compare the value of the half-life you obtained in Calculation 10 with the known value of 8.04 days for radioactive iodine by finding the percent error. How does the value calculated from λ compare with the one read directly from the decay curve in Calculation 8?

QUESTIONS

1. In the case of a short-lived radioisotope, the time needed to complete an experiment may be long enough to decrease appreciably the total number of radioactive atoms present, so that the intensity of radiation will be decreased. How long could you work with a given sample of P^{32} before it would have decayed sufficiently to make a 0.5% difference between the initial activity and the activity after a certain time t? The half-life of P^{32} is 14.3 days.

2. A source containing 10 microcuries of radioactive cobalt (Co^{60}) is stored away. What percentage of the cobalt will disintegrate in one year? The half-life of radioactive cobalt is 5.27 years.

3. A solution containing 10 microcuries of radioactive phosphorus (P^{32}) is obtained on the first of the month. What fraction of the radioactive phosphorus remains after 10 days? The half-life of radioactive phosphorus is 14.3 days.

4. A certain radioactive isotope can decay in either of two different ways characterized by decay constants (probabilities of disintegration per unit time) λ_1 and λ_2, respectively. Derive an expression for the half-life $t_{\frac{1}{2}}$ of this isotope in terms of λ_1 and λ_2.

5. A radioactive isotope has a half-life $t'_{\frac{1}{2}}$ and decays to a new isotope, which is also radioactive. This latter isotope has a half-life of $t''_{\frac{1}{2}}$, decaying in turn to a third (stable) isotope. At an initial instant $t = 0$ a certain sample contains only N_0 atoms of the original isotope. Find an expression for the number of atoms of the stable isotope present as a function of time.

The Hydrogen Spectrum. 49
The Rydberg Constant

The research work done in the field of spectroscopy, both experimental and theoretical, has contributed a great deal to our knowledge of the physical nature of things, a knowledge not only of the earth, but of the sun, of the stars, and of interstellar space. This experiment contains a discussion of the Bohr theory of the hydrogen atom, which marked the beginning of a new era in spectroscopy and atomic structure. The object of this experiment is to measure the wavelengths of the first four lines of the Balmer series of the hydrogen spectrum by means of a diffraction grating spectrometer, and from these measurements to calculate the value of the Rydberg constant.

THEORY

The so-called *bright-line* spectrum of an element is produced by exciting that element in the gaseous state by means of an electric discharge or by heating. Such a spectrum consists of bright lines, each corresponding to light of a definite color and hence wavelength. The number and relative positions of the lines depend on the element and on the method of excitation employed. Wavelengths of visible light are usually expressed in angstrom units, although in recent years there has been an attempt to make the nanometer the standard unit of wavelength for infrared, visible, and ultraviolet electromagnetic radiation. One angstrom unit is defined as 10^{-10} meter and is therefore 10^{-8} centimeter. A nanometer is 10^{-9} meter and hence 10^{-7} centimeter. Clearly there are ten angstrom units in a nanometer.

The frequencies of the light emitted by an excited atom of a gas or vapor at low pressure are characteristic of that particular atom, and the spectrum produced is the bright-line spectrum of that element. When an atom absorbs radiant energy, it absorbs the same frequencies of light it would emit if it were excited. The sun's spectrum is an absorption spectrum. The main body of the sun emits a continuous spectrum, but the relatively cooler gases surrounding the sun absorb the energy corresponding to the line spectra of the elements present in the sun's atmosphere. These absorption lines appear as dark lines or missing lines in the continuous spectrum of the sun. They were first observed by Fraunhofer and are called *Fraunhofer lines*.

It is found that among the many lines appearing in the spectrum of an element, certain ones may be picked out that are clearly related to each other, following a simple law and forming a series, and there may be many different series for the same element. Many attempts were made, over a period of years, to explain the origin of spectral lines and the various relationships found. It was thought at first that the frequencies of the light emitted by a particular element might be related like the fundamental and the harmonics of a vibrating system. But this explanation turned out to be unsuccessful. Even the one visible series in hydrogen, the simplest element and therefore the element with the simplest spectrum, defied explanation until finally in 1885 Johann Jakob Balmer (1825–1898), a Swiss school teacher and lecturer, found an empirical formula that correctly gave the wavelengths of the lines of this series. The Balmer formula may be stated as

$$\frac{1}{\lambda} = R\left(\frac{1}{2^2} - \frac{1}{n^2}\right) \qquad (168)$$

where λ is the wavelength of a visible line in the hydrogen spectrum, R is a constant called the Rydberg constant, and n takes on the integral values 3, 4, 5, etc., corresponding to the various lines. Thus setting $n = 3$ gives the wavelength of the longest wavelength (red) line H_α in the visible spectrum of hydrogen; $n = 4$ the next longest (H_β); $n = 5$ the next (H_γ); etc. For very large n the wavelength approaches a limiting short value $4/R$ called the *series limit*. Note that in Equation 168 the Rydberg constant R must have the dimensions of reciprocal length and that the unit of length chosen dictates the units in which the wavelength λ will come out.

A theoretical derivation of the Balmer formula was introduced in 1913 by Niels Bohr (1885–1962) and this proved to be the beginning of a vast development in atomic theory. Bohr took as his starting point the nuclear atomic model already proposed by Ernest Rutherford (1871–1937). This model of atomic structure postulates a heavy nucleus containing most of the atomic mass and all of the atom's positive charge and around which one or more negative electrons circulate in orbits under the influence of the electrostatic attraction between the un-

363

like charges. The situation is very simple for the case of one-electron atoms such as hydrogen, singly ionized helium, doubly ionized lithium, etc. The attractive force between the electron and the nucleus is then simply

$$F = \frac{1}{4\pi\epsilon_0} \cdot \frac{(Ze)e}{r^2} = \frac{Ze^2}{4\pi\epsilon_0 r^2}$$

where F is the force in newtons, Z is the atomic number (1 for hydrogen, 2 for helium, etc.), e is the charge on the electron in coulombs, ϵ_0 is the permittivity of free space in the rationalized MKS system, and r is the distance between the electron and the nucleus in meters. Assuming a circular orbit, we note that r is the radius of this orbit and F provides the centripetal force mv^2/r necessary to maintain it. We therefore require

$$\frac{Ze^2}{4\pi\epsilon_0 r^2} = \frac{mv^2}{r}$$

or
$$mv^2 = \frac{Ze^2}{4\pi\epsilon_0 r} \qquad (169)$$

where m is the mass of the electron in kilograms and v its speed in meters per second.

Equation 169 is a necessary relation between the speed of the electron and the radius of its orbit in the Rutherford model, but it does not predict that any particular orbit will be preferred. Moreover the model itself contains a serious flaw in that, according to classical electromagnetic theory, the radial acceleration of the charged electron should cause electromagnetic energy to be radiated continuously until the electron falls into the nucleus and the atom collapses.

Since actual hydrogen atoms neither radiate continuously nor collapse, Bohr postulated the existence of certain so-called *stationary orbits* in which the electron did not radiate classically. These orbits are specified by the *quantum condition*

$$mvr = n\frac{h}{2\pi} \qquad (170)$$

which states that the electron's angular momentum mvr cannot have any value but only certain allowed ones. These values must all be integral multiples of a basic package or *quantum $h/2\pi$* of angular momentum that cannot be subdivided. Here h is Planck's constant, and the fact that the angular momentum must be an integral multiple of $h/2\pi$ is expressed in Equation 170 by requiring that n be an integer (1, 2, 3, etc.). An integer that tells how many quanta are to be taken is called a *quantum number,* and n, the first such number in atomic theory, is called the *principal quantum number.* Notice that h must have the same dimensions as angular momentum, namely kilogram-meter²-seconds⁻¹ or joule-seconds.

Equations 169 and 170 are two simultaneous equations in v and r and may be solved simultaneously for these quantities. The result is

$$r = n^2\frac{\epsilon_0 h^2}{\pi m Z e^2} \qquad \text{and} \qquad v = \frac{1}{n} \cdot \frac{Ze^2}{2h\epsilon_0} \qquad (171)$$

These relations show that only certain values of r and v are allowed, namely those obtained when the positive integers are successively substituted for n.

Following in the footsteps of the German physicist Max Planck (1858–1947), Bohr further assumed that in transferring from one allowed orbit to another, an electron had to make the change instantaneously (in order not to be "caught" in a forbidden position) and accomplished this by either absorbing or giving up the energy difference between these two states (orbits) in a single package, that is, an energy quantum. To determine the size of this package we must investigate the energy the electron has when moving in any given orbit. Its kinetic energy is $mv^2/2$, and Equation 169 shows that this is just $Ze^2/8\pi\epsilon_0 r$. The electron also has potential energy — $Ze^2/4\pi\epsilon_0 r$ by reason of its position in the electric field of the nucleus considered as a point charge Ze. Note that the negative sign comes from assuming that the zero of potential energy is at infinity and that since the electron and the nucleus attract each other, work is done by the electric field on the electron rather than work being done on the field when the electron is brought into its orbit from infinity.

The total energy of the electron is the sum of its kinetic and potential energies and is thus given by

$$E = \frac{Ze^2}{8\pi\epsilon_0 r} - \frac{Ze^2}{4\pi\epsilon_0 r} = -\frac{Ze^2}{8\pi\epsilon_0 r}$$

Notice that it is zero when the electron is stopped at infinity, that is, when it has been just removed from the atom to form a hydrogen ion, a doubly ionized helium ion, or a triply ionized lithium ion.

Insertion of the allowed values of r from Equation 171 yields allowed values of the energy, one corresponding to each allowed orbit. These are given by

$$E_n = -\frac{1}{n^2} \cdot \frac{mZ^2e^4}{8\epsilon_0^2 h^2} \qquad (172)$$

The difference between any two of these energies may be written

$$\Delta E = E_{n_2} - E_{n_1} = \frac{mZ^2e^4}{8\epsilon_0^2 h^2}\left(\frac{1}{n_1^2} - \frac{1}{n_2^2}\right) \qquad (173)$$

and if in a change between the orbits specified by the quantum numbers n_1 and n_2 this energy difference must be absorbed or emitted in a single package, one way to do this is to have the electron absorb or emit a train of electromagnetic waves whose frequency is given by the Planck formula

$$\Delta E = h\nu \qquad (174)$$

where ΔE is the energy difference between two allowed states (orbits) in joules, h is Planck's constant in joule-seconds, and ν is the frequency of the radiation in hertz (seconds^{-1}). Substitution of Equation 174 in 173 yields

$$\nu = Z^2 \frac{me^4}{8\epsilon_0^2 h^3}\left(\frac{1}{n_1^2} - \frac{1}{n_2^2}\right)$$

which may be stated in terms of the radiation wavelength as

$$\frac{1}{\lambda} = Z^2 \frac{me^4}{8\epsilon_0^2 ch^3}\left(\frac{1}{n_1^2} - \frac{1}{n_2^2}\right) \qquad (175)$$

where c is the velocity of light and use has been made of the usual relation $\nu = c/\lambda$ between frequency and wavelength.

Equation 175 shows that if a hydrogen atom is excited to any high energy state, that is, if it has its electron placed in an orbit of relatively large radius specified by the value n_2 of the quantum number n, it may make transitions to any lower energy state (smaller radius orbit) specified by the smaller quantum number value n_1 by radiating electromagnetic waves whose wavelength is given by this equation. Various series in hydrogen are predicted by taking $Z = 1$ and setting $n_1 = 1$ while n_2 successively takes on the values 2, 3, 4, . . . for the first possible series; $n_1 = 2$ and $n_2 = 3, 4, 5, . . .$ for the next; $n_1 = 3$ and $n_2 = 4, 5, 6, . . .$ for the next; etc. The situation is illustrated in Fig. 129, which depicts the Bohr circular orbits schematically (not to scale) and shows various possible transitions arranged according to series. Notice that the Balmer formula, Equation 168, is identical with Equation 175 when $Z = 1$, $n_1 = 2$, and the Rydberg constant is defined as

$$R = \frac{me^4}{8\epsilon_0^2 ch^3}$$

Agreement between accurately measured values of the Rydberg constant and values calculated from this relation using the best available numerical data for the fundamental physical constants is remarkably close. This is particularly true if account is taken of the fact that the nucleus is not truly stationary but rotates along with the electron about the atom's common center of mass. To do this, a so-called "reduced mass" is calculated for the electron, which compensates for the assumption that the nucleus is a fixed point. The effect decreases for heavier nuclei, since the atom's center of mass then comes closer and closer to coinciding with the nucleus, but there is nevertheless an observable shift in the value of the Rydberg constant as applied to hydrogen, deuterium, singly ionized helium, etc., as shown in the following table:

Rydberg constant for hydrogen	$R_H = 109,677.58$ cm^{-1}
Rydberg constant for deuterium	$R_D = 109,707.42$ cm^{-1}
Rydberg constant for ionized helium	$R_{He} = 109,722.27$ cm^{-1}
Rydberg constant for infinitely heavy nucleus	$R_\infty = 109,737.31$ cm^{-1}

Notice that the change occurs in the fifth significant figure and thus affects only very precise measurements.

In addition to the excellent agreement between the theoretical and experimental values of the Rydberg constant, Bohr's prediction of other series besides that of Balmer has been verified. Thus the Lyman series ($n_1 = 1$; $n_2 = 2, 3, . . .$), the Paschen series ($n_1 = 3$, $n_2 = 4, 5, . . .$), the Brackett series ($n_1 = 4$, $n_2 = 5$, 6, . . .), and even the Pfund series ($n_1 = 5$, $n_2 = 6$, 7, . . .) have all been observed since Balmer's time and are found to follow Equation 175 with great precision. Special equipment is required to observe these series,

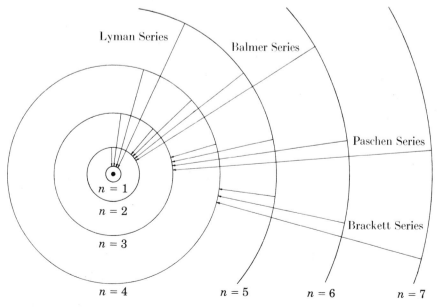

Figure 129 Bohr Circular Orbits of Hydrogen with Transitions for Different Series of Lines

Energy in ergs, *W*   Quantum number, *n*

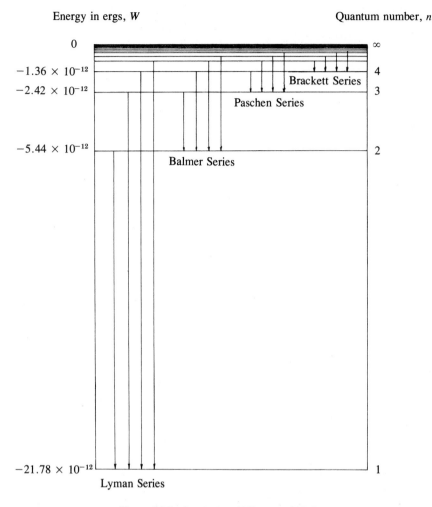

Figure 130   Energy Level Diagram of Hydrogen

however, since the Lyman lines are all in the ultraviolet region of the spectrum and the other series are all in the infrared. Thus the Balmer series forms the only visible part of the hydrogen spectrum, and even its shorter wavelength lines fall in the ultraviolet. In the present experiment only the four longest wavelength Balmer lines will be studied. In order of decreasing λ, these are often designated H_α, H_β, H_γ, and H_δ.

The presentation of energy states and the transitions between them can be made much more quantitative than is possible with a diagram like Fig. 129 by using an *energy level diagram*. Figure 130 shows the energy level diagram for hydrogen. The allowed energies are plotted on a vertical scale, a horizontal line being drawn at the appropriate level for each such energy and accordingly called an energy level. The levels are labeled both with the appropriate energies computed from Equation 172 and with the corresponding quantum numbers. When transitions are indicated by vertical lines drawn between the levels involved, the length of each line is proportional to the frequency of the emitted radiation and hence inversely proportional to the wavelength. Fig. 130 clearly indicates why the Lyman series should be in the

ultraviolet and the Paschen and Brackett series in the infrared. The energy level diagram greatly simplifies the description of the spectrum. It must be remembered that each series is an aggregate of infinitely many lines and that the whole spectrum contains an infinite number of series. An energy level diagram allows the whole spectrum to be simply represented by only one aggregate of infinitely many levels. From these levels, all the lines of all the series can be derived.

In the present experiment a diffraction grating will be used to measure the hydrogen wavelengths. The theory and operation of such a grating is fully covered in Experiment 44, which should be reviewed in preparation for the present work. If you have not done that experiment, read the theory and procedure sections, noting Equation 159 in particular. You should also note, however, that the setup in Experiment 44 was an extremely elementary one that enabled you to concentrate your attention on the operation of the grating. In particular, the method used for determining the diffraction angle θ was not capable of great precision. We shall, therefore, now use a spectrometer similar to that studied in Experiment 42 with the transmission grating of Experiment 44 sub-

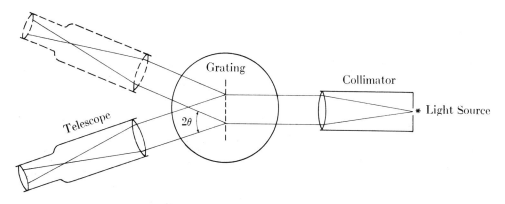

Figure 131 The Diffraction Grating Spectrometer

stituted for the prism, as shown in Fig. 131. Here again the collimator contains a slit through which light from the source enters the spectrometer and a lens whose function is to render the rays coming from the slit parallel. Compare this arrangement with that used in Experiment 44 to direct approximately parallel rays at the grating. The grating is mounted on the rotating table in the middle of the spectrometer, which is adjusted so that the plane of the grating is at right angles to the incoming rays. The diffracted image of the slit is then viewed with the telescope, which is rotated until the image falls on the cross hairs in the eyepiece. The angle θ can then be read directly from the spectrometer's divided circle rather than having to be calculated from length measurements, as in Experiment 44. Note that with the grating, an absolute measure of the wavelength λ forming a particular image of the slit can be obtained if the grating constant is known. No calibration curve is required as it was with the prism in Experiment 42. A grating spectrometer is used in the present work both for this reason and to give the opportunity for experience with this important instrument.

APPARATUS

1. Spectrometer, for prism or grating
2. Diffraction grating
3. Grating holder
4. Atomic hydrogen discharge tube
5. Discharge-tube power supply
6. Tube holder

PROCEDURE

1. The spectrometer should be already adjusted by the instructor, with the collimator and telescope properly focused and aligned, and the spectrometer table leveled so that its axis of rotation is the same as the axis of the spectrometer. CAUTION: Do not handle the spectrometer at all until the instructor has demonstrated its operation to you, for it is very easy to upset the adjustments. Then, be extremely careful in operating it.

2. Place the diffraction grating in its mounting at the center of the spectrometer table. The unruled side of the grating should face the collimator, and the plane of the grating should be made as nearly perpendicular to the axis of the collimator as can be judged with the eye. The lines on the grating should be parallel to the slit of the collimator, and the center of the grating itself should be as nearly as possible in the center of the spectrometer table.

3. Place the stand containing the hydrogen discharge tube in front of the spectrometer, with the discharge tube just outside the collimator slit (see Fig. 132). Connect the leads from the tube holder terminals to the high-voltage power supply. Adjust the spectrometer so that the collimator points directly at the discharge tube's center. Close the power supply switch. CAUTION: Do not touch the discharge tube or the high-voltage wires while the current is on.

4. Observe the central image produced by the diffraction grating. To do this, turn the telescope into line with the collimator and view this image through the eyepiece. It will be an image of the collimator slit and will have the same color as the hydrogen discharge. The slit should be made as narrow as possible and yet remain visible in the telescope. Make sure the telescope is adjusted so that when the eyepiece is focused on the cross hairs, the image of the slit is also in focus.

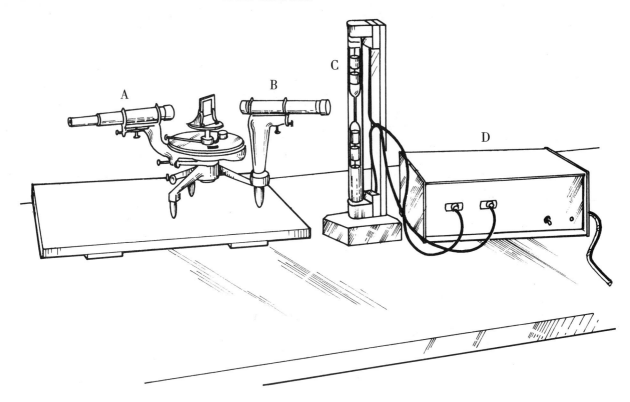

Figure 132   Arrangement of Apparatus for the Diffraction Grating Spectrometer: (A) Telescope, (B) Collimator, (C) Discharge Tube, (D) Power Supply

5. Move the telescope to the right so as to view the first-order hydrogen spectrum. Adjust the position of the telescope until the red hydrogen line falls exactly on the intersection of the cross hairs of the telescope.

6. Directly under the arm supporting the telescope is a clamping device that rotates with the arm and permits the telescope to be clamped by means of a radial screw. A tangent screw on the arm allows a fine adjustment of the setting of the telescope; however, the tangent screw is used only after the telescope has been clamped with the radial screw. Clamp the telescope by means of the radial screw and locate the position of the red hydrogen line by setting the intersection of the cross hairs exactly on the red line by means of the tangent screw. Record the setting of the telescope to the nearest minute of arc, reading the divided circle with the help of the vernier.

7. Repeat Procedures 5 and 6 for the blue-green, the blue, and the violet hydrogen lines. Record the setting of the telescope to the nearest minute of arc for each of the lines measured.

8. Move the telescope to the left of the central image, so as to view the left-hand first-order hydrogen spectrum. Adjust the position of the telescope until the red hydrogen line falls exactly on the intersection of the cross hairs of the telescope. Record the setting of the telescope to the nearest minute of arc.

9. Repeat Procedure 8 for the blue-green, the blue, and the violet hydrogen lines. Record the setting of the telescope to the nearest minute of arc for each of the lines measured.

D A T A

Number of lines per inch on the grating	_____	Accepted value of Rydberg constant	_____
Grating constant in centimeters	_____	Percent error	_____
Calculated value of Rydberg constant	_____		

Hydrogen Lines		Setting of Telescope		Calculated Values			
Wavelength Angstrom Units	Color	Right	Left	θ	Sin θ	Computed Wavelength	Percent Deviation
6562.8	red						
4861.3	blue-green						
4340.5	blue						
4101.7	violet						

CALCULATIONS

1. Compute the grating constant, that is, the distance between the lines on the grating in centimeters, from the number of lines per inch stated on the grating.

2. Compute the angle of diffraction θ for each of the four lines measured. The angle of diffraction is one half of the angle between the two settings of the telescope, for each line.

3. Compute the value of sin θ for each of the four lines measured.

4. Compute the wavelength of each line, using the value of the grating constant previously calculated and applying Equation 159. The wavelength should be expressed in angstrom units, evaluated to four significant figures.

5. Calculate the percent deviation of the computed values of the wavelengths from the known values.

6. Calculate the value of the Rydberg constant from the computed wavelength of each line.

7. Find the average value of the computed Rydberg constant.

8. Calculate the percent deviation of the computed value of the Rydberg constant from the known value.

QUESTIONS

1. Calculate the angular separation between the D lines of sodium for a diffraction grating of 15,000 lines per inch in the first-order spectrum. Calculate the separation in the second-order spectrum. The wavelengths of the lines are 5890 and 5896 Å respectively.

2. Using the Balmer formula, calculate the wavelength of the fifth line of the Balmer series of hydrogen lines.

3. Explain the significance of the Fraunhofer lines in the spectrum of the sun.

4. Calculate the wavelength of the first line of the Balmer series in deuterium, or heavy hydrogen. Use the Balmer formula, which gives the reciprocal wavelength, $1/\lambda$, and calculate the wavelength in angstroms to five significant figures. For the Rydberg constant, use the value appropriate for deuterium. Also find the wavelength difference in angstroms between this line and the corresponding hydrogen line by using the known value of the red hydrogen line. It was the accurate measurement of these lines, and the interpretation given by the Bohr theory, that led to the discovery of heavy hydrogen by Urey, Brickwedde, and Murphy in 1932.

5. A famous constant called the *fine structure constant* may be defined as the ratio of the speed of the electron in the lowest ($n = 1$) Bohr orbit to the speed of light. Derive an expression and calculate the numerical value of the fine structure constant. Also calculate the reciprocal of this value. It is a famous number which you may already have heard of.

6. Suppose an electron makes a transition between two energy states in hydrogen both of which have very high quantum numbers n_1 and n_2 but which are close together, that is, for which $n_2 - n_1$ is a small integer like 1, 2, or 3. Show that the frequency of the light radiated is approximately an integral multiple or harmonic of the fundamental frequency of revolution of the electron in its lowest orbit. This is an example of Bohr's *correspondence principle,* which states that at high quantum numbers the quantum picture must become identical to the classical one.

7. (a) What is meant by an atom's *ionization energy?* (b) What is the ionization energy of hydrogen? (c) Show that the *series limit* of the Lyman series involves an energy identical to the hydrogen ionization energy.

8. What lines in the spectrum of ionized helium lie in the visible range (4000 to 7000 Å)?

The Ratio e/m for Electrons 50

The physical nature of cathode rays is revealed by the deflections they undergo as they pass through electric and magnetic fields. The direction of these deflections shows that these rays consist of particles that have a negative electric charge and a definite mass. The investigations of J. J. Thomson in 1897 led to a measurement of the ratio of charge to mass for these particles, which have been called electrons. The results of these measurements showed that the atom is not indivisible, that the electron is part of an atom, and that it is much smaller than the smallest atom. The object of this experiment is to determine the value of the ratio of charge to mass of an electron. From the known value of the charge on the electron and the value of e/m obtained with this experiment, the mass of the electron can be computed.

THEORY

One of the simplest methods of measuring the ratio e/m, or the *ratio of charge to mass* of an electron, was devised by Lenard, and consists in measuring the deflection of a beam of electrons in a known magnetic field. The beam of electrons is bent into a circular path. From the known values of the accelerating potential, the strength of the magnetic field, and the radius of curvature of the path, the value of e/m can be calculated.

When an electron or any other charged particle moves in a magnetic field in a direction at right angles to the field, it is acted upon by a force perpendicular to the direction of the field and the direction of motion of the particle. The force on an electron traveling perpendicular to a magnetic field is given by

$$F = Bev$$

where B is the strength of the magnetic field, e is the charge on the electron, and v is the velocity of the electron.

Since the force is always perpendicular to the electron's direction of motion, it makes the electron move in a circular path whose plane is perpendicular to the direction of the magnetic field. The force required to keep a body moving in a circle is

$$F = m \frac{v^2}{r}$$

where F is the force in newtons, m is the body's mass in kilograms, v is the speed in meters per second, and r is the radius of the circle in meters. This required centripetal force is provided by the force exerted on the electron by the magnetic field. Therefore, we have $mv^2/r = Bev$ or

$$\frac{mv}{r} = eB \tag{176}$$

Note that the *momentum mv* of the electron determines the radius of its path in a given magnetic field. Note also that in our MKS system of units the field strength B must be measured in webers/meter², or (as they are now called) teslas.

When an electron is accelerated by a potential difference V, it gains a kinetic energy equal to the work done on it by the field. Because the potential difference is by definition the work done per unit charge, the work done on the electron is Ve. Therefore, the kinetic energy $\frac{1}{2}mv^2$ of the electron is given by

$$\tfrac{1}{2} mv^2 = Ve$$

where m is the mass of the electron in kilograms, v is its velocity in meters per second, V is the accelerating potential in volts, and e is the electron's charge in coulombs. Solving, we find that the velocity acquired by the electron is

$$v = \sqrt{2Ve/m}$$

and that substitution in Equation 176 yields

$$\frac{e}{m} = \frac{2V}{B^2 r^2} \tag{177}$$

This expression gives the *ratio of charge to mass* of an electron in terms of the accelerating potential, the strength of the magnetic field, and the radius of the circular path of the beam of electrons. When these quantities are known, the value of e/m can be computed.

The apparatus to be used for measuring the ratio of e/m is based on a design originally proposed by K. T. Bainbridge. It consists of a large vacuum tube mounted at the center of a pair of Helmholtz coils that provide the uniform magnetic field needed for the operation of the tube. The arrangement is shown in Fig. 133. The beam of electrons is produced by an electron gun composed of a straight filament parallel to the axis of the coils and surrounded by a coaxial anode containing a single slit parallel to the filament. Electrons emitted from the heated

373

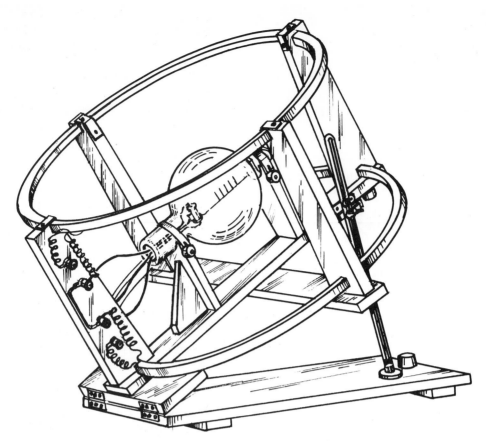

Figure 133   *e/m* Vacuum Tube and Helmholtz Coils

filament are accelerated by the potential difference applied between the filament and the anode. Some of the electrons come out as a narrow beam through the anode slit. The tube contains a trace of mercury vapor that serves to render the path of the electrons visible. When the electrons collide with the mercury atoms, some of the atoms will be ionized; as these ions recombine with stray electrons, the characteristic mercury light is emitted. Since the recombination and the emission of light occur very near the point where ionization took place, the path of the beam of electrons is visible as a bluish-white streak.

A pair of Helmholtz coils is a particular arrangement of two coaxial, circular coils of radius *a* with their planes parallel and separated by a distance *a* equal to the radius of the coils. This arrangement is very useful because it produces an almost uniform magnetic field over a fairly large region near the center of the coils. The magnetic field at the center is parallel to the axis of the coils, and its magnitude is given by

$$B = \frac{8\mu_0 NI}{\sqrt{125}a} \qquad (178)$$

where *B* is the magnetic field strength in teslas, *N* is the number of turns of wire on each coil, *I* is the current through the coils in amperes, and *a* is the mean radius of each coil in meters.

Each coil of the pair of Helmholtz coils used in this experiment has 72 turns of copper wire with a resistance of approximately 1 ohm. The mean radius of the coil is 0.33 meters. The coils are supported in a frame which can be adjusted with reference to the angle of dip so that the magnetic field produced by the coils will be parallel to the earth's magnetic field, but in the opposite direction. A graduated scale indicates the angle of tilt.

The magnetic field of the Helmholtz coils causes the beam of electrons to move in a circular path whose radius decreases as the magnetic field increases. The field can be adjusted until the sharp outer edge of the beam coincides with the outer edge of one of the five bars spaced at different distances from the filament. The vacuum tube used in this experiment has been so designed that the radius of the circular path can be conveniently measured. In its manufacture, five cross bars have been attached to the staff wire, and this assembly is attached to the filament assembly in such a way that the distance between each cross bar and the filament is accurately known. The distance in centimeters between the filament and the outer edge of each bar is given below. These distances are the diameters of the circles which the electron beam will be made to describe.

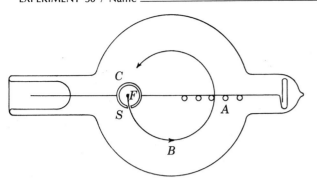

Figure 134 Sectional View of *e/m* Tube and
Filament Assembly

Crossbar Number	Distance to Filament
1	6.48 cm
2	7.75 cm
3	9.02 cm
4	10.30 cm
5	11.54 cm

Fig. 134 is a sectional view of the tube and filament assembly. *A* represents the five cross bars attached to the staff wire; *B* shows a typical path of the beam of electrons; *C* represents the cylindrical anode, with the slit *S*; and *F* represents the filament.

The field circuit for the Helmholtz coils is shown in

Fig. 135. The voltage supply E_3 consists of a 12-volt storage battery or a power supply capable of delivering 5 amperes of filtered direct current continuously. Resistance R_2 is the tubular rheostat of about 200 ohms resistance. R_3 is the tubular rheostat of about 10 ohms resistance. *A* is a D.C. ammeter (0–5 amperes).

The circuit for the *e/m* tube is shown in Fig. 136. The voltage supply E_1 consists of a "B" battery that will deliver 22.5 volts or a 30-volt D.C. power supply. The voltage supply E_2 consists of a 6-volt storage battery or power supply capable of delivering 5 amperes of filtered direct current continuously. Resistance R_1 is the tubular rheostat of about 5 ohms resistance. *A* is a D.C. ammeter (0–10 amperes).

The action of the *e/m* tube and Helmholtz coils can be used in explaining the principle of the *mass spectrometer*. Equation 177 for *e/m* shows that charged particles having the same ratio of charge to mass will follow the same circular path when they are accelerated by a given voltage and then acted upon by a given magnetic field. However, if the particles all have the same charge but different masses, they will follow circular paths of different radii. Thus, by measuring the radius of a particular path, the mass of the particles in that path can be determined very accurately. This is essentially the principle on which the mass spectrometer operates.

The accepted values of *e*, *m*, and *e/m* for the electron are as follows:

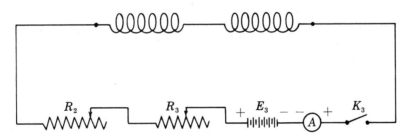

Figure 135 Field Circuit for the Helmholtz Coils

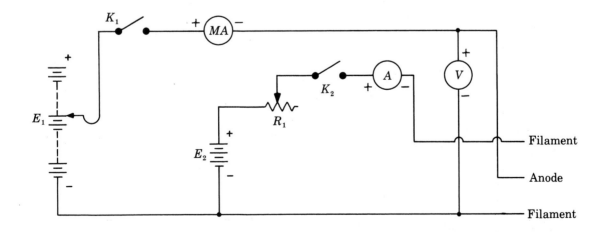

Figure 136 Circuit for the *e/m* Tube

Charge on the electron:	$e = 1.602 \times 10^{-19}$ coulombs
Mass of the electron:	$m = 9.11 \times 10^{-31}$ kilograms
Ratio of charge to mass for the electron:	$e/m = 1.759 \times 10^{11}$ coulombs/kilogram

APPARATUS

1. e/m tube mounted between Helmholtz coils
2. 6-volt D.C. power supply or storage battery
3. 12-volt D.C. power supply or storage battery
4. Tubular rheostat of about 200 ohms resistance
5. Tubular rheostat of about 10 ohms resistance
6. Tubular rheostat of about 5 ohms resistance
7. "B" battery ($22\frac{1}{2}$ and 45 volts), or 30-volt D.C. power supply

8. D.C. voltmeter (0–50 volts)
9. D.C. ammeter (0–10 amperes)
10. D.C. ammeter (0–5 amperes)
11. D.C. milliammeter (0–10 milliamperes)
12. Three single-pole, single-throw switches
13. Compass
14. Dip needle

PROCEDURE

1. The Helmholtz coils will be already connected in the circuit as shown in Fig. 135. The e/m tube will be already connected in the circuit as shown in Fig. 136. These connections should be checked and approved by the instructor. All switches should be open.

2. The Helmholtz coils have been designed to provide the uniform magnetic field required in the operation of the tube. The axis of the coils may be inclined to the dip angle for quantitative measurements in such a way that the magnetic field produced by the coils may be adjusted to cancel the earth's magnetic field. This is done by making the magnetic field produced by the coils parallel to the earth's magnetic field, but oppositely directed. First orient the Helmholtz coils so that the e/m tube will have its long axis in a magnetic north-south direction, as determined by a compass. Then measure the magnetic dip at this location by using a good dip needle. Finally tip the coils up until the plane of the coils makes an angle with the horizontal equal to the complement of the angle of dip. The axis of the coils will now be parallel to the earth's magnetic field. For example, in New York State the magnetic dip is about 74° and thus the plane of the coils should make an angle of about 16° with the horizontal. A graduated scale connected to the apparatus indicates the angle of tilt of the plane of the coils. If the dip needle is now placed in a north-south direction on the inclined board supporting the coils, the needle will point along the axis of the coils.

3. Make sure that the axis of the coils has been set so that it points along the earth's magnetic lines of force. Now check to see if the magnetic field produced by each of the coils is in the direction opposing that of the earth's field. To do this, send a small current through the Helmholtz coils. Make sure that there is maximum resistance in the field circuit by checking on the setting of both rheostats R_2 and R_3, and then close switch K_3. Check the direction of the magnetic field produced by the current in each coil by placing a compass very close to the coil, to see in what direction the needle is deflected when the current is turned on. From this, the direction of the magnetic field produced by the coils can be determined. The direction of the magnetic field produced by each coil should be upward, along the axis of the coils. This will be opposite to the direction of the earth's magnetic field, which is downward along the axis. If one of the coils produces a field in the wrong direction, the current through that coil should be reversed. Now open switch K_3.

4. Apply the accelerating potential to the anode of the tube by closing switch K_1. If a "B" battery is used, this will be 22.5 volts. If you have an adjustable power supply, set it between 25 and 30 volts.

5. Make sure that there is maximum resistance in the filament circuit by checking on the setting of rheostat R_1. Now close switch K_2 in the filament circuit. Start with a low filament current and increase it very slowly and carefully by gradually decreasing resistance R_1 until the proper electron emission is obtained. This is checked by watching the milliammeter in the anode circuit, which reads the plate current. As soon as anode current begins to appear, the filament current should be increased extremely slowly and carefully, while the value of the anode current is continually watched. When the anode current reaches a value of 6 milliamperes, do not increase the filament current any further. This will avoid overheating the filament and possibly burning it out. Record the value of the anode current and of the filament current.

6. The thin path of light produced in the path of the beam of electrons as they pass through the mercury vapor should now be clearly visible. We shall merely call this path the electron beam. If the tube has been properly rotated so that the crossbars extend upward from the staff wire, the electron beam should be horizontal. But the electron beam will be deflected slightly toward the base of the tube by the earth's magnetic field. To balance out the effect of the earth's field, a small current should be sent through the Helmholtz coils. Make sure that there is maximum resistance in the field circuit by checking on the setting of both rheostats R_2 and R_3. Now close switch K_3 and send a small current through the coils. If the magnetic field of the coils tends to straighten the beam, the coils are properly connected. If the field increases the deflection toward the base, the field current is flowing in the wrong direction.

7. Adjust the value of the field current until the electron beam is straight. Slowly increase the value of the current by gradually decreasing resistance R_2. Keep watching the electron beam and slowly increase the current until the beam is perfectly straight, and perpendicular to the long axis of the tube. This adjustment should be made very carefully and accurately. When the electron beam is straight, the magnetic field of the coils is just equal to the earth's magnetic field, so that the resultant field at the electron beam is zero. Record this value of the current as I_1. This current will be constant throughout the experiment, since the accelerating voltage is kept constant.

8. Increase the field current slowly until the electron beam describes a circle. The current is increased by gradually decreasing resistance R_2. Keep on decreasing only resistance R_2 until it is completely out of the circuit. Then begin decreasing resistance R_3 and from then on, use only resistance R_3 to control the current. This is done to prevent overloading R_2, which cannot carry too big a current. Adjust the value of the current carefully until the bright, sharp outside edge of the beam just clears the outside edge of crossbar number 5, the one that is farthest away from the filament of the tube. Record this value of the current as I_2 in the table. Also record the accelerating voltage indicated by the voltmeter.

The outside edge of the beam is used because it contains the electrons with the greatest velocity. The electrons emitted from the negative end of the filament fall through the greatest potential difference between filament and anode, and thus have the greatest velocity. This is the potential difference that the voltmeter measures, and it is the one to be used in the calculations. The electron beam spreads as it goes around the tube, but all of the spreading is toward the inside of the circle. This spreading is produced by electrons that have a smaller velocity and by electrons that have a component of velocity parallel to the magnetic field.

9. Keep on increasing the field current slowly and adjust its value carefully until the bright, sharp outside edge of the electron beam just clears the outside edge of crossbar number 4. Again record this value of the current as I_2 in the table. Record the accelerating voltage to be sure there has been no change. Repeat the same procedure for crossbars 3, 2, and 1 and record the value of the current I_2 and the accelerating voltage in each case.

DATA

Number of turns on each coil	_____	e/m calculated value		_____
Mean radius of each coil	_____	e/m accepted value		_____
Filament current	_____	Percent error		_____
Anode current	_____			

Crossbar Number	Radius of Circle	Anode Voltage	I_1	I_2	I	B	e/m
1							
2							
3							
4							
5							

CALCULATIONS

1. Calculate the *radius* of the circle of the electron beam for each crossbar distance.

2. For each value of the field current I_2 subtract the current I_1 to obtain the value of the current I, which is the current needed to produce the magnetic field that deflects the electrons.

3. For each value of the field current I, calculate the field strength B of the magnetic field produced within the Helmholtz coils.

4. For each circular path described by the electron beam, compute the value of *e/m* by using the appropriate value of the accelerating potential, the magnetic field strength, and the radius of the circle.

5. Find the average of the values of *e/m* obtained in Calculation 4 and record this as the calculated value of *e/m*. Compare your result with the accepted value by finding the percent error.

QUESTIONS

1. The oil-drop experiment gives a value of 1.60×10^{-19} coulombs of charge on the electron. Using this value and your calculated value for *e/m*, compute the mass of the electron.

2. Name two sources of error in your measurements which are most likely to affect the accuracy of your results.

3. Compute the velocity of an electron which has been accelerated through a difference of potential of 100 volts. Express your result in centimeters per second.

4. Calculate the strength of the magnetic field required to bend electrons that have been accelerated through a potential difference of 100 volts into a circle 10 centimeters in diameter. Express your result in gauss, a commonly used unit of magnetic field strength. 10^4 gauss = 1 tesla.

5. Show how the action of the *e/m* tube and Helmholtz coils can be used in explaining the principle of a mass spectrometer. In a mass spectrometer, isotopes of the same element can be separated. Charged particles carrying the same charge, but differing in mass, can be produced and studied in this type of tube.

6. Calculate the expected spread in the beam at crossbar number 5 due to the voltage drop across the filament. Do this by noting that the accelerating voltage between the anode and the negative end of the filament is the voltage read on the voltmeter, whereas the voltage between the anode and the positive end is the voltmeter reading minus the filament voltage. Estimate the filament voltage by taking an approximate value for the setting of R_1, finding the drop across this resistance due to your recorded filament current, and subtracting this drop from the battery voltage E_2.

7. Show that if the magnetic field is held constant, the time T required for an electron to make a complete circle in your *e/m* tube and return to the anode is independent of the accelerating voltage by deriving an expression for this time. The reciprocal $f = 1/T$ is called the electron's *cyclotron frequency*.

The Photoelectric Effect. 51
Planck's Constant

The quantum theory has revolutionized our ideas of the nature of radiation, the nature of atoms and molecules, and the interaction between radiation and atomic particles. One of the cornerstones of modern physics, the quantum theory, was introduced by Max Planck (1858–1947) in an attempt to derive a law of blackbody radiation that would agree with experimental observations. Planck made some entirely new and radical assumptions about the nature of radiation. He assumed that any radiation of a definite frequency cannot be emitted or absorbed in arbitrary amounts but must instead be always emitted or absorbed in whole multiples of a discrete quantity of energy. This quantity of energy, called a quantum, is proportional to the frequency of the radiation. The constant of proportionality is a universal constant, now known as Planck's constant.

Making use of the quantum idea, Albert Einstein (1879–1955) developed the law of the *photoelectric effect*. He assumed that light of a definite frequency can transfer its entire quantum of energy to a single electron, even though the electron may be far from the source of radiation. This assumption extended the quantum theory to the transfer of radiant energy through space by means of light quanta. The purpose of this experiment is to verify Einstein's theory of the photoelectric effect and to determine experimentally the value of Planck's constant.

THEORY

One of the earliest applications of the quantum theory was in the explanation of the photoelectric effect. This effect was discovered by Heinrich Hertz in 1887, when he observed that an electric spark started more readily when the electrodes of a spark gap were exposed to ultraviolet light. The effect was investigated one year later by Hallwachs, who found that a freshly polished zinc plate that was insulated and charged negatively would lose its charge when exposed to ultraviolet light; he also noted that there was no effect if the charge on the plate was positive. Ten years later, both J. J. Thomson and P. Lenard showed that the action of the light caused emission of free negative charges from the metal surface. These charges were found to be the same as other electrons, but they were called *photoelectrons*. Usually, electrons are liberated from substances only by ultraviolet light. This is true for nearly all of the known metals. However, a few elements, namely the alkali metals lithium, sodium, potassium, rubidium, and cesium, are exceptions, for they will eject photoelectrons even when visible light falls on them.

The photoelectric effect is the emission of electrons from a metallic surface by the incidence of a beam of light. The number of electrons emitted per second is directly proportional to the intensity of the light. The energy of the electrons emitted depends only on the frequency of the light and the kind of metal used. It does not depend on the intensity of the light. Early in the study of photoelectricity, certain characteristics of the photoelectric effect were clearly established:

1. The photoelectric current is directly proportional to the intensity of the light falling on the emitting surface.
2. The maximum kinetic energy of the ejected photoelectrons does not depend on intensity of light, but it does depend on frequency; for a monochromatic beam of light, the maximum kinetic energy of the electrons increases with the frequency of the light.
3. For a particular metal, there is a definite cutoff frequency ν_0, below which no photoelectric effect occurs.
4. There is no detectable time lag between the impinging of light on the surface of the metal and emission of the photoelectrons.

For physicists of the nineteenth century, the phenomena of interference and diffraction had firmly established the wave theory of light. Similarly, the discovery of electromagnetic waves by Heinrich Hertz in 1886, and his experimental work with these waves, further confirmed Maxwell's electromagnetic (wave) theory of light. But physicists found that the photoelectric effect could not be explained on the basis of this theory. According to the wave theory, the kinetic energy of the photoelectrons should have increased as the intensity of the light increased. However, the kinetic energy of the

381

photoelectrons was independent of the intensity of the light, depending only on the frequency. According to this theory also, the photoelectric effect should have occurred for any frequency of light, provided that the light had sufficient intensity. However, for each metal there was a characteristic cutoff frequency below which the photoelectric effect did not occur, no matter how intense the illumination. According to the wave theory, the energy of light waves was distributed equally over the wave front. Hence, if an electron obtained its energy by the ordinary process of absorption, there should have been a considerable time lag between the beginning of illumination and the start of the photoelectric current. But there was no such detectable time lag. Thus, it could only be concluded that the electromagnetic theory of light failed to explain the photoelectric effect.

To remedy this situation, Einstein introduced a fundamental change in the concept of the nature of light. Making use of Planck's quantum idea, Einstein postulated that radiation is not a smooth, continuous flow of energy as it would be according to the wave theory. He viewed it instead as a series of discontinuous, concentrated packages of energy, called *photons*. Thus a photon was a single quantum of electromagnetic radiation. The energy of a single photon was determined as

$$E = h\nu$$

where E is the energy of the quantum in joules, ν is the frequency of the radiation, and h is Planck's constant. Einstein further assumed that the energy in each photon was so concentrated that the photon must either transfer all its energy to a single electron or "miss" the electron and transfer none whatever. These ideas of Einstein concerning the nature of light extended the quantum theory to the transfer of radiant energy through space by means of light quanta.

By using the extension of the quantum theory just described, Einstein arrived in 1905 at a satisfactory explanation of the photoelectric effect. He assumed that in the photoelectric process a whole quantum of radiant energy was absorbed by a single electron. He assumed further that a certain amount of energy was required to liberate the electron from the metal, and that any additional energy the electron absorbed was used to give it its kinetic energy. Applying the principle of conservation of energy, Einstein obtained the expression

$$h\nu = W + \tfrac{1}{2}mv^2 \; + \tfrac{1}{2}mv^2 \tag{179}$$

This is the *Einstein photoelectric equation* and represents energy transfer between a photon and an electron. The first term, $h\nu$, is the total energy content of a single quantum of light, the incoming photon, where h is Planck's constant and ν is the frequency of the light. The term W is the energy needed to get the electron free from the atoms and away from the metal surface; it is called the *work function* of the metal. The last term, $\tfrac{1}{2}mv^2$, is the kinetic energy of the ejected electron, where m is the mass of the electron and v is its maximum velocity.

Einstein's application of the quantum theory to the photoelectric effect explains in a simple way all the characteristics that could not be explained by the wave theory of light. According to Einstein's theory, the high light intensity means a large number of quanta passing a point per second in a light ray. Thus, the photoelectric current per second in a light ray. Thus, the photoelectric current is directly proportional to the intensity of the light, because the number of photoelectrons emitted (current) is proportional to the number of photons reaching the surface. The maximum kinetic energy of the electrons ejected from a surface by a monochromatic beam of light increases with the frequency of the light. This is explained simply by the original assumption of the quantum theory — that the energy in a quantum of radiation is directly proportional to the frequency — and by Einstein's assumption that a whole quantum of energy is absorbed by a single electron. The theory accounts for the existence of a low frequency limit for the photoelectric effect, because the quantity of energy W is the least amount of energy that an electron can receive and still escape from the surface of the metal. The energy that an electron receives is proportional to the frequency of the light; if the frequency is less than a certain lower limit, the energy the electron receives is less than the amount needed for the electron to escape. Thus the theory predicts what is observed — that the work function W is a constant characteristic of the metal, while Planck's constant h is a universal constant that depends on the nature of light. Finally, there is no measurable time lag between the arrival of the first photons and the emission of the photoelectrons. This confirms the assumption of the *photon theory*, that the energy in each photon is so concentrated that the photon can transfer its whole energy content to a single electron. Thus Einstein's application of the quantum theory to the photoelectric effect not only explains the effect completely but gives additional insight into the nature of light.

Experimental investigations of the photoelectric effect were carried out in 1912 by A. L. Hughes, and by O. W. Richardson and K. T. Compton. Their experiments showed that the energy of the photoelectrons increased proportionately with the frequency of the light, and that the constant of proportionality was approximately equal to Planck's constant h. This was the first experimental confirmation of Einstein's photoelectric equation. In 1916, R. A. Millikan carried out many experiments that established the photoelectric equation completely and led to a very accurate determination of the value of Planck's constant. Millikan carefully planned and executed an extensive series of experiments. He systematically studied the photoelectric effect by varying separately the intensity of the light, the wavelength, and the metal illuminated. In the first set of experiments, Millikan varied the intensity of the light, while keeping the metal surface and the wavelength of the light unchanged.

When monochromatic light falls on a metal plate, it will liberate photoelectrons, which can be detected as a current if a potential difference V is applied between the metal plate, which is the cathode, and a collecting wire, which is the anode. When the anode is positive with respect to the cathode, it attracts the negative electrons, and as V is increased the resulting photoelectric current reaches a certain limiting value at which all the photoelectrons ejected from the cathode are collected by the anode. This current was found by Millikan to be directly proportional to the intensity of the light over an extremely wide range of values. If, on the other hand, V is made negative, the photoelectric current continues to flow in the same direction as before but decreases as the negative value of V is increased until a value V_s of V is reached at which the current drops to zero. V_s is called the *stopping potential* and is a measure of the kinetic energy of the fastest-ejected photoelectrons. This is because when the anode is negative, the electric field opposes the motion of the electrons, but for small values of V some still get through to the anode, showing that they are emitted from the cathode with a finite velocity. As V is made more and more negative, only the faster-moving electrons reach the anode, until when $V = V_s$ even the fastest ones are stopped before being collected, hence the term "stopping potential." From measurements of the stopping potential, Millikan found that the maximum velocity and thus the maximum kinetic energy of the ejected photoelectrons is independent of the intensity of the light for light of a particular wavelength.

If monochromatic light of different wavelengths is used, the stopping potential is found to have a definite value for each wavelength. If the stopping potential is plotted as a function of the frequency of the light, in the case of an individual metal, a linear relation is found to exist between these variables. Millikan discovered that for an individual metal, such as sodium, a straight line of definite slope and definite intercept is obtained. These results can be simply explained in terms of Equation 179. Here $\frac{1}{2}mv^2$ is the kinetic energy of the photoelectrons emitted by the cathode, and we recall that the stopping potential V_s is the *retarding* voltage required to just stop the fastest electron before it gets to the anode. Therefore, in going to the anode from the cathode this fastest electron has done work eV_s against the electric field and has given up its entire kinetic energy $\frac{1}{2}mv^2$ in doing so. Had $\frac{1}{2}mv^2$ been greater than eV_s, the electron would still have been moving when it got to the anode and would thus have been collected there. Had $\frac{1}{2}mv^2$ been less than eV_s, the electron would have stopped a long way from the anode. Because V was adjusted to just barely cut off the collected current, we know that the fastest electrons stop just before getting to the anode, and we can write

$$\frac{1}{2}mv^2 = eV_s$$

The photoelectric equation can then be written as

$$h\nu = W + V_s e$$

Division by the electronic charge e and rearrangement of terms leads to

$$V_s = \frac{h}{e}\nu - \phi \qquad (180)$$

where V_s is the stopping potential, ν is the frequency of the light, e is the charge on the electron, and $\phi \equiv W/e$ is the work function of the cathode material expressed in volts. Note that ϕ is the work required to bring a coulomb's worth of electrons (6.25×10^{18} electrons) through the cathode surface, whereas W is the work required to extract a single electron. Since work functions are usually listed in volts, it is the quantity ϕ that is of particular interest.

Equation 180 is the equation of a straight line with slope h/e and intercept $-\phi$. A plot of corresponding values of frequency and stopping potential should therefore be a straight line from whose slope Planck's constant h may be obtained. Moreover, the intercept will give the work function ϕ of the cathode material. Notice that if a frequency ν_0 is defined such that $\phi = h\nu_0/e$, then Equation 180 can be written as

$$V_s = \frac{h}{e}\nu - \frac{h}{e}\nu_0 \qquad (181)$$

which says that the stopping potential just becomes zero when $\nu = \nu_0$. In other words, ν must be equal to or greater than ν_0 for the photocathode to emit any electrons at all. The quantity ν_0 is therefore called the *photoelectric threshold frequency*.

In his final set of experiments, Millikan varied the metal exposed to the light. He found that if the stopping potential were plotted as a function of the frequency of the light, a straight line plot could be obtained for each metal. All of these lines were noted to have the same slope as the one obtained for sodium, but lines for different metals had different intercepts. Such results meant that h was a constant and the same for all metals, or a universal constant, while the work function ϕ was a constant characteristic of the metal. The value of the photoelectric threshold frequency ν_0 lay in the ultraviolet for most metals, but for the alkali metals and for barium and strontium it lay in the visible region, for potassium it lay in the red, and for cesium, in the infrared. From the results of his systematic series of experiments, Millikan was able to verify Einstein's photoelectric equation completely, in every detail.

This experiment involves an investigation of some characteristics of the photoelectric effect. A monochromatic beam of light of constant intensity is sent into a vacuum photoelectric cell and made to strike the cell's cathode. Photoelectrons are emitted from the cathode surface and are collected by the anode (collecting wire) of the cell. This flow of electrons constitutes the photoelectric current. We shall study the variation of this current as the voltage between the cathode and the anode is varied. When the potential of the anode is positive with respect to the cathode, the photoelectric current flows. Even when the potential is zero, some current flows because the photoelectrons are emitted with enough kinetic energy to be able to reach the anode. When the potential is made negative with respect to the cathode, the current decreases until a large enough value of negative potential is reached, at which point the photoelectric current drops to zero. This is the stopping potential V_s. We shall measure V_s for different wavelengths of light.

The best light source for our purpose is a high-intensity mercury arc lamp. This source will produce very intense mercury lines that are far enough apart to be separated and used as an intense monochromatic beam of light. To accomplish this, two types of light filters are used. One type transmits only light of a desired wavelength. The other, known as a cutoff filter, transmits light of wavelengths down to a certain limit but no light from below this limit. Both types can be used in this experiment, but those given preference in the Apparatus Notes (after Appendix Tables) are of the kind that transmit only a desired wavelength. If a filter transmits more than one line, the intense line of shortest wavelength is the one to be used, because the shortest wavelength controls the stopping potential. This is the way the filters are labeled. In measuring the stopping potential for different wavelengths of light, we use the appropriate filter for each wavelength.

For each filter used and for light of a definite wavelength, a curve is plotted to show the relation be-

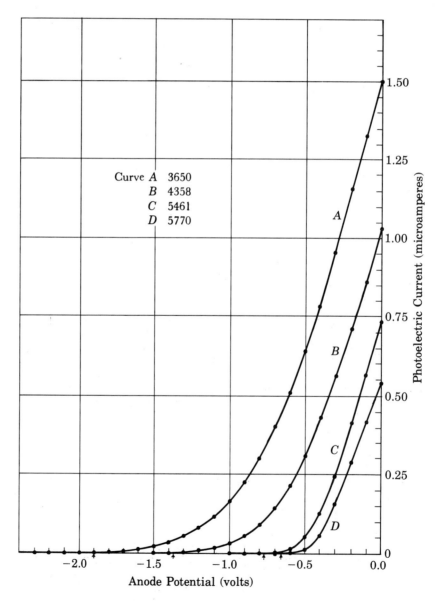

Curve A	3650
B	4358
C	5461
D	5770

Figure 137   Variation of Photoelectric Current with Anode Potential for Light of Different Wavelengths

tween anode voltage and photoelectric current. The general appearance of these curves is shown in Fig. 137, in which the curves obtained for four different wavelengths are plotted from data obtained under actual laboratory conditions. From curves such as these, the value of the stopping potential corresponding to each wavelength can be read off. Another curve is then plotted showing the relation between the frequency of the light and the stopping potential. This curve turns out to be a straight line. Such a curve is shown in Fig. 138, which was plotted from the data obtained from Fig. 137. Such a straight line can easily be shown to correspond to one plotted from the final form of Einstein's photoelectric equation (Equation 180). Therefore, from the slope of this straight line, the value of Planck's constant h can be determined. How this is done can best be explained by the following

example, which uses the straight line of Fig. 138. The slope m of the straight line is given by

$$m = \frac{V_1 - V_0}{\nu_1 - \nu_0} \qquad m = \frac{2.00 - 0.00}{(8.45 - 3.60) \times 10^{14}}$$

$$m = 0.4124 \times 10^{-14}$$

The value of Planck's constant h is given by

$$h = m \times 1.602 \times 10^{-19}$$

$$h = 0.4124 \times 10^{-14} \times 1.602 \times 10^{-19}$$

$$h = 6.60 \times 10^{-34} \text{ joule-sec}$$

This is the calculated value of Planck's constant, which in this case comes to within less than 1% of the accepted value.

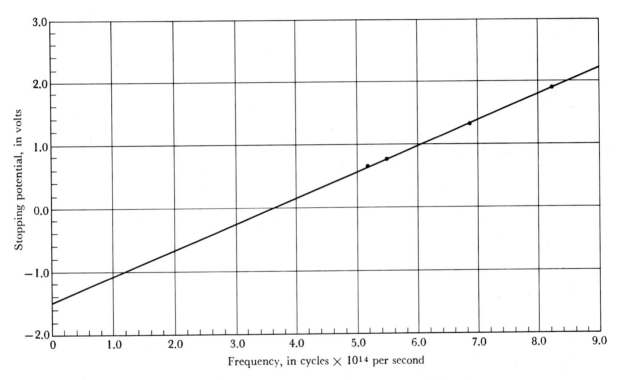

Figure 138   Variation of Stopping Potential with Frequency of the Light

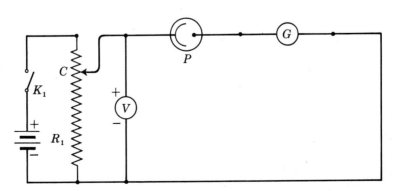

Figure 139   Circuit Diagram for Planck's Constant Experiment

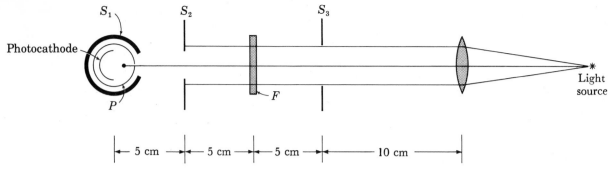

Figure 140 Optical Arrangement for Measurements on the Photoelectric Effect

The circuit and components used in this experiment are shown in the schematic diagram of Fig. 139. The voltage supply consists of two good dry cells of 1.5 volts each, connected in series. Resistance R_1 is a 5000-ohm rheostat of a type often called a "volume control" because it is used for this purpose in radio and television sets. Its sliding contact is moved by a control shaft on which a large knob can be mounted to allow accurate adjustment. V is a D.C. voltmeter (0–3 volts), P is a standard vacuum phototube (type 929), and G is a very sensitive current meter. An electronic instrument called a picoammeter (because it can read currents as low as a few times 10^{-12} ampere) will be used.

The optical arrangement for this experiment is shown in Fig. 140. The light source is a high-intensity mercury arc lamp, which is enclosed in a metal housing provided with an adjustable projection lens. The lamp is mounted on an adjustable support stand. The lens should be adjusted to produce a parallel beam of light.

The remainder of the apparatus is mounted on the auxiliary optical bench, which is 50 cm long and provides an easy and convenient means of lining up the optical parts. The vacuum phototube P is an RCA type 929, which has an octal base like the type 6AU4GT discussed in Experiment 35. The octal base is described in Procedure 1 of that experiment, and the connection diagram for the type 929 is given in Fig. 141. Note that this is a *bottom* view of the base. The socket into which the phototube is plugged is mounted on a wooden block provided with a metal rod that fits the carrier supports on the optical bench. Care should be taken in making connections to the tube socket to remember that when the socket is viewed from above, the pin numbering proceeds counterclockwise and is thus the *reverse* of that shown in Fig. 141. Be careful to select the right terminals and avoid short circuits to adjacent terminals. The phototube is covered by a cardboard light shield S_1 that admits light only through a 1.5 × 2.5 cm aperture. The shield rests on the wooden block and is oriented so that the aperture is in front of the anode and thus directs light to the active cathode surface.

S_2 is an iris diaphragm mounted on one of the carrier supports. Its purpose is to regulate the amount of light entering the phototube and also to prevent the entrance of any light that has not passed through the filter F. The latter is mounted on another carrier support by means of a lens holder. S_3 is a screen having an aperture whose purpose is to regulate the amount of light delivered to the light filter from the source.

The pieces of apparatus on the optical bench should all be lined up accurately so that the central ray of the light beam entering the phototube passes through the centers of all the apertures. Similarly, the height of the mercury arc lamp should be properly adjusted so that the beam of light coming from it is accurately centered on the optical system's axis.

When the optical arrangement described above is used, the experiment does not have to be performed in a dark room. It can be performed in a dimly lighted room with very good results, since not enough stray light can enter the phototube to spoil the accuracy of the measurements. However, you should realize that this experiment is an excellent test of your abilities as an experimenter; good results are in direct proportion to the care exercised in setting up the apparatus and taking the necessary readings.

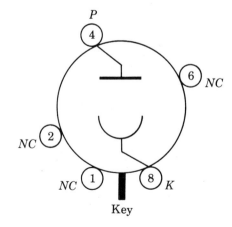

P = Plate or Anode
K = Cathode
NC = No Connection

Figure 141 Base Connection Diagram for the Type 929 Phototube

APPARATUS

1. Photoelectric cell (vacuum phototube type 929)
2. Mercury arc lamp, high-intensity
3. Light filters for the mercury lines: yellow, 5770; green, 5461; blue, 4358; ultraviolet, 3650
4. Picoammeter
5. Auxiliary optical bench, 50 cm long
6. Iris diaphragm
7. Two lens holders
8. Two screen holders
9. Four optical bench carrier supports
10. Mounted socket, 8-pin (octal) for phototube
11. Screen with aperture for S_3
12. Volume-control-type potentiometer (5000 ohms, wirewound)
13. D.C. voltmeter (0–3 volts)
14. Two good dry cells of 1.5 volts each
15. Single-pole, single-throw switch

PROCEDURE

1. Record the type of photoelectric cell supplied to you and the type of light filter to be used for each of the four mercury lines. Turn on the picoammeter, allow it to warm up, and check the zero setting by placing the range switch in the zero position and adjusting the zero control for zero meter reading. The range switch should be kept in this position except when actual measurements of the photoelectric current are being made.

2. Begin the experiment by putting the 5770 filter in place on the optical bench. The optical apparatus should already have been arranged on the bench as shown in Fig. 140 and lined up so that the center of the light beam entering the phototube passes through the center of all the apertures. The mercury arc lamp should be placed and its height adjusted to line it up with the axis of the optical system.

3. Close the switch in the phototube circuit (K_1 in Fig. 139). The circuit connections are such that a negative voltage is applied to the anode of the phototube. The cathode is connected to the positive terminal. The voltmeter reads the positive voltage on the cathode, which is also the value of the negative voltage on the anode. Adjust the rheostat until the voltage on the anode of the phototube is zero as read on the voltmeter.

4. Plug the cord connected to the mercury arc lamp into the 115-volt A.C. line. Allow about 3 minutes for the lamp to reach high intensity. Its lens should now be adjusted to produce a parallel beam of bluish-white light aimed straight down the centerline of the apertures. This beam will be limited by screen S_3 so that a beam of rectangular cross section reaches the filter. This is a cutoff light filter that does not allow any wavelengths below 5770 Å to pass through it. The light passing through the filter and reaching the second screen and finally, the phototube, will be yellow in color. Open wide the aperture of the iris diaphragm S_2. The entire system should be well centered so that the maximum amount of light reaches the cathode of the phototube. The room should be dimly lighted so that there will not be enough stray light entering the phototube to spoil the accuracy of the results.

   With the anode voltage set at zero, switch the picoammeter to a range on which you can read 2 microamperes (2×10^{-6} ampere). Then slowly adjust the iris diaphragm aperture until a reading of between 1.5 and 2.0 microamperes is obtained. Record this reading to the nearest tenth of a scale division, estimating the tenth of a division carefully.

5. Do not disturb the setting of the iris diaphragm or move any of the apparatus while making the remainder of the measurements with the 5770 filter. Set the anode voltage at exactly 0.10 volt by adjusting the rheostat. The picoammeter reading should now be lower than before if the connections to the phototube have been properly made. Record the new picoammeter reading to the nearest tenth of a scale division.

6. Set the anode voltage at exactly 0.20 volt and record the picoammeter reading obtained. Proceed to increase the anode voltage in 0.10-volt steps until a reading of 1.5 volts is reached. Record the indicated photoelectric current to the nearest tenth of a scale division for each setting of the anode voltage. You will find that as the anode voltage is increased, the photoelectric current decreases until it becomes zero for some definite value of the anode voltage. Switch the picoammeter to a more sensitive scale as required and make sure to take the reading near the zero value of the current carefully and accurately to the nearest tenth of a division. *Note:* You may find that the photoelectric current does not approach zero as a limit but approaches some other value close to zero. This is usually due to some spurious current produced by emission from the anode. Some spurious current can be detected in almost all cases if the sensitivity of the picoammeter is turned up far enough. Do not attempt to use a range over a factor of 10 more sensitive than the one you used to observe the photoelectric current with no anode voltage in Procedure 4.

7. Turn the picoammeter range switch to its zero position. Remove the 5770 filter from the optical bench and replace it with the 5461 filter. This is a filter that lets through only the green mercury line at 5461 Å, or else it does not allow any wavelengths below 5461 Å to pass through it. Make sure that the filter is properly lined up so that the center of the light beam passes through the center of the filter. With the anode voltage still at 1.50 volts, set the picoammeter to a range on which you can read 2 microamperes. Slowly reduce the anode voltage to zero while observing the picoammeter reading. With the anode voltage set at zero, the iris diaphragm aperture should be carefully adjusted until a reading of between 1.5 and 2.0 microamperes is obtained. Record this reading to the nearest tenth of a scale division.

8. Do not disturb the setting of the iris diaphragm or move any of the apparatus while making the remainder of the measurements with the 5461 filter. Set the anode voltage at exactly 0.10 volt and record the new picoammeter reading to the nearest tenth of a scale division. Then increase the anode voltage in 0.10-volt steps until a maximum reading of 1.50 volts has been reached. Record the picoammeter reading to the nearest tenth of a scale division for each setting of the anode voltage. Make sure to take the readings near the zero value of the current very carefully and accurately to the nearest tenth of a scale division. Switch the picoammeter to a more sensitive range if necessary to get good readings of the small current values, but do not increase the meter sensitivity by more than a factor of ten.

9. Repeat Procedures 7 and 8 with the 5461 filter replaced by the filter for the blue mercury line at 4358 Å. The iris diaphragm aperture should be set to give a photoelectric current between 2.0 and 2.5 microamperes with zero anode voltage when the 4358 filter is used, and the anode voltage should be increased in 0.10-volt steps until a maximum of 2.00 volts has been reached.

10. Repeat Procedures 7 and 8 using the 3650 filter. This is the ultraviolet filter UV-1 that transmits the mercury line at 3650 Å. The iris diaphragm aperture should be set to give a photoelectric current between 2.0 and 2.5 microamperes with zero anode voltage, and the anode voltage should be increased in 0.10-volt steps until a maximum of 2.50 volts has been reached.

11. Return the picoammeter range switch to the zero position. Shut off the picoammeter and the mercury arc and disconnect the battery by opening switch K_1.

DATA

Type of photoelectric cell used _____

Type of light filters used for the mercury lines:

Yellow, 5770 _____

Green, 5461 _____

Blue, 4358 _____

Ultraviolet, 3650 _____

Work function of the photocathode _____

Calculated value of Planck's constant _____

Accepted value of Planck's constant _____

Percent error _____

Mercury Lines		Frequency	Stopping Potential
Wavelength, Angstrom Units	Color	Cycles Per Second	Volts
5770	Yellow	5.20×10^{14}	
5461	Green	5.50×10^{14}	
4358	Blue	6.88×10^{14}	
3650	Ultraviolet	8.22×10^{14}	

Anode Voltage	Photoelectric Current, Microamperes			
	Filter 5770	Filter 5461	Filter 4358	Filter 3650
0.00				
−0.10				
−0.20				
−0.30				
−0.40				
−0.50				
−0.60				
−0.70				
−0.80				
−0.90				
−1.00				
−1.10				
−1.20				
−1.30				
−1.40				
−1.50				
−1.60				
−1.70				
−1.80				
−1.90				
−2.00				
−2.10				
−2.20				
−2.30				
−2.40				
−2.50				

CALCULATIONS

1. Plot a curve to show the relation between the anode voltage and the photoelectric current for light of a definite wavelength. Using the data for the ultraviolet mercury line at 3650 Å, plot the anode voltage along the x axis and the corresponding values of the photoelectric current along the y axis. Use a volt for the unit of voltage and a microampere for the unit of current. Choose the scale on the graph paper in such a way that the curve will cover practically the whole page (see Fig. 137.). Keep in mind that the anode voltages are negative. Each plotted point should be a small dot surrounded by a small circle, so that the point will be clearly visible on the graph. A smooth curve should be drawn through the plotted points in such a way as to fit them as closely as possible. Plot a second curve on the same piece of graph paper, with the same coordinates scale as before, using the data for the green mercury line at 5461 Å. Again plot the anode voltage

along the x axis and the corresponding values of the photoelectric current along the y axis. Draw a smooth curve through the plotted points. The two curves will be distinct, so that it will be easy to notice the effect of a change of wavelength on the stopping potential.

2. Plot a third curve on a separate piece of graph paper with the same coordinates and the same scale as before. For this plot, the data for the blue mercury line at 4358 Å are to be used. As before, plot the anode voltage along the x axis and the corresponding values of the photoelectric current along the y axis. Draw a smooth curve through the plotted points. Plot a fourth curve, using the same piece of graph paper, the same coordinates, and the same scale as for the third curve, but using the data for the yellow mercury line at 5770 Å. Plot the anode voltage along the x axis and the corresponding values of the photoelectric current along the y axis as before. Draw a smooth curve through the plotted points. Again, the two curves will be distinct, and the effect of a change of wavelength on the stopping potential will be easily seen.

3. From each of the four curves obtained in Calculations 1 and 2, read off and record the value of the stopping potential corresponding to each wavelength. The cutoff voltage, or the stopping potential, is the voltage at which the photoelectric current becomes zero, for light of a given wavelength. It is the voltage needed to prevent the fastest-moving electrons produced by light of a given frequency from reaching the anode. This voltage can be read from any curve that shows the relation between the anode voltage and the photoelectric current. *Note:* If the photoelectric current does not approach zero as a limit, but some other value close to zero, the stopping potential is taken as the point at which the curve is tangent to the direction of the x axis for each individual curve. This point should be read as carefully as possible to the nearest hundredth of a volt.

4. Plot another curve to show the relation between the frequency of the light and the stopping potential. For each wavelength used, plot the value of the frequency of the light corresponding to that wavelength along the x axis, and the corresponding value of the stopping potential along the y axis. Choose the scale on the coordinates in such a way that the value of the frequency can be plotted from 0 to 9.0×10^{14} cycles per second, and the stopping potential from -2.0 to $+3.0$ volts. Notice that in this curve the stopping potential is plotted as a positive voltage, which is the voltage actually read by the voltmeter. This voltage is a measure of the kinetic energy of the fastest-moving photoelectrons ejected by light of a given frequency. Draw a smooth curve through the four plotted points, extending it to both the x and y intercepts (see the curve shown in Fig. 138.).

5. From the curve obtained in Calculation 4, read the value at the point of intercept of the straight line with the vertical ($\nu = 0$) axis. According to Equation 180, this value is the negative of the work function of the photocathode material in volts. Record your result for the work function so obtained.

6. Compute the value of Planck's constant h from the slope of the straight line obtained in Calculation 4. To do this pick two widely separated points (ν_1, V_1) and (ν_2, V_2) on the line. The slope m is then given by

$$m = \frac{V_2 - V_1}{\nu_2 - \nu_1}$$

Notice that, as shown by Equation 180, this slope is h/e and in fact has the dimensions of volt-seconds or joule-seconds/coulomb. Multiply your result by the known value 1.602×10^{-19} coulomb for the electronic charge to get Planck's constant in joule-seconds. Use at least three significant figures in your calculations and record the resulting value of h.

7. The accepted value of Planck's constant is $h = 6.626 \times 10^{-34}$ joule-seconds. Compare your calculated value with the accepted value by finding the percent error.

QUESTIONS

1. Show how the frequencies listed for each of the mercury lines in the data table were obtained from the respective wavelengths, and check that the given values are correct.

2. What do you think is the biggest source of error in determining the stopping potential for photoelectrons produced by light of a given wavelength? Explain.

3. Using the value of the work function found in this experiment, calculate the corresponding photoelectric threshold frequency. Use the accepted value of Planck's constant. Note that Equation 181 indicates that $\nu = \nu_0$ when the stopping potential is zero, so that ν_0 can also be read from your plot in Calculation 4 as the value of the frequency at the intersection of your graph and the horizontal line at $V = 0$. Compare the two values of the photoelectric threshold frequency so obtained.

4. The work function for a metal used for the cathode of a photoelectric cell is 1.5 volts. Calculate the longest wavelength a beam of incident light can have if it is to have just enough energy to cause the emission of photoelectrons from the surface of this metal. What is the color of this light?

5. Calculate the amount of energy contained in a photon, that is, a quantum of visible light, of wavelength 5900 Å. Suppose that a yellow lamp is sending out radiation of this one wavelength at the rate of one watt (one joule per second). How many quanta of radiation are emitted each second?

6. Suppose the classical rather than the quantum theory were true. How long would it take for an electron to get through the surface of the photocathode of Question 3 held 10 centimeters away from a 100-watt light bulb? Assume that the light bulb is a point source radiating a total of 100 watts uniformly in all directions. This means that the radiated energy is distributed uniformly over spherical wave fronts. Assume also that an electron in the metal surface collects only the energy it intercepts and that it is a tiny sphere 10^{-13} centimeters in diameter.

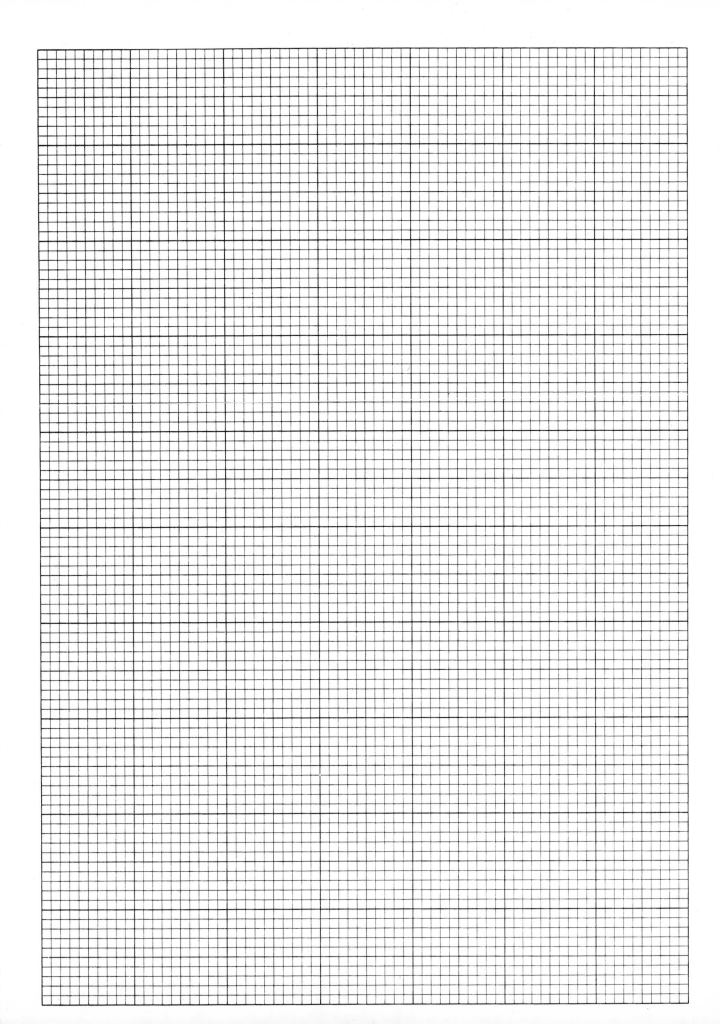

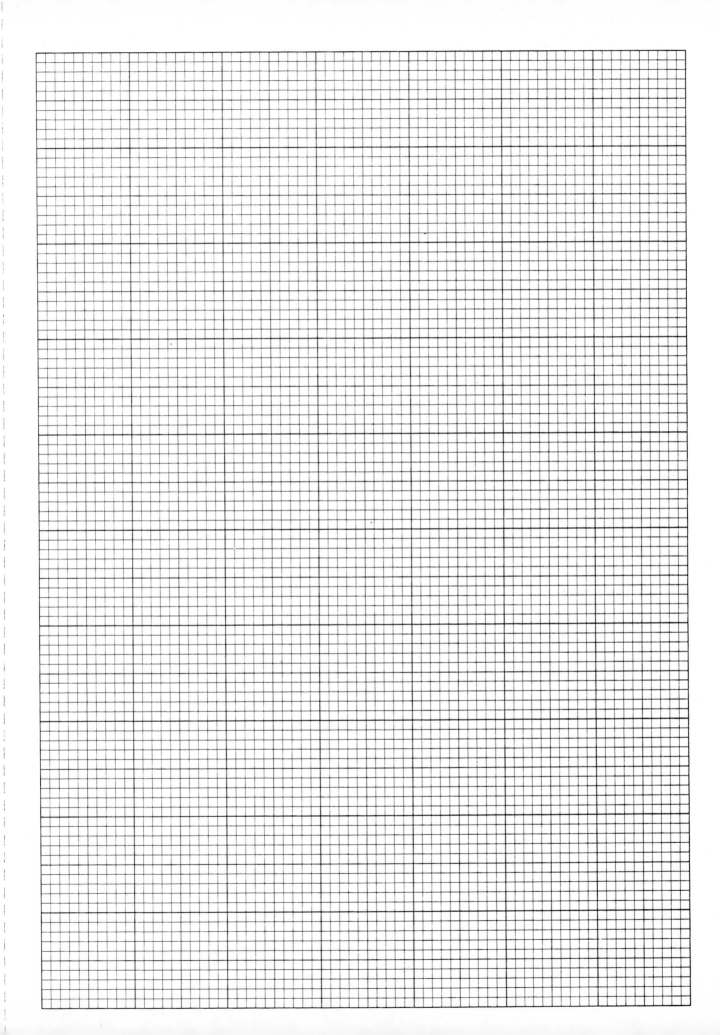

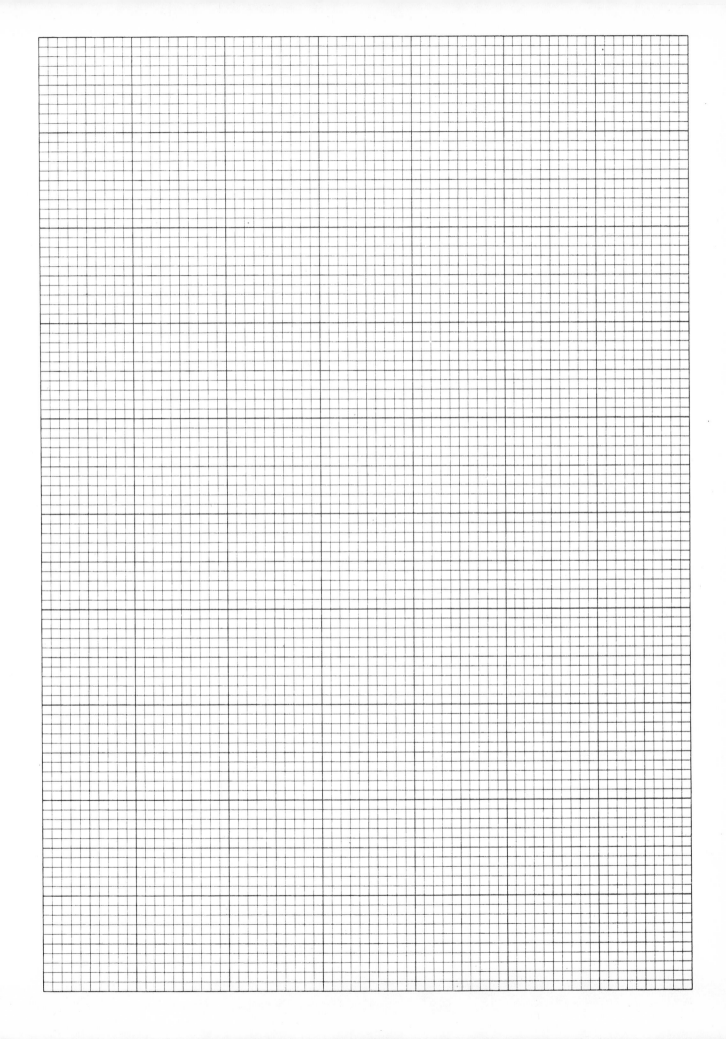

The Charge of the Electron. 52
Millikan's Oil Drop Experiment

The charge of the electron is one of the most important and fundamental physical constants. R. A. Millikan's determination of the value of this charge in his oil drop experiment proved to be the simplest and most direct method of finding this fundamental constant. The experiment showed that electric charges are not infinitely divisible, it provided the evidence for the existence of an elementary electric charge, and it gave us the means of determining the value of this charge very accurately. The object of this experiment is to determine the charge of the electron using Millikan's oil drop experiment.

THEORY

The charge of the electron is a constant that is fundamental in modern physics. Some early attempts to measure this charge were made during the study of the charge on gaseous ions such as are produced in a gas by X rays. To measure these charges, Townsend in 1897 used the clouds that form about the ions in saturated air. By observing the rate of fall of a cloud, which is made up of small droplets of water, and applying Stoke's law for the free fall of spheres through a viscous medium, he could determine the size of the droplets. From a measurement of the total amount of water in the cloud, he could calculate the number of droplets it contained. By assuming that each droplet contained one ion, and measuring the total charge in the cloud, he could then calculate the charge on a single ion. For this charge he obtained a value of 1.13×10^{-19} coulomb. In 1903, H. A. Wilson used the same method, with the exception of the addition of an electric field. His method was thus somewhat similar to Millikan's oil drop method, except that Wilson used water drops and made his measurements on the entire cloud instead of on a single drop. For the charge on a single ion, Wilson obtained the value of 1.03×10^{-19} coulomb. The values of both Townsend and Wilson are of the right order of magnitude for the charge of the electron, but because of the errors involved in the measurements, neither is very accurate. One of the errors introduced was the assumption that each droplet contained one electron. Another error resulted from the loss of water droplets in evaporation during the course of the experiment.

The most accurate method yet devised for measuring the charge of the electron was developed by Millikan during the years 1909 to 1913. He used oil droplets to make the first accurate measurement of the value of this constant and to show that the charge of the electron has a single, discrete value. Hundreds of times he determined the electrical charge on individual drops of oil located singly between the two horizontal parallel plates of an electrical capacitor. By measuring the velocity of fall of the drop under gravity and the velocity of rise when the plates were at a high electrical potential difference, he obtained data from which the charge on a drop could be calculated. He found that the charge was always an integral multiple of a certain small amount of charge, which he presumed to be the charge of a single electron.

In our experiment, fine watch oil is used; there will then be no loss of weight of the droplet by evaporation. The oil droplets are charged as they are sprayed from an atomizer, and then allowed to fall through a small hole in the center of the top plate of the capacitor. Air currents are avoided by enclosing the chamber with two glass plates, which also permit the viewing of the droplets. In the measuring process, a single oil drop is observed with a measuring microscope, and the velocity of fall determined by means of a scale contained in the eyepiece of the microscope. This method is extremely sensitive and very accurate, since the effect of a single electron on the motion of the oil drop can be detected and measured.

During the experiment, the oil drop is subject to three different types of forces: gravitational force, viscous resistance of the air, and electrical force. For very accurate measurements even the buoyant force of the air has to be taken into account, but we can neglect the effect of this force. By analyzing the effect of the other forces on the motion of the oil drop, we can derive an expression that allows us to calculate the charge on the oil drop and eventually to determine the charge on the electron.

When there is no electric field present, the oil drop will fall slowly, acted upon by the downward pull of the force of gravity and the retarding frictional force that contact with air molecules provides. Whenever a body is moving through the air or any other viscous material, the motion of the body is opposed by a frictional force owing to the viscous resistance of the medium. This resisting force is known experimentally to be proportional to the velocity of the moving body. Thus, when the oil drop falls through the air, the frictional force of the air quickly increases as the velocity of the drop increases, and the frictional force soon becomes equal to the force of gravity on the oil drop. The net force acting on the oil drop is then zero, and the oil drop no longer accelerates but moves with a constant velocity. This constant, or terminal, velocity is quickly reached by a very small droplet, and its value depends on the radius of the droplet.

The resistance of a viscous fluid, such as air, to the steady motion of a moving sphere can be obtained from Stokes' law, which gives the retarding frictional force acting on the sphere as

$$f = 6\pi r \eta v \qquad (182)$$

where r is the radius of the sphere, η is the coefficient of viscosity of the fluid, and v is the velocity of the sphere.

For an oil drop that has reached a constant, or terminal, velocity, the upward retarding force is equal to the downward gravitational force. We write

$$f_1 = mg = 6\pi r \eta v_1 \qquad (183)$$

where f_1 is the frictional force, m is the mass of the oil drop, g is the value of the acceleration of gravity, mg represents the gravitational force, r is the radius of the oil drop, η is the coefficient of viscosity of the air, and v_1 is the terminal velocity of the oil drop for free fall under gravity.

If an electric field exists between the plates of a capacitor on both sides of an oil droplet, the field can be applied in such a direction and be of such magnitude as to make the oil drop move upward with a new terminal velocity. The frictional force again opposes the motion, and is thus in this case directed downward. The electric force upward now balances the frictional force and the gravitational force, which are both downward. Taking the direction of these forces into account, we see that the retarding frictional force is given by

$$f_2 = Eq - mg = 6\pi r \eta v_2 \qquad (184)$$

where v_2 is the upward terminal velocity of the oil drop under the influence of the electric field E. In all three expressions for the retarding frictional force, this force is equal to some constant times the velocity of the oil drop, and the constant in all cases is the one given by Stokes's law. In the last expression, f_2 is the frictional force; E is the value of the electric field; q is the charge on the oil drop; and Eq represents the electric force on the oil drop, since the force on a charge q placed in an electric field E equals Eq. Putting the value of mg from Equation 183 into Equation 184 yields

$$Eq - 6\pi r \eta v_1 = 6\pi r \eta v_2$$

Simplifying, and solving for q, we obtain

$$q = \frac{6\pi r \eta}{E}(v_2 + v_1) \qquad (185)$$

This expression gives the value of the charge q in terms of the electric field E, the radius of the oil drop r, the coefficient of viscosity η, the velocity v_1 of the oil drop for free fall under gravity, and the upward velocity v_2 of the oil drop under the influence of the electric field E.

The value of the electric field can be easily computed. The intensity of the electric field between two parallel plates is given by

$$E = \frac{V}{d} \qquad (186)$$

where E is the electric field; V is the voltage, or the potential difference, between the plates; and d is the distance between the plates.

The radius of the oil drop may be calculated by noting that if this drop is truly a sphere, then its volume is $4\pi r^3/3$ and its mass m is given by

$$m = \rho \frac{4}{3}\pi r^3$$

where ρ is the density of the oil. If this expression for m is substituted in Equation 183 and the result is solved for r, we get

$$r = \sqrt{\frac{9\eta v_1}{2g\rho}} \qquad (187)$$

Now the charge on the oil drop can be calculated from Equation 185. Remember that all the data used for calculating the charge on an oil drop must be for the same oil drop. The velocity of fall under gravity will be the same for the same oil drop. But the upward velocity under the influence of the electric field will be different if the electric charge on the oil drop has changed during the course of the experiment. Each different upward velocity will correspond to a different value of the charge on the oil drop.

When many values for charge on the oil drops have been determined, all of these charges will be integral multiples of a certain small amount of charge, which is the smallest charge that can be found. If this smallest charge is assumed to be the charge of the electron, each charge found may be represented by the expression

$$q = ne$$

where q is the charge, n is a small integer, and e is the charge of the electron. After the different values of the charge have been determined, the value of the charge e can be found by dividing each value of q by the appropriate n, which is an integer between 1 and about 20. These values of e will be the calculated values, obtained from the experiment, of the charge on the electron.

When Millikan determined the value of the charge on the electron, he found it to depend somewhat on the size of the oil drop used. But experimental evidence had shown that the value of the charge on the electron was a constant, since an oil drop gains or loses charges of a definite amount regardless of the size of the drop. Millikan suspected that the discrepancy was due to Stokes's law, which he found did not hold accurately for very small drops. The motion of very small droplets is actually affected by the bombardment of air molecules, and this affects the predicted value of the velocity. Thus, he had to make a correction to Stokes's law in expressing the velocity of the oil drop, a correction that depended on the air pressure and on the size of the oil drop. The correction involved replacing the velocity v, given by Stokes's law, with the expression

$$\frac{v}{(1 + (b/pr))}$$

Since substitution of Equation 187 in 185 gives an expression for q that contains the velocity to the three-halves power, the correction for the charge on the electron can be put simply in the form

$$e = \frac{e'}{(1 + (b/pr))^{3/2}} \qquad (188)$$

where e is the corrected value of the charge on the electron, e' is the value of the charge on the electron obtained from the experiment, b is a constant of numerical value 0.000617, p is the barometric pressure in centimeters of mercury, and r is the oil-drop radius in centimeters.

The apparatus used in this experiment is an adaptation of that used by Robert A. Millikan. The design was developed by the Sargent–Welch Scientific Company, based upon the recommendations of J. B. Hoag, and thus the apparatus is called the Hoag–Millikan Oil Drop Apparatus.

The essential components are a specially designed capacitor, a measuring microscope, a light source, and a toggle-switch assembly, all mounted on a common sup-

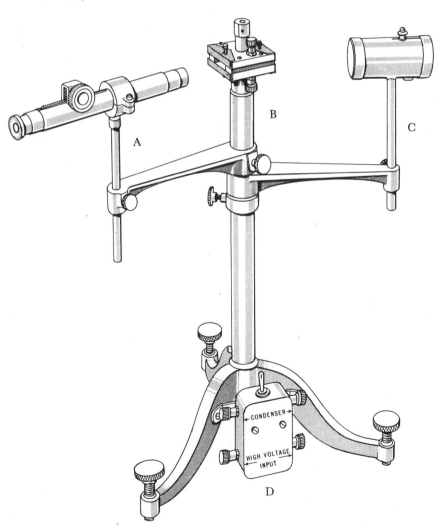

Figure 142   Hoag-Millikan Oil Drop Apparatus: (A) Measuring Microscope, (B) Condenser Unit, (C) Light Source, (D) Switching Assembly

port. The general appearance of the apparatus is shown in Fig. 142.

The capacitor unit consists of two horizontal metal plates, 5 cm square, separated by bakelite strips of equal and uniform thickness. The plates are about 4.5 mm apart, and are mounted on the central support shaft, completely insulated from the shaft and from each other. A single small hole in the center of the upper plate allows the oil droplets to fall from the spray chamber above the plate into the observation chamber. The spray chamber in the more recent models is about 6 cm high and slips over the viewing chamber. Oil from the atomizer is sprayed into a small hole near the top of the spray chamber. The oil droplets fall freely in the spray chamber, and many of them pass through the small hole into the viewing chamber. Two glass windows, one at the front and one at the rear of the chamber, permit the illuminating and viewing of the oil droplets, while eliminating air currents within the chamber. The glass windows are removable to allow measurements of the plate separation to be made with calipers. A binding post is provided on each plate for high-voltage connection.

The light source is an incandescent lamp, which should be operated on 6.3 volts A.C. but may also be operated from a 6-volt battery. The light can be focused to a narrow beam to illuminate the center of the viewing chamber. This is done by varying slightly the position of the bulb in its housing and by carefully adjusting the orientation and the elevation of the light source.

The measuring microscope has a magnification of approximately 22X. Measurements are made by means of a graduated scale contained in the microscope, which first has to be calibrated with the calibration scale supplied. The depth of focus is sufficiently short that the calibration remains unchanged so long as the object is in sharp focus. The microscope is provided with a rack-and-pinion fine-focusing arrangement. Focusing involves inserting a small pin through the hole in the center of the upper plate of the viewing chamber; the sharp point of the pin is then put into focus. In the focusing process, both elevation and orientation of the microscope are carefully adjusted.

The switching assembly is mounted at the base of the apparatus. It contains four binding posts — two to be connected to the high-voltage supply, and two to the capacitor plates. It also has a toggle switch that reverses the polarity of the plates; when placed in the vertical position, this switch disconnects the power source and short-circuits the plates, thus allowing free fall of the oil droplets.

The accepted values for e, m, and e/m for the electron are the following:

$e = 1.602 \times 10^{-19}$ coulomb	(charge on the electron)
$m = 9.11 \times 10^{-31}$ kilogram	(mass of the electron)
$e/m = 1.759 \times 10^{11}$ coulomb/kilogram	(ratio of charge to mass for the electron)

APPARATUS

1. Oil drop apparatus, Hoag-Millikan
2. Power supply, 0 to 300 volts D.C.
3. D.C. voltmeter (0–300 volts)
4. Watch oil, $\frac{1}{4}$-oz bottle
5. 6-volt storage battery or 6.3-volt filament transformer (if not provided in the power supply unit)
6. Stop watch or stop clock
7. One single-pole, single-throw switch

PROCEDURE

1. Record the reading of the laboratory barometer in centimeters of mercury.

2. Record the spacing of the capacitor plates in centimeters. Spacing should already have been determined by the instructor.

3. Record the calibration of the scale in the eyepiece of the microscope in centimeters per scale division. This should also have been determined previously by the instructor.

4. The microscope and light source should already have been focused by the instructor, but if this has not been done, focusing instructions will be found in the Apparatus Notes for this experiment (see Appendix). With all the adjustments properly made, the microscope will be focused on the region through which the oil drops will come so that these drops will appear as bright stars on a dark background. They will be sharply in focus, and the background of light will just permit the scale to be clearly visible. The oil drops will appear to fall right along the scale. Do not disturb the focusing of the microscope or the adjustments of either the microscope or the light source.

5. Make certain that the pin has been removed from the small hole in the upper plate of the capacitor and that the toggle switch is in the vertical position. Replace the spray chamber on top of the upper plate. Plug the line cord of the high-voltage power supply into a 115-volt, 60-cycle A.C. outlet. Turn the power switch on and adjust the voltage to 200 volts. Close the switch to light the lamp on the light source. CAUTION: Do not touch the capacitor plates or the high-voltage wires while the supply is on.

6. Now the apparatus is ready for operation. Gently spray some oil into the spray chamber by placing the nozzle of the atomizer close to the small hole near the top and directing the spray toward the top of the chamber. Do not spray toward the small hole in the upper plate of the condenser. Being small, the hole could easily be clogged with oil. The oil spray should be fine and uniform within the chamber, a condition that indicates that the droplets are of the proper size. The oil drops will become charged in the atomizing process.

7. As soon as the oil has been sprayed, look through the microscope. At first, a diffused light will be seen. This will thin out and small individual bright spots will appear, which are reflections of light from the oil drops and appear like stars of different magnitudes. The oil drops are falling under the force of gravity, but note that they are moving at diverse speeds. The large ones fall fastest and quickly disappear from the field of view; they are too large to use. One suitable for our purpose will be moving slowly and can be kept in the field of view for a long time by properly manipulating the toggle switch. When the switch is in the vertical position, the oil drop falls freely under gravity because no voltage is being applied to the plates. When the switch is placed to the right, one plate is made positive and the other negative. When the switch is placed to the left, the polarity of the plates is reversed. Determine the position of the switch that makes the upper plate positive by noticing when the oil drop moves upward. When the switch is placed in the opposite position, the oil drop moves downward in the same direction as when it is under gravity alone but much faster. The switch is not used in this position to make any measurements because of the possibility that the oil drop will quickly disappear from view and be lost. It is better to rely only on gravity for the downward pull. Remember, however, that the downward motion of the oil drop is seen as upward motion in the microscope, and upward as downward—the image in the microscope is inverted.

8. Begin by observing the motion of a slow-moving droplet falling freely under gravity, the toggle switch being in the vertical position. By manipulating the switch, practice making the oil drop move up and down. Be sure not to let the oil drop fall out of the field of view in either direction. Rely on gravity for the downward motion, and apply the electric field only for the upward motion. Remember that directions are reversed in the microscope and that only two of the three possible positions of the toggle switch are to be used.

Choose two points on the scale such as 2.00 and 8.00, and record their value. With a stop watch, measure the transit time for the oil drop between these two points for both upward and downward motions. Select an oil drop that requires about 20 seconds for the downward motion and about 30 seconds for the upward motion. Keep the oil drop under observation for as long as possible—30 minutes or longer. Determine and record the transit time alternately for the downward motion, or free fall, and for the upward motion under the influence of the electric field. Do this until you have made five observations of the downward motion and five of the upward motion. Thereafter, record only the transit time for the upward motion for the same oil drop, and observe it as long as possible. The values of the transit time for the downward motion will be very nearly equal. Those for the upward motion will differ but are likely to fall into groups, in which intragroup values will differ by only a fraction of a second and intergroup values by several seconds. For the downward motion, the time of transit will remain the same for the same oil drop, since velocity depends only on the size of the drop and the force of gravity, which remain the same. For the upward motion, the time of transit will be different if the charge on the oil drop changes by one or two electrons, because the velocity of the oil drop depends on the electric force, which in turn depends on the electrical charge on the drop. Thus in measuring the transit time for the upward motion, one is likely to observe a sudden change in the velocity of the oil drop when it gains or loses an electron. It is better not to record that particular reading but to wait for the next one when the velocity has settled down to a new value.

Note: It is advisable to have two observers make the measurements. One manipulates the toggle switch, watches the oil drop continuously, and handles the stop watch. The other reads the stop watch and records all the readings. Make a total of at least 30 readings for the upward transit time, whether for a single oil drop or for several. For the downward motion, a total of 5 readings for each oil drop is sufficient.

9. The table (see Data) for recording downward motion of the oil drop has three oil drops listed. This is done to provide space for recording data if it becomes necessary to make measurements on three different oil drops. It is quite possible, however, to make all the measurements needed for the experiment on a single oil drop. The only condition is that there should be several different groupings of the velocities for upward motion of the oil drop, representing different values of the electric charge on the same oil drop. This condition will be met if the oil drop gains or loses an electron under the influence of radiation or through collisions. If this is the case, all of the data can be taken on a single oil drop.

The reason for listing three oil drops is that sometimes an oil drop may be accidentally lost after a series of measurements have been made, but not enough measurements to be very conclusive. Then another oil drop must be used and a new set of measurements made. But in calculations for determining the value of the electric charge on an oil drop, all the data used must be for the same oil drop.

When sufficient observations have been made, the power supply should be turned off. Open the switch to shut off the light source. Turn the power switch off. Pull out the plug of the high-voltage supply unit. Leave the toggle switch in its vertical position to short-circuit the capacitor plates.

DATA

Barometer reading _____

Separation of the capacitor plates _____

Calibration of the scale in the eyepiece _____

Density of the watch oil _____

Distance traversed by oil drop, in scale readings, from_____ to_____ _____

Distance traversed by oil drop, in centimeters _____

Voltage applied to the plates of the capacitor _____

Electric field between the plates of the capacitor _____

Calculated value of the charge on the electron _____

Accepted value of the charge on the electron _____

Percent error _____

Downward Motion of Oil Drop Free Fall Under Gravity		
Oil Drop No. 1 Time of Transit	Oil Drop No. 2 Time of Transit	Oil Drop No. 3 Time of Transit
1.	1.	1.
2.	2.	2.
3.	3.	3.
4.	4.	4.
5.	5.	5.
Average Time _____	Average Time _____	Average Time _____
Velocity of Oil Drop _____	Velocity of Oil Drop _____	Velocity of Oil Drop _____
Radius of Oil Drop _____	Radius of Oil Drop _____	Radius of Oil Drop _____

Upward Motion of Oil Drop Under Influence of Electric Field		
Oil Drop No. 1 Time of Transit	Oil Drop No. 2 Time of Transit	Oil Drop No. 3 Time of Transit

Upward Motion of Oil Drop Under Influence of Electric Field		
Oil Drop No. 1 Time of Transit	Oil Drop No. 2 Time of Transit	Oil Drop No. 3 Time of Transit

Upward Motion of Oil Drop Under Influence of Electric Field		
Oil Drop No. 1 Average Time	Oil Drop No. 2 Average Time	Oil Drop No. 3 Average Time
Group I _____	Group I _____	Group I _____
Group II _____	Group II _____	Group II _____
Group III _____	Group III _____	Group III _____
Velocity of Oil Drop	Velocity of Oil Drop	Velocity of Oil Drop
Group I _____	Group I _____	Group I _____
Group II _____	Group II _____	Group II _____
Group III _____	Group III _____	Group III _____

Calculated Values of Electrical Charge on Oil Drop	Assumed Values of Small Integer n to Obtain $q = ne$	Calculated Values of Charge on the Electron

CALCULATIONS

1. Calculate the distance traversed by the oil drop from the value of the distance expressed in scale readings and the calibration of the scale in the eyepiece. Express the distance in centimeters.

2. Calculate the average time of transit for free fall under gravity for oil drop No. 1. If measurements have been made on oil drops No. 2 and No. 3, calculate the average transit time for these drops also.

3. Calculate the velocity of free fall for oil drop No. 1, using the value of the distance traversed and the value of the average transit time for free fall. Express the velocity in centimeters per second. If measurements have been made on oil drops No. 2 and No. 3, calculate the velocity of free fall for these drops also.

4. Calculate the average time of transit for the upward motion of oil drop No. 1. This is the motion under the influence of the electric field. Notice that the observed values of the transit time listed in the table are likely to fall into groups in which the values belonging to any one group do not differ by more than a fraction of a second. The values belonging to different groups will be found to differ by several seconds. From the transit times for oil drop No. 1, make a list of those belonging to the same group and call it Group I. Calculate the average time of transit for Group I and record it in the table. Make a separate list of the transit times for oil drop No. 1 belonging to a different group and call it Group II. Calculate the average time of transit for Group II and record it in the table. If there is another group present, make another list of transit times for oil drop No. 1 belonging to this group and call it Group III. Calculate the average transit time for Group III and record it in the table. If more than one oil drop has been used in making measurements of the transit time for upward motion under the influence of the electric field, calculate the average time for each of these by the method just described. Do this even if there is only one group present for an individual oil drop.

5. Calculate the velocity of the upward motion of oil drop No. 1 under the influence of the electric field by using the values of distance traversed and average transit time for each group listed. Express the velocity in centimeters per second and record it in the table. If measurements have been made on oil drops No. 2 and No. 3, calculate the velocity of upward motion for these oil drops also, for each group listed. Record the data in the table.

6. Calculate the radius of the oil drop from Equation 187. Use 980 cm/sec² for the acceleration of gravity; 0.890 gram/cm³ for the density of Nye's watch oil; and 0.000182 poise (dyne-sec/cm²) for the viscosity of air. The radius of the oil drop will then come out in centimeters. If measurements have been made on oil drops No. 2 and No. 3, calculate the radii of these drops also. Record the resulting values in the table of the downward motion of the oil drop.

7. Compute the value of the electric field used in your measurements from Equation 186. Express the distance between the capacitor plates in meters so that the electric field will come out in newtons/coulomb.

8. Calculate the value of the electrical charge on oil drop No. 1 from Equation 185. In these calculations the MKS system of units will be used. All mechanical quantities must therefore be converted to this system, and the value of the charge will then come out in coulombs.

Calculate the value of the electrical charge on oil drop No. 1 from the value of the upward velocity of the oil drop, found for Group I, as listed in the table. This is the term v_2 appearing in Equation 185. The term v_1 is the velocity of oil drop No. 1 for free fall under gravity as given in the table for the downward motion. Both of these velocities must be converted to meters/sec by multiplying their values in cm/sec by 10^{-2}. The factor r is the value of the radius of oil drop No. 1 as found in Calculation 6 but multiplied by 10^{-2} to convert it to meters. E is the value of the electric field intensity found in Calculation 7, and η is the coefficient of viscosity of air, which may be taken as 0.000182 poise. A poise is a dyne-sec/cm² and must be converted to newton-sec/meter² for use in Equation 185.

Similarly, calculate the value of the electrical charge on oil drop No. 1 from the value of the upward velocity of the oil drop, under the influence of the electric field, found for Group II, and also for Group III, as listed in the table.

9. If measurements have been made on oil drops No. 2 and No. 3, calculate the electrical charge on each of these oil drops, as described in Calculation 8. Do this even if there is only one group of velocities present for an individual oil drop.

10. List all the values of the electrical charge found on all of the oil drops measured, for the different groups represented. Assume that each charge found may be represented by the expression $q = ne$, where q is the charge, n is a small integer, and e is roughly equal to the known value of the charge on the electron. Next to each charge, list the small interger n, between 1 and about 20, that best fulfills the condition of the above equation. Assuming the relation to hold, calculate the value of e that will be obtained for each value of the charge listed, and record the value of e thus found. These values of e represent the calculated values of the charge on the electron, obtained from the experiment.

11. Find the average of the calculated values of the charge on the electron, obtained in Calculation 10, and record this in the data as the calculated value of the charge on the electron. Compare your result with the accepted value by finding the percent error.

QUESTIONS

1. Name at least three possible sources of error in this experiment that are likely to affect the accuracy of the results.

2. (a) By using your calculated value for the radius of oil drop No. 1, as found in Calculation 6, compute the volume of this oil drop. Express the volume in cubic centimeters. (b) Compute the mass of the oil drop, if the density of the oil is 0.890 gram per cubic centimeter.

3. Assume that oil drop No. 1 used in the experiment was suspended in the electric field of the capacitor plates, held stationary under the influence of the electric field and the force of gravity. Calculate the value of the voltage that would have to be applied to the plates of the capacitor to accomplish this, assuming that the charge on the oil drop was equal to that produced by five electrons. For the mass of the oil drop, use the value obtained in Question 2.

Under these conditions when the oil drop is suspended, the gravitational force is equal to the electric force acting on the oil drop. The relation between the two forces is $mg = Eq$, where mg is the force of gravity, m is the mass of the oil drop in kilograms, and g is the value of the acceleration of gravity in meters/sec^2. The term Eq represents the electric force on the oil drop, since the force on a charge q placed in an electric field E is equal to Eq. Use the MKS system of units as before.

4. In determining the charge on the electron, Millikan found that a small error was introduced if the oil drop used was too small. This is due to a failure of Stokes' law to predict accurately the velocity of very small droplets. If a correction to Stokes' law is introduced, the velocity can be predicted much more accurately. When this is done, Equation 188 gives a corrected value for the charge on the electron. Use this expression to correct your result. For e', use your calculated value of the charge on the electron as recorded in the data. For the radius of the oil drop use the one that corresponds to oil drop No. 1, expressed in centimeters. Compare the corrected value of the charge on the electron with the accepted value by finding the percent error.

5. The accepted value for the ratio of charge to mass for the electron is $e/m = 1.759 \times 10^{11}$ coulomb/kilogram. Using this value of e/m and your calculated value for the charge on the electron, compute the mass of the electron. Express your result in kilograms and in grams.

6. A hydrogen atom consists of a single proton (its nucleus) and an electron. (a) What fraction of the mass of the hydrogen atom is due to the electron? (b) Calculate the ratio of the proton mass to the electron mass.

The First Excitation 53
Potential in Mercury

This experiment is a modern version of the famous Franck-Hertz experiment, which first demonstrated that an atom appeared to have discrete energy states not only in radiative interactions but in any interaction involving an energy exchange. Prior to this work, Planck's radiation quantum $h\nu$ and Bohr's stationary states of the hydrogen atom looked a little like theoretical machinations cooked up to fit the experimental data. In the Franck-Hertz experiment, however, the atom receives energy by being bombarded, not by a somewhat mysterious photon, but by electrons, which everybody thought of as

"real" particles whose kinetic energy could be readily established. Thus the experiment provides convincing evidence that no matter how the energy is delivered, an atom will accept it only in certain amounts, namely those that bring the total energy in the atom to one of its allowed levels. In the present experiment, the method of Franck and Hertz will be applied to mercury, and the first excitation potential, that is, the energy required to raise a mercury atom from its ground state to its first excited state, will be measured.

THEORY

Shortly after the turn of the century it was recognized that atoms both accepted and emitted radiation of particular frequencies corresponding to energies $h\nu$ which apparently represented differences between allowed energy levels in which the atoms could exist. Spectroscopic studies could then be used to determine where on an energy scale these allowed levels lay. In 1914 J. Franck and G. Hertz provided further evidence for the existence of discrete atomic energy levels by showing that excitation to the same levels already determined spectroscopically could be achieved by having a bombarding electron instead of incident radiation deliver the requisite energy. A diagram of their original apparatus is shown in Fig. 143. Electrons are emitted thermionically from the hot filamentary cathode and are accelerated towards the grid by the adjustable potential V_a, whose polarity is such as to make the grid positive with respect to the

filament so that it (the grid) functions as the anode. A collector electrode is mounted as shown to receive the electrons that get through the wire mesh of which the grid is made. A small potential V_r of such a polarity as to oppose the electron advance towards the collector is connected between the collector and the grid. If the tube is highly evacuated, all the electrons emitted from the filament get to the collector with energy $e(V_a - V_r)$, provided $V_a > V_r$, which is the usual case.

Now if some gas is introduced into the tube, the electrons will collide with the gas atoms on their way to the collector. Provided the electrons have less energy than that required to raise these atoms from their ground state to the first excited level above the ground state, quantum theory tells us that no energy transfer between a colliding electron and atom can take place. In other words, the collisions must all be perfectly elastic. Thus,

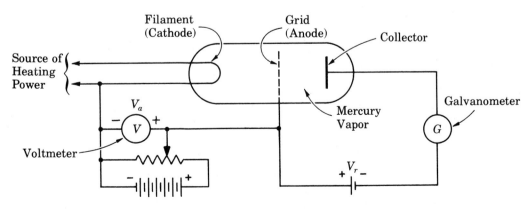

Figure 143   Schematic Diagram of the Apparatus of Franck and Hertz

407

although an electron may bounce around from atom to atom a good deal, it will not lose any energy in the process and will thus still have eV_a joules when it gets to the grid and be collected at the collector with energy $e(V_a - V_r)$. If, however, V_a is raised until eV_a becomes equal to the energy difference between the ground and first excited levels, then an energy transfer can take place in a collision, the electron losing all its kinetic energy to the atom, which is accordingly raised to its first excited state. The electron now no longer has enough energy to make it to the collector, since it must overcome the potential barrier eV_r to do so, and is thus not collected. This is revealed by a decrease in the collector current as indicated by the galvanometer. A further increase in V_a results in an electron having (or, if it has not reached the grid, gaining) kinetic energy beyond the amount lost to an atom in a collision. Therefore, as we continue to raise V_a, the collected current rises from the dip observed when eV_a was just equal to the difference between the ground and first excited state energy (V_a equal to the first excitation potential, as it is called) since the electrons have enough energy to surmount the retarding potential barrier even after losing the first excitation energy in the collision. However, if V_a is raised still further, a point will come at which an electron makes a collision in which it loses an energy equal to the atom's first excitation energy and then goes on with enough energy to make a second inelastic collision in which it loses another identical package of energy. After losing two such packages (one to each of two different atoms) the electron will again find itself without enough energy to

get to the collector and a second dip in the collector current will be observed. Thus, as V_a is increased we will find a dip in the collector current whenever V_a is equal to an integral multiple of the first excitation potential of the gas used in the tube. A typical plot of galvanometer current against V_a is shown in Fig. 144. Note that the proper value of the first excitation potential is obtained by measuring between adjacent dips rather than between the first dip and $V_a = 0$ because the actual potential between grid and filament must include the contact potential between these electrodes. In determining the excitation potential as the difference in V_a between adjacent dips we are in effect subtracting the potential required for n inelastic collisions from that required for $(n + 1)$ of them. Because the contact potential is present as an additive constant in both these quantities, it subtracts out.

At this point the reader may well ask why the gas atoms get raised to only their first excited state. Why, when V_a is large, cannot an electron gain sufficient energy to raise one atom to a higher state rather than raising each of two to their first excited levels? The answer is that such a process is indeed possible, and the chance of its occurring depends on the gas pressure in the tube. If this is relatively high, the probability of a given electron's colliding with an atom is high and the *mean free path,* defined as the average distance the electron goes before making a collision (the average distance between collisions), is short. In this case the electron never gets much more than the first excitation level's worth of kinetic energy between collisions and thus, with high V_a, raises several atoms to their first excited levels in several

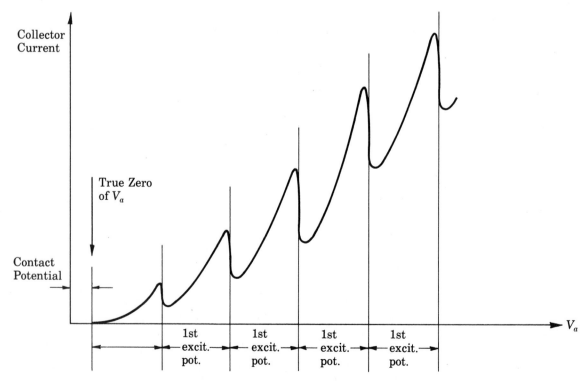

Figure 144   Typical Collector Current Curve

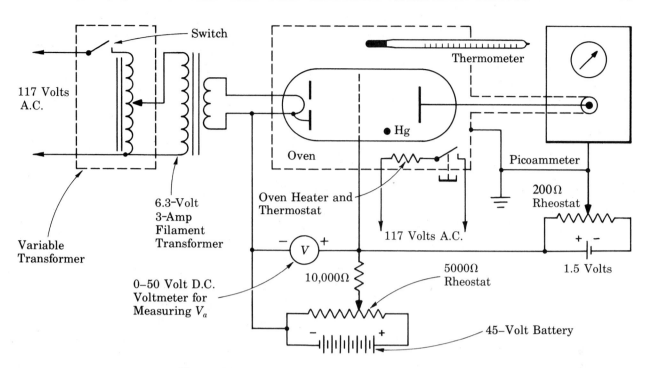

Figure 145 Wiring Diagram for the Franck-Hertz Experiment

collisions rather than going a long distance without hitting any atom and then giving all the kinetic energy thus acquired to one atom, thereby raising it to a higher level. In the present experiment we shall use mercury vapor as the gas and adjust its pressure by controlling the temperature of the tube, into which some liquid mercury has been introduced. We shall use a sufficiently high pressure so that negligibly few collisions resulting in atoms being raised to excited states higher than the first will occur. Adjustment of the pressure to allow observation of excitation to the higher levels is very critical and will not be attempted.

The experimental setup consists of a special, highly evacuated tube containing a filament, a grid, a collector electrode, and some mercury as described above mounted in an oven enclosure equipped with a thermostatically controlled heater. This assembly is wired to an external circuit as shown in the diagram of Fig. 145, which is seen to resemble closely the simplified drawing of the Franck-Hertz apparatus already given but shows all the ancillary equipment used in our version of the experiment. The galvanometer is seen to be a picoammeter, which is needed because the collected currents are small. Because "floating" the picoammeter case at a potential other than ground is undesirable, this point in the circuit is chosen as ground. The dry cell supplies a maximum retarding potential of 1.5 volts, the actual

value of V_r (which is usually of the order of 1 volt) being set by the associated rheostat. A large 45-volt battery or a power supply capable of delivering 30 to 50 volts provides the accelerating grid-to-filament potential V_a. If a battery or fixed power supply is used, a 5000-ohm rheostat is wired as shown to make V_a adjustable. If, however, a variable power supply is available, this rheostat becomes unnecessary. Note, however, that neither side of the power supply is returned to the common ground point, so that only a supply whose negative and positive sides are both isolated from its chassis may be substituted for the 45-volt battery.

V_a is measured with a 50-volt D.C. voltmeter that, because of the high resistance of the circuit, should have a sensitivity of at least 20,000 ohms per volt. An electronic voltmeter (EVM) may be used here, but only if its "low" side is not internally grounded to its chassis. A high-sensitivity voltmeter is needed for V_a because of the 10,000-ohm resistor connected between the source of V_a and the tube grid. This resistor protects the tube against excessive grid current should a discharge occur in the mercury vapor for any reason. In normal operation the grid draws so little current that the voltage drop across this resistor is negligibly small. The meter used to measure V_a must not, however, itself draw an appreciable current.

APPARATUS

1. Franck-Hertz tube
2. Oven for the above tube
3. Thermometer (-10 to $+225°C$)
4. Small variable line transformer
5. Filament transformer, 6.3 volts, 3 amperes
6. Tubular rheostat (about 200 ohms)
7. Volume-control-type potentiometer (5000 ohms, wirewound)
8. 10,000-ohm, 2-watt carbon resistor
9. 45-volt battery or power supply capable of delivering 30 to 50 volts
10. 1.5-volt dry cell
11. Picoammeter
12. D.C. voltmeter (0–50 volts)

PROCEDURE

1. Plug in the oven heater power cord and insert the thermometer in the hole provided for it in the enclosure. A few turns of tape will prevent the thermometer's falling all the way in. Place the tape so that the thermometer bulb is even with the midsection of the tube. The proper oven temperature is in the range $170°C$ to $190°C$, $180°C$ being about right. The thermostat should be adjusted by means of the knob on the side of the enclosure so that the oven temperature stabilizes at this value. On the Klinger apparatus, the thermostat knob turns through $360°$, after which, if it is being turned *away* from you as you face the apparatus (the direction for higher temperatures), there will be a click indicating that the thermostat has shifted from its highest to its lowest setting. Continued rotation brings the setting up again from the minimum for another revolution. CAUTION: Although the thermostat knob may be turned either way to make small adjustments in the temperature, do *not* attempt to turn it backwards (towards lower temperatures) through the end of its range.

2. Assemble and wire the apparatus while the oven is warming up. Fig. 145 is the wiring diagram and Fig. 146 is a sketch of a complete setup using the Heath EUW-17 power supply for V_a, the Simpson model 260 volt–ohm–milliammeter as the voltmeter, and the picoammeter described in the *American Journal of Physics* article mentioned in the Apparatus Notes (see Appendix). Make sure that the picoammeter range switch is set to its zero position and the rheostat or power supply controlling V_a adjusted to make this voltage zero. Do not plug in or turn on the picoammeter, the variable transformer, or the power supply (if used) until your circuit has been checked by the instructor. Then plug in and turn on the picoammeter so as to allow it to warm up.

3. When the oven temperature has stabilized at around $180°C$, set the retarding potential V_r at about 1.3 volts. If a multirange meter is being used to measure V_a, it may be temporarily disconnected and used to measure V_r after being switched to a low voltage range. CAUTION: Don't forget to return it to the 50-volt range before reconnecting it to read V_a. If a multirange meter is not available, the value of V_r (which is not critical) may be judged from the position of its controlling rheostat. Record the temperature at which you are running the experiment.

4. Set V_a to its maximum value (45 volts), make sure the picoammeter is zeroed, and then switch it to its 10^{-8} ampere full-scale range. Plug in and turn on the variable transformer and advance it slowly and carefully until the picoammeter reads approximately 2×10^{-9} amperes. *Note:* The setting required to do this varies widely with different Franck–Hertz tubes. In some cases it will be found necessary to turn the variable transformer almost all the way up, and even then a collector current much above 1×10^{-9} amperes may not be obtainable. The tube is acceptable providing the collector current is not less than this value. Slight reduction of V_r and/or the temperature may be tried in an effort to raise the collector current under these circumstances, and you should also vary V_a a little to make sure it is not set by chance at the bottom of a dip. On the other hand, you may find that a 2×10^{-9}-ampere collector current is obtained with relatively low filament power. Again, this may be normal for the tube you are using. However, if the picoammeter starts reading much more than 2×10^{-9} ampere before the variable transformer has been turned halfway up, shut off the filament at once and call the instructor.

5. Reduce V_a to zero and switch the picoammeter to the 10^{-9}-ampere full-scale range. Slowly increase V_a while watching the picoammeter to see when the collector current goes through its first peak. Set V_a accurately on the peak and record the values of V_a and the peak current.

6. Continue to advance V_a until the collector current goes through the minimum following the peak found in Procedure 5. Set V_a accurately on this minimum and record V_a and the current.

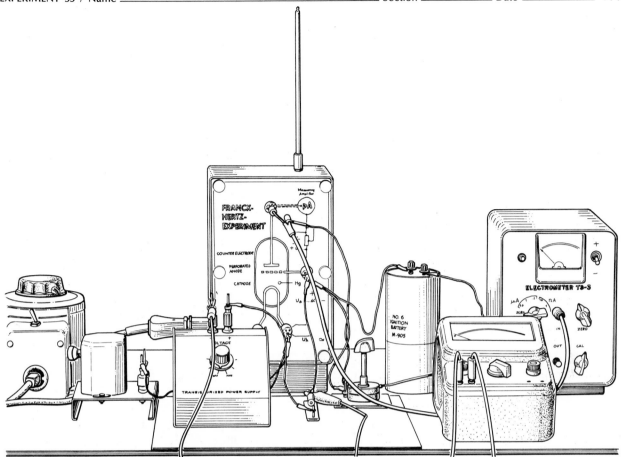

Figure 146   Typical Arrangement of Components for the Franck–Hertz Experiment. The Oven Containing the Special Tube is in the Center with the Power Supply for V_a in Front of It. The Picoammeter is at the Right.

7. Advance V_a until a second peak is found and record the voltmeter and picoammeter readings. Continue to the following minimum and again record the collector current and accelerating voltage.

8. Repeat Procedure 7 for the subsequent peaks and dips. At least six well defined dips should be identifiable in the accelerating voltage range from zero to 45 volts.

9. Return V_a to zero and take data for a plot of collector current as a function of accelerating voltage. Do this by advancing V_a in 2-volt steps and recording the collector current observed at each step. Continue this process until you reach 42 volts or the maximum voltage available from your power supply, whichever occurs first. Return V_a to zero, turn the variable transformer all the way down, and reset the picoammeter to its zero position when you have finished the experiment.

DATA

Oven operating temperature   _____

Peaks	Accelerating Voltage, Volts	Collector Current, Amperes	Minima	Accelerating Voltage, Volts	Collector Current Amperes
First			First		
Second			Second		
Third			Third		
Fourth			Fourth		

Peaks	Accelerating Voltage, Volts	Collector Current, Amperes	Minima	Accelerating Voltage, Volts	Collector Current Amperes
Fifth			Fifth		
Sixth			Sixth		
Seventh			Seventh		

Accelerating Voltage, Volts	Collector Current, Amperes	Accelerating Voltage, Volts	Collector Current Amperes
0		22	
2		24	
4		26	
6		28	
8		30	
10		32	
12		34	
14		36	
16		38	
18		40	
20		42	

Average value of the first excitation potential   _____

Calculated wavelength of the mercury resonance line   _____

Percent deviation from the spectroscopically observed value   _____

CALCULATIONS

1. Plot a graph of collector current as a function of accelerating voltage using all your data. Use the collector current values as ordinates and the accelerating voltage values as abscissas. Choose your scales so as to cover as much of a full sheet of graph paper as possible. Note that since you have measured the actual voltages and currents for all the maxima and minima in your curve, you know exactly where these maxima and minima occur and do not have to guess at the points where the curve turns around.

2. The first excitation potential of mercury in volts (joules per coulomb) is the difference in voltage between any two adjacent minima on your curve. A good average value can be obtained from your data by subtracting the value of V_a corresponding to the first well defined minimum from the value for the last one and dividing by the number of intervening spaces between dips. Thus in Fig. 144 an average value of the excitation potential could be obtained by subtracting V_a for the first minimum from V_a for the last (the fifth) and dividing by four. Record your result.

3. The energy actually delivered to each mercury atom struck by an electron is the kinetic energy of the electron, and this is equal to the electron accelerating voltage multiplied by the electronic charge. Thus $\Delta V_a e$ is the excitation energy in joules, ΔV_a being the voltage difference between adjacent minima obtained in Calculation 2. If a mercury atom were in its first excited state and made a transition to the ground state, this is the energy it would have to give up. If it gave it up by radiating a Planck quantum $h\nu$, then $h\nu$ should be equal to $\Delta V_a e$ and the frequency of the radiated light would be given by

$$\nu = \frac{\Delta V_a e}{h}$$

Since $\lambda = c/\nu$, the wavelength of this light is

$$\lambda = \frac{ch}{\Delta V_a e}$$

where λ is the wavelength in meters, c is the velocity of light in meters per second, h is Planck's constant in joule-seconds, ΔV_a is the first excitation potential of mercury in volts, and e is the charge on the electron in coulombs. Use this expression to find the wavelength of the mercury resonance line (the light emitted by a mercury atom in making a transition from its first excited state to its ground state). Express your result in angstroms by multiplying the wavelength in meters by 10^{10}.

4. The resonance wavelength for mercury is 2536 Å. Compare this value with that obtained in Calculation 3 by finding the percent error. *Note:* The first excited state of mercury is a triplet state consisting of three closely spaced energy levels. The 2536 Å wavelength corresponds to a transition to the ground state from the middle one of these, radiative transitions from the other two being forbidden. Transitions to any of the three states from the ground state as a result of electron bombardment are permitted, however, hence the value of the excitation potential obtained in this experiment is an average of three unresolved excitation potentials corresponding to excitations to each of the three states of the triplet. This is slightly higher than the excitation to the middle state alone, hence you should expect your calculated wavelength to be a little shorter than the one given for the mercury resonance line.

QUESTIONS

1. In this experiment, would you expect to see radiation coming from the mercury vapor in the Franck-Hertz tube as you increased the accelerating voltage through its range? Explain your answer.

2. (a) Define a wave number. (b) Express the mercury resonance line in wave numbers.

3. Why are the transitions between the ground and first excited states of atoms such as mercury, sodium, or hydrogen called "resonance lines"?

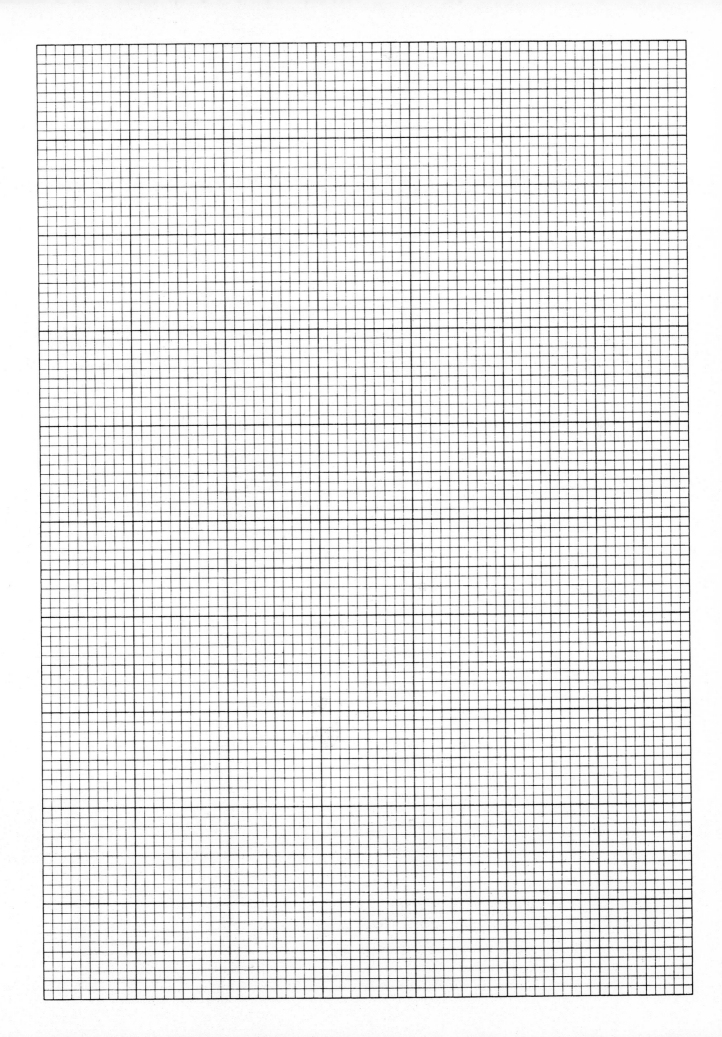

Appendix Tables

TABLE I. Physical Constants

Velocity of light in vacuum	$\begin{cases} = 3.00 \times 10^8 \text{ meters per second} \\ = 186{,}000 \text{ miles per second} \end{cases}$
Velocity of sound in air at 0°C	$\begin{cases} = 331.4 \text{ meters per second} \\ = 1087 \text{ feet per second} \end{cases}$
Acceleration of gravity	$\begin{cases} = 9.8 \text{ meters per second per second} \\ = 32.2 \text{ feet per second per second} \end{cases}$
1 standard atmosphere	$\begin{cases} = 76 \text{ centimeters of mercury} \\ = 1.013 \times 10^5 \text{ newtons per square meter} \\ = 14.7 \text{ pounds per square inch} \end{cases}$
1 horsepower	$\begin{cases} = 550 \text{ foot-pounds per second} \\ = 746 \text{ watts} \end{cases}$
1 faraday	$= 96{,}500$ coulombs
Avogadro's number	$= 6.022 \times 10^{23}$ molecules per mole
Mechanical equivalent of heat	$= 4.186$ joules per calorie
Absolute zero	$= -273.18$°C
Electron charge	$= 1.602 \times 10^{-19}$ coulomb
Electron mass	$= 9.109 \times 10^{-31}$ kilogram
Proton-electron mass ratio	$= 1836$
Boltzmann's constant	$= 1.381 \times 10^{-23}$ joule/°K
Planck's constant	$= 6.626 \times 10^{-34}$ joule-second
Permittivity of free space	$= (1/36\pi) \times 10^{-9}$ farad per meter
Permeability of free space	$= 4\pi \times 10^{-7}$ weber per ampere-meter
Heat of fusion of water at 0°C	$= 79.6$ calories per gram
Heat of vaporization of water at 100°C and 1 atmosphere pressure	$= 539$ calories per gram

TABLE II. Metric and English Equivalents

1 inch	= 2.540 centimeters	1 centimeter	= 0.3937 inch
1 foot	= 30.48 centimeters	1 meter	= 39.37 inches
1 mile	= 1.609 kilometers	1 kilometer	= 0.6214 mile
1 ounce	= 28.35 grams	1 gram	= 0.03526 ounce
1 pound	= 453.6 grams	1 kilogram	= 2.205 pounds
1 quart	= 0.9463 liter	1 liter	= 1.057 quarts

TABLE III.   Densities
(in grams per cubic centimeter)

Solids				Liquids		Gases *(at 0°C, 760 mm pressure)*	
Aluminum	2.7	Glass, crown	2.5 −2.7	Alcohol, ethyl	0.79	Air	0.001293
Brass	8.4	Glass, flint	3.0 −3.6	Alcohol, methyl	0.81	Carbon dioxide	0.001977
Copper	8.90	Granite	2.7	Carbon tetrachloride	1.60	Helium	0.0001785
Gold	19.3	Marble	2.6 −2.8	Ether	0.74	Hydrogen	0.00008988
Iron	7.85	Wood		Gasoline	0.68	Nitrogen	0.001251
Lead	11.3	Balsa	0.11−0.13	Mercury	13.6	Oxygen	0.001429
Nickel	8.75	Cork	0.22−0.26	Olive oil	0.92		
Platinum	21.5	Ebony	1.11−1.33	Turpentine	0.87		
Silver	10.5	Maple	0.62−0.75	Water, pure	1.00		
Steel	7.8	Oak	0.60−0.90	Water, sea	1.025		
Zinc	7.1	Pine	0.35−0.50				

TABLE IV.   Young's Modulus
(in dynes per square centimeter)

Aluminum	7.0×10^{11}
Brass	9.2×10^{11}
Copper	$12 \ \times 10^{11}$
Phosphor bronze	$12 \ \times 10^{11}$
Steel	19.2×10^{11}

TABLE V.   Coefficient of Linear Expansion
(per degree Celsius)

Aluminum	0.000024	Lead	0.000029
Brass	0.000018	Nickel	0.000014
Copper	0.000017	Platinum	0.000009
German silver	0.000018	Silver	0.000019
Glass	0.000009	Steel	0.000011
Gold	0.000014	Tin	0.000027
Iron	0.000012	Zinc	0.000026

TABLE VI.   Specific Heat

Aluminum	0.22	Mercury	0.033
Brass	0.092	Nickel	0.11
Copper	0.093	Steel	0.12
Glass	0.16	Tin	0.054
Iron	0.11	Water	1.00
Lead	0.031	Zinc	0.093

TABLE VII.   Resistivity of Metals
(in ohm-cm)

Aluminum	$2.6 \ \times 10^{-6}$	German silver	$33.0 \ \times 10^{-6}$
Brass	$6.4–8.4 \ \times 10^{-6}$	Iron	$10 \ \times 10^{-6}$
Constantan	$44.1 \ \times 10^{-6}$	Manganin	$44.0 \ \times 10^{-6}$
Copper	1.72×10^{-6}	Nichrome	$100 \ \times 10^{-6}$
		Silver	1.63×10^{-6}

TABLE VIII.   Electrochemical Data

Element	Atomic Mass	Valence	Electrochemical Equivalent, Grams Per Coulomb
Aluminum	27.1	3	0.0000936
Copper	63.6	2	0.0003294
Copper	63.6	1	0.0006588
Gold	197.2	3	0.0006812
Hydrogen	1.008	1	0.00001046
Iron	55.8	3	0.0001929
Iron	55.8	2	0.0002894
Lead	207.2	2	0.0010736
Nickel	58.7	2	0.0003040
Oxygen	16.0	2	0.00008291
Silver	107.9	1	0.0011180

TABLE IX.   Vapor Pressure of Water

Temperature, °C	Vapor Pressure, mm of Hg	Temperature, °C	Vapor Pressure, mm of Hg
0	4.58	55	118.0
5	6.54	60	149.4
10	9.21	65	187.5
15	12.79	70	233.7
20	17.54	75	289.1
25	23.76	80	355.1
30	31.82	85	433.6
35	42.18	90	525.8
40	55.32	95	633.9
45	71.88	100	760.0
50	92.51		

TABLE X.   Dielectric Constant and Strength

Material	Dielectric Constant	Dielectric Strength*, Kilovolts per Millimeter
Vacuum	1.000 00	∞
Air	1.000 34	0.8
Paper	3.5	14
Ruby Mica	5.4	160
Amber	2.7	90
Porcelain	6.5	4
Fused Quartz	3.8	8
Pyrex Glass	4.5	13
Polyethylene	2.3	50
Polystyrene	2.6	25
Teflon	2.1	60
Neoprene	6.9	12
Pyranol Oil	4.5	12

*This is the maximum potential gradient that may exist in the dielectric without the occurrence of electrical breakdown.
Table from D. Halliday and R. Resnick, "Fundamentals of Physics," John Wiley & Sons, Inc., New York, 1970, p. 495.

TABLE XI.   Standard Wire Sizes
(Resistance for Standard Copper Wire)

A.W.G. No.	Diameter, inches × 10⁻³	Resistance of 1000 feet, ohms at 25°C	A.W.G. No.	Diameter inches × 10⁻³	Resistance of 1000 feet, ohms at 25°C
12	80.8	1.619	22	25.3	16.46
14	64.1	2.575	24	20.1	26.17
16	50.8	4.094	26	15.9	41.62
18	40.3	6.510	28	12.6	66.17
20	32.0	10.35	30	10.0	105.2

TRIGONOMETRIC FUNCTIONS

ANGLE	SINE	COSINE	TANGENT	COTANGENT	ANGLE
0	0.0000	1.0000	0.0000	∞	90
1	0.0175	0.9998	0.0175	57.2900	89
2	0.0349	0.9994	0.0349	28.6363	88
3	0.0523	0.9986	0.0524	19.0811	87
4	0.0698	0.9976	0.0699	14.3007	86
5	0.0872	0.9962	0.0875	11.4301	85
6	0.1045	0.9945	0.1051	9.5144	84
7	0.1219	0.9925	0.1228	8.1443	83
8	0.1392	0.9903	0.1405	7.1154	82
9	0.1564	0.9877	0.1584	6.3138	81
10	0.1736	0.9848	0.1763	5.6713	80
11	0.1908	0.9816	0.1944	5.1446	79
12	0.2079	0.9781	0.2126	4.7046	78
13	0.2250	0.9744	0.2309	4.3315	77
14	0.2419	0.9703	0.2493	4.0108	76
15	0.2588	0.9659	0.2679	3.7321	75
16	0.2756	0.9613	0.2867	3.4874	74
17	0.2924	0.9563	0.3057	3.2709	73
18	0.3090	0.9511	0.3249	3.0777	72
19	0.3256	0.9455	0.3443	2.9042	71
20	0.3420	0.9397	0.3640	2.7475	70
21	0.3584	0.9336	0.3839	2.6051	69
22	0.3746	0.9272	0.4040	2.4751	68
23	0.3907	0.9205	0.4245	2.3559	67
24	0.4067	0.9135	0.4452	2.2460	66
25	0.4226	0.9063	0.4663	2.1445	65
26	0.4384	0.8988	0.4877	2.0503	64
27	0.4540	0.8910	0.5095	1.9626	63
28	0.4695	0.8829	0.5317	1.8807	62
29	0.4848	0.8746	0.5543	1.8040	61
30	0.5000	0.8660	0.5774	1.7321	60
31	0.5150	0.8572	0.6009	1.6643	59
32	0.5299	0.8480	0.6249	1.6003	58
33	0.5446	0.8387	0.6494	1.5399	57
34	0.5592	0.8290	0.6745	1.4826	56
35	0.5736	0.8192	0.7002	1.4281	55
36	0.5878	0.8090	0.7265	1.3764	54
37	0.6018	0.7986	0.7536	1.3270	53
38	0.6157	0.7880	0.7813	1.2799	52
39	0.6293	0.7771	0.8098	1.2349	51
40	0.6428	0.7660	0.8391	1.1918	50
41	0.6561	0.7547	0.8693	1.1504	49
42	0.6691	0.7431	0.9004	1.1106	48
43	0.6820	0.7314	0.9325	1.0724	47
44	0.6947	0.7193	0.9657	1.0355	46
45	0.7071	0.7071	1.0000	1.0000	45
ANGLE	COSINE	SINE	COTANGENT	TANGENT	ANGLE

LOGARITHMS

N	0	1	2	3	4	5	6	7	8	9
10	0000	0043	0086	0128	0170	0212	0253	0294	0334	0374
11	0414	0453	0492	0531	0569	0607	0645	0682	0719	0755
12	0792	0828	0864	0899	0934	0969	1004	1038	1072	1106
13	1139	1173	1206	1239	1271	1303	1335	1367	1399	1430
14	1461	1492	1523	1553	1584	1614	1644	1673	1703	1732
15	1761	1790	1818	1847	1875	1903	1931	1959	1987	2014
16	2041	2068	2095	2122	2148	2175	2201	2227	2253	2279
17	2304	2330	2355	2380	2405	2430	2455	2480	2504	2529
18	2553	2577	2601	2625	2648	2672	2695	2718	2742	2765
19	2788	2810	2833	2856	2878	2900	2923	2945	2967	2989
20	3010	3032	3054	3075	3096	3118	3139	3160	3181	3201
21	3222	3243	3263	3284	3304	3324	3345	3365	3385	3404
22	3424	3444	3464	3483	3502	3522	3541	3560	3579	3598
23	3617	3636	3655	3674	3692	3711	3729	3747	3766	3784
24	3802	3820	3838	3856	3874	3892	3909	3927	3945	3962
25	3979	3997	4014	4031	4048	4065	4082	4099	4116	4133
26	4150	4166	4183	4200	4216	4232	4249	4265	4281	4298
27	4314	4330	4346	4362	4378	4393	4409	4425	4440	4456
28	4472	4487	4502	4518	4533	4548	4564	4579	4594	4609
29	4624	4639	4654	4669	4683	4698	4713	4728	4742	4757
30	4771	4786	4800	4814	4829	4843	4857	4871	4886	4900
31	4914	4928	4942	4955	4969	4983	4997	5011	5024	5038
32	5051	5065	5079	5092	5105	5119	5132	5145	5159	5172
33	5185	5198	5211	5224	5237	5250	5263	5276	5289	5302
34	5315	5328	5340	5353	5366	5378	5391	5403	5416	5428
35	5441	5453	5465	5478	5490	5502	5514	5527	5539	5551
36	5563	5575	5587	5599	5611	5623	5635	5647	5658	5670
37	5682	5694	5705	5717	5729	5740	5752	5763	5775	5786
38	5798	5809	5821	5832	5843	5855	5866	5877	5888	5899
39	5911	5922	5933	5944	5955	5966	5977	5988	5999	6010
40	6021	6031	6042	6053	6064	6075	6085	6096	6107	6117
41	6128	6138	6149	6160	6170	6180	6191	6201	6212	6222
42	6232	6243	6253	6263	6274	6284	6294	6304	6314	6325
43	6335	6345	6355	6365	6375	6385	6395	6405	6415	6425
44	6435	6444	6454	6464	6474	6484	6493	6503	6513	6522
45	6532	6542	6551	6561	6571	6580	6590	6599	6609	6618
46	6628	6637	6646	6656	6665	6675	6684	6693	6702	6712
47	6721	6730	6739	6749	6758	6767	6776	6785	6794	6803
48	6812	6821	6830	6839	6848	6857	6866	6875	6884	6893
49	6902	6911	6920	6928	6937	6946	6955	6964	6972	6981
50	6990	6998	7007	7016	7024	7033	7042	7050	7059	7067
51	7076	7084	7093	7101	7110	7118	7126	7135	7143	7152
52	7160	7168	7177	7185	7193	7202	7210	7218	7226	7235
53	7243	7251	7259	7267	7275	7284	7292	7300	7308	7316
54	7324	7332	7340	7348	7356	7364	7372	7380	7388	7396

LOGARITHMS

N	0	1	2	3	4	5	6	7	8	9
55	7404	7412	7419	7427	7435	7443	7451	7459	7466	7474
56	7482	7490	7497	7505	7513	7520	7528	7536	7543	7551
57	7559	7566	7574	7582	7589	7597	7604	7612	7619	7627
58	7634	7642	7649	7657	7664	7672	7679	7686	7694	7701
59	7709	7716	7723	7731	7738	7745	7752	7760	7767	7774
60	7782	7789	7796	7803	7810	7818	7825	7832	7839	7846
61	7853	7860	7868	7875	7882	7889	7896	7903	7910	7917
62	7924	7931	7938	7945	7952	7959	7966	7973	7980	7987
63	7993	8000	8007	8014	8021	8028	8035	8041	8048	8055
64	8062	8069	8075	8082	8089	8096	8102	8109	8116	8122
65	8129	8136	8142	8149	8156	8162	8169	8176	8182	8189
66	8195	8202	8209	8215	8222	8228	8235	8241	8248	8254
67	8261	8267	8274	8280	8287	8293	8299	8306	8312	8319
68	8325	8331	8338	8344	8351	8357	8363	8370	8376	8382
69	8388	8395	8401	8407	8414	8420	8426	8432	8439	8445
70	8451	8457	8463	8470	8476	8482	8488	8494	8500	8506
71	8513	8519	8525	8531	8537	8543	8549	8555	8561	8567
72	8573	8579	8585	8591	8597	8603	8609	8615	8621	8627
73	8633	8639	8645	8651	8657	8663	8669	8675	8681	8686
74	8692	8698	8704	8710	8716	8722	8727	8733	8739	8745
75	8751	8756	8762	8768	8774	8779	8785	8791	8797	8802
76	8808	8814	8820	8825	8831	8837	8842	8848	8854	8859
77	8865	8871	8876	8882	8887	8893	8899	8904	8910	8915
78	8921	8927	8932	8938	8943	8949	8954	8960	8965	8971
79	8976	8982	8987	8993	8998	9004	9009	9015	9020	9025
80	9031	9036	9042	9047	9053	9058	9063	9069	9074	9079
81	9085	9090	9096	9101	9106	9112	9117	9122	9128	9133
82	9138	9143	9149	9154	9159	9165	9170	9175	9180	9186
83	9191	9196	9201	9206	9212	9217	9222	9227	9232	9238
84	9243	9248	9253	9258	9263	9269	9274	9279	9284	9289
85	9294	9299	9304	9309	9315	9320	9325	9330	9335	9340
86	9345	9350	9355	9360	9365	9370	9375	9380	9385	9390
87	9395	9400	9405	9410	9415	9420	9425	9430	9435	9440
88	9445	9450	9455	9460	9465	9469	9474	9479	9484	9489
89	9494	9499	9504	9509	9513	9518	9523	9528	9533	9538
90	9542	9547	9552	9557	9562	9566	9571	9576	9581	9586
91	9590	9595	9600	9605	9609	9614	9619	9624	9628	9633
92	9638	9643	9647	9652	9657	9661	9666	9671	9675	9680
93	9685	9689	9694	9699	9703	9708	9713	9717	9722	9727
94	9731	9736	9741	9745	9750	9754	9759	9763	9768	9773
95	9777	9782	9786	9791	9795	9800	9805	9809	9814	9818
96	9823	9827	9832	9836	9841	9845	9850	9854	9859	9863
97	9868	9872	9877	9881	9886	9890	9894	9899	9903	9908
98	9912	9917	9921	9926	9930	9934	9939	9943	9948	9952
99	9956	9961	9965	9969	9974	9978	9983	9987	9991	9996

Apparatus Notes

These notes are intended to give the specifications of some of the commercial apparatus appearing in various figures and around which many of the experiments in this book were developed. In addition, special considerations in the operation of some of the experiments and suggestions for optional equipment that might prove useful are presented. It was felt that this material, largely of interest to the instructor, should not be included in the individual experiment writeups where it would lengthen the writeups unduly and possibly confuse the student. Interested students are, of course, encouraged to read this material if they wish, but it is chiefly aimed at the instructor whose job it is to install new experiments and get them to work properly.

The major suppliers of the laboratory equipment used for the experiments in this book are the following:

Central Scientific Company
11222 Melrose Avenue
Franklin Park
Chicago, Illinois 60131
Telephone: (312) 451-0150

Sargent-Welch Scientific Company
7300 North Linder Avenue
Skokie, Illinois 60076
Telephone: (312) 677-0600 or 267-5300

Klinger Educational Products Corporation
83-40 Parsons Boulevard
Jamaica, New York 11432
Telephone: (212) 297-8080 or 657-0536

The Central Scientific Company is often known as "Cenco" and will be so referred to hereafter. The other two will be referred to as "Sargent-Welch" and "Klinger," respectively. Suppliers of special equipment will be mentioned as they come up.

EXPERIMENT 1

The triple-beam balance is a simple Ohaus type supplied by both Cenco and Sargent-Welch. The equal-arm balance may be any of a number of two-pan analytical balances with varying sensitivities. As a rule, expensive high-sensitivity enclosed units should not be selected for a large undergraduate laboratory. The Sargent-Welch Cat. No. S-3478 with a capacity of 2000 grams and a sensitivity of 50 milligrams has proven satisfactory and has the advantage of including a tray for supporting a beaker under one of the arms as required for Experiment 12 (see Fig. 30).

EXPERIMENT 2

The force table shown in Fig. 13 may be either the Cenco Cat. No. 74285-000 or the Sargent-Welch Cat. No. 0740. Each includes four pulleys that mount on the circumference of the table, four weight hangers, a center ring and pin, and a set of slotted weights. The Cenco unit is machined from solid aluminum and is the one used in the development of this experiment and Experiment 11. However, it is more expensive than the Sargent-Welch apparatus, which has a wooden base under the thin aluminum table top.

EXPERIMENT 3

This experiment was designed around the Sargent-Welch meter-stick balance set, Cat. No. 0748. This set includes the meter stick, the three knife-edge clamps, the support stand, and the set of hooked weights specified in the apparatus list. The equal-arm balance may be the same as the one used in Experiment 1.

EXPERIMENT 4

The Atwood machine shown in Fig. 18 is a lecture-demonstration setup incorporating special features such as a brake and an automatic timing device that are unnecessary for the student laboratory. A complete system of this sort is listed by Sargent-Welch as

Cat. No. 0846. The experiment can, however, be done with the relatively inexpensive equipment called for in the apparatus list. The only critical item is the pulley, which must have as little friction as possible and be light and not too large in diameter, as its moment of inertia is to be neglected. It cannot be too small, however, as the rising and falling weights must be able to pass each other without colliding. A deluxe but expensive example is Cenco's Cat. No. 75655-000. Less expensive and extremely satisfactory is the Sargent-Welch Cat. No. 0778A.

EXPERIMENT 5

This experiment is designed for, and Fig. 19 shows, the ballistic pendulum apparatus supplied by Cenco under Cat. No. 75425-000. The spring gun is demountable and can be used separately, *e.g.*, in Experiment 11. The gun assembly and the steel ball (with hole for loading on the gun plunger) may be purchased separately under Cat. Nos. 00067-495 and 00060-516, respectively.

EXPERIMENT 6

Fig. 20 illustrates the Cenco Cat. No. 74470-000 centripetal force apparatus. The unit must be spun by a motor-driven rotator, which can maintain a constant speed and is equipped with a counter so that the revolutions may be timed. The rotator intended for use with the above-mentioned centripetal force apparatus is the Cenco Cat. No. 74350-001. Sargent-Welch supplies a less expensive unit that nevertheless meets all the requirements under Cat. No. S-76558. A still less expensive rotator is the Sargent-Welch Cat. No. S-62420. It is quite satisfactory but requires an adaptor chuck to take the centripetal force apparatus and has no counter, so that some separate means of measuring its shaft rpm must be provided. Sargent-Welch Cat. No. S-23305 revolution counter or Cat. Nos. S-79490 or S-79494-30 tachometers (that make the stop watch unnecessary) will serve this purpose.

EXPERIMENT 7

Any smooth, dry board will work satisfactorily in this experiment. The Cenco Cat. No. 75680-000 pulley is arranged for convenient mounting at the board's end; Cat. No. 75675-000, which is a pulley provided with a clamp, may be used if a non-permanent mounting is desired.

EXPERIMENT 8

The board and pulley listed for this experiment may be the same as that used in Experiment 7. Both Cenco and Sargent-Welch supply pulley sheaves of the type shown in Fig. 23. Suitable wheel-and axle combinations are listed by Cenco (Cat. No. 75746-000) and by Sargent-Welch (Cat. No. 0752). These units have a choice of wheel diameters, but only the smallest and largest need be used in the present experiment. The moment of inertia apparatus to be specified for Experiment 10 can also be used here, thus reducing the number of different items that must be stocked if both experiments are to be done.

EXPERIMENT 9

The Young's modulus apparatus shown in Fig. 24 is the Sargent-Welch Cat. No. 0572. It includes the three-legged optical lever and some weights, but the 1-kg weights called for in the apparatus list must be obtained separately. A less expensive apparatus set is the Sargent-Welch Cat. No. 0578. Cenco also lists a Young's modulus apparatus under Cat. No. 74000-000. In this case the optical lever (Cat. No. 74002-000) must be ordered separately. The telescope and scale may be the Sargent-Welch Cat. No. 2738, which is also specified for use with the current balance in Experiment 23 (see Fig. 43). The Cenco number for their telescope and scale is Cat. No. 82382-000.

EXPERIMENT 10

The moment of inertia apparatus illustrated in Fig. 25 is Cenco Cat. No. 75290-000. The current version of this apparatus is made to take spark recording paper for use with a spark timer, but this refinement is neither needed nor called for in Experiment 10's apparatus list or procedure instructions. Alternate units are the Sargent-Welch Cat. Nos. 0570J or 0574.

EXPERIMENT 11

The conservation of angular momentum apparatus illustrated in Figs. 27 and 29 was developed at Boston University. Details are available by calling the Department of Physics, (617) 353-2600. Briefly, the apparatus consists of a steel bar 20 inches long, 2 inches wide, and $\frac{1}{4}$ inch thick. Holes are drilled at its center for the supporting shaft and $\frac{3}{4}$ inch in from each end for the catcher cup and counterweight. The catcher cup is made from a piece of $2\frac{1}{4}$-inch O.D. aluminum tube $3\frac{3}{4}$ inches long. The open ends are closed with plastic inserts, one of which has a central hole through which the ball enters the cup. The aluminum tube is epoxied to a small aluminum block fitted with a stud by which the cup can be mounted on the bar. The sides of the block are accurately aligned with the cup axis so as to allow the cup to be set at angles of 45° or 90° to the bar with the drafting triangle. Three disks cut from 2-inch diameter brass stock serve as a nut into which the cup mounting stud is screwed, a nut that screws onto the threaded portion of the supporting shaft that sticks through the central hole in the bar, and a counterweight that balances the cup assembly, respectively. The cup mounting nut is $\frac{7}{16}$ inch thick, the supporting shaft nut $1\frac{1}{16}$ inch thick, and the counterweight $\frac{3}{4}$ inch thick. The supporting shaft is carried in ball bearings that are pressed into a brass tube attached to a base plate that in turn is bolted to the center of the force table used in Experiment 2. The experiment was developed with the Cenco force table, which was modified by having four holes for mounting the angular momentum apparatus drilled and tapped in it. A small pointer is attached to the rotating bar and points to the scale along the table's circumference so that the bar's angular position may be read. The spring gun is taken from the Blackwood ballistic pendulum (Experiment 5) and mounted on a two-by-four block so that it can be clamped to a support rod as shown in Fig. 29. The table must be leveled so that the rotating bar is in a truly horizontal plane; leveling screws are provided in the table legs for this purpose. Fig. 29 shows how the apparatus is assembled. The table is placed as close to the gun as possible while still allowing the catcher cup to clear the extended plunger of the spring gun as it comes around. The gun is aimed directly into the cup by eye, and a few trial shots are fired to ensure that the ball goes into the cup and stays there. Remarkably consistent timed runs can then be made and excellent results obtained.

EXPERIMENT 12

A balance equipped with a platform to support the beaker is listed by Sargent-Welch as Cat. No. S-3478. This unit has already been mentioned in connection with Experiment 1 and can in fact be used in all experiments in this book requiring an equal-arm balance.

EXPERIMENT 13

No special equipment is needed for this experiment, since all items on the apparatus list are common laboratory supplies.

EXPERIMENT 14

The vibrator pictured in Fig. 33 is supplied by Sargent-Welch as Cat. No. 3256A. Attention is called to the paragraph in the theory section of the experiment, which states that the vibration frequency is 120 hertz rather than the actual line frequency of 60 hertz. This is because the blade of the vibrator is not magnetized and is thus attracted to the electromagnet pole no matter which way the current flows through the coil. The blade, therefore, moves toward the pole as the current rises on both the positive and negative half-cycle and springs back as the current falls, giving a frequency of twice the line frequency. Arrangements of this sort are sometimes deliberately used as frequency doublers.

A good example of the rod-mounted pulley is Cenco's Cat. No. 75660-000.

EXPERIMENT 15

An example of a sonometer turned by weights as shown in Fig. 34 is the Sargent-Welch Cat. No. 3352. Both Cenco and Sargent-Welch supply a wide selection of tuning forks.

EXPERIMENT 16

The resonance tube apparatus needed for this experiment is available from either Cenco or Sargent-Welch. Representative catalog numbers are, for Cenco, 84930-000, and for Sargent-Welch, 3308 or 3309.

EXPERIMENT 17

The basic apparatus is available from Cenco under Cat. No. 77411-000 or from Sargent-Welch under Cat. No. 1633. The Cenco unit employs a 6-volt light bulb to indicate micrometer contact instead of a buzzer, and the bulb and associated wiring are built into the base. The experimenter need only apply 6 volts from a suitable source (such as the filament transformer to be specified for Experiments 33 through 37) to the terminals provided. The Sargent-Welch apparatus requires the separate buzzer, dry cell, and associated wiring called for in the apparatus list.

The Sargent-Welch setup includes a steel test rod, so that only the aluminum and copper rods need be ordered separately. The

Cenco equipment comes without rods, so that all three must be ordered. The following table gives the respective catalog numbers:

Rod Material	Cenco Cat. No.	Sargent-Welch Cat. No.
Steel	77418-000	1645
Aluminum	77420-000	1635
Copper	77424-000	1637

Several kinds of boilers (steam generators) and associated tripod stands are supplied by both Sargent-Welch and Cenco. The Sargent-Welch Cat. No. 1625 is a recommended example. Representative Cenco Cat. Nos. are 70728-000, 77936-000 and 77915-000. Because the last-named is made of glass, it is not recommended for student laboratories.

EXPERIMENT 18

Any of various boilers and calorimeters with which the average physics laboratory is usually well supplied may be used. If new equipment is to be ordered, attention is called to the Sargent-Welch Cat. No. 1625 steam generator and Cat. No. 1689 calorimeter. Cat. No. 1630C is a complete calorimeter set, which includes the above-mentioned steam generator and calorimeter along with some other minor accessories. If gas or alcohol burners are not available in the laboratory, the Sargent-Welch Cat. No. 1626 steam generator is provided with a built-in 450-watt electric heater, and the Cat. No. 1630D calorimeter set includes this boiler rather than No. 1625. Cenco lists a set of metal sample cylinders as Cat. No. 78125-000 and a suitable calorimeter as Cat. No. 77970-000.

EXPERIMENT 19

See the apparatus notes for the previous experiment for a discussion of appropriate boilers and calorimeters. In addition, a water trap (Apparatus item 3) is required. Either the Sargent-Welch Cat. No. 1629 or the Cenco Cat. No. 77945-000 will be satisfactory.

EXPERIMENT 20

Cenco Cat. No. 76365-000 is a Boyle's law apparatus similar to that illustrated in Fig. 37. It can be adapted for the air thermometer work by the addition of a bulb such as the Sargent-Welch Cat. No. 1601 or 1083B. However, a complete setup similar to that shown in Fig. 38 and capable of performing all the work in this experiment is available from Sargent-Welch under Cat. No. 1083. A somewhat more deluxe but much more expensive version is listed as Cat. No. 1083H.

EXPERIMENT 21

The original version of the ''mechanical equivalent of heat'' apparatus shown in Fig. 39 was made by Leybold-Heraeus Company in Germany and sold under their Cat. No. 38814. The U.S. distributor is the LaPine Scientific Company with offices in Chicago, Illinois; Irvington, New York; and Berkeley, California. Currently an improved modern version is available from Klinger as Cat. No. KH2040. In addition to the basic apparatus designated by this number, a special calorimeter (Cat. No. KH2042K), expanded scale thermometer (Cat. No. KH2042J), and 5-kg weight (Cat. No. KH2041) are needed.

EXPERIMENT 22

Klinger supplies a complete setup for Rüchhardt's method of measuring γ, the large (10-liter) flask or aspirator complete with rubber stoppers and stopcock tube being listed as Cat. No. KH2001 and the precision glass tube (60 cm length) with matching steel ball as Cat. No. KH2002. The precision tube is expensive and easily broken, so that attempts have been made to obtain such tubes in bulk from glass suppliers and to get matching ball bearings. The laboratory staff at Boston University has been reasonably successful in this endeavor, and further details may be obtained by calling the Physics Department at (617) 353-2600. It must be emphasized, however, that the conflicting requirements of low friction and good sealing between the ball and the tube bore make the choice and cleaning of the ball and tube extremely critical. The clearance must be between 0.0001 and 0.0003 inch and must be maintained within this tolerance for the entire length of the tube. Needless to say, the slightest speck of lint or dust on the ball or in the tube will suffice to stop the ball's motion and abort the run. Moreover, no lubricant has ever been found that improves the operation of the experiment, the best results being obtained when the ball and bore are absolutely dry. When necessary, cleaning is accomplished with crystal ammonia. Detergent household ammonia and the usual laboratory solvents such as trichlorethylene, carbon tetrachloride, acetone, and alcohol must not be used as they leave enough of a deposit to jam the ball. After cleaning with the clear ammonia, the tube and ball may be rinsed with distilled water and must then be *thoroughly* dried. Lintless gun-cleaning cloth patches or lintless

"Kimwipe" tissues have been used successfully. *It is imperative* that the instructor warn all students to maintain the required high degree of cleanliness and to be careful with the precision tubing. The success of the experiment depends on good technique.

Argon is available in the usual gas cylinders from local welding suppliers. The cylinder must be fitted with a pressure-reducing regulator. An output pressure setting of 3 to 5 psig with the connecting hose in place and the stopcock on the apparatus closed will result in an adequate gas flow when the stopcock is opened.

EXPERIMENT 23

This experiment was designed around the Sargent-Welch Cat. No. 2353 current balance. The Cat. No. 2738 telescope and scale and the Cat. No. S-4253 fractional weight set are also required. The remainder of the equipment called for in the apparatus list is available in most physics laboratories, but special mention should be made of the power supply, which must be capable of delivering 10 amperes continuously. Use of a 12-volt storage battery in the circuit of Fig. 44 gives good results. Sargent-Welch lists a recommended D.C. power supply as Cat. No. S-30967-95. This supply will deliver up to 20 amperes, so that additional data may be taken at higher currents, but its output voltage is only 6 volts. This is adequate for the experiment, but the 1-ohm current-limiting resistor should be omitted, as the wiring will introduce sufficient minimum resistance to prevent the flow of excessive current when no more than 6 volts is applied. In addition, high-current rheostats are often difficult to adjust with any precision, and an improved power circuit might be obtained by omitting both the rheostat and the 1-ohm resistor in Fig. 44 and operating the power supply from a variable line transformer, *e.g.,* a "Variac" or a "Powerstat." See the Apparatus Notes for Experiments 37 and 53 for a discussion of such line transformers, which will be needed for those experiments anyway. For the present application, however, a larger unit will be required than the Superior model 10B recommended in those cases. The Superior "Powerstat" Model 116B appearing in Fig. 146 would be quite satisfactory. A considerable saving in expense can be achieved by building one's own power supply, which can consist simply of an appropriately rated filament transformer and a couple of solid-state rectifier diodes. Details can be obtained from the Boston University Department of Physics by calling (617) 353-2600. Again, use of a variable line transformer provides a convenient, smooth, and precise way of adjusting the output current.

EXPERIMENT 24

The apparatus list for this experiment calls for rather common electrical equipment available from many suppliers, and so no particular catalog numbers need be quoted. A variety of meters is available from both Cenco and Sargent-Welch, but in the interest of keeping costs down, the possibility of using panel meters from local electronic supply houses should be investigated. In addition, Heathkit supplies a number of items for electrical and electronic experiments at reasonable prices. For example, although there are many small power supplies with output voltage variable from 0 to 15 volts and suitable current ratings, the Heathkit model IP-2728 is an excellent choice. The older (and unregulated) Model EUW-17 is also quite satisfactory and relatively inexpensive. Appropriate tubular rheostats are listed by Cenco under Cat. No. 82910. The present experiment was developed using the 90-ohm (Cat. No. 82910-011) and 180-ohm (Cat. No. 89210-010) units.

EXPERIMENT 25

This experiment uses the Sargent-Welch Cat. No. 2809 Wheatstone bridge. A complete set, Cat. No. S-68780, includes, in addition to the Cat. No. 2809 slide-wire, a set of resistors, a 1000-ohm decade box, a galvanometer, a dry cell, and a contact key; in short, almost all the apparatus items listed for the experiment. A slide-wire bridge is also available from Cenco under Cat. No. 83191-000. If the sample wire board called for in Apparatus item 8 cannot be made up in the laboratory, an excellent substitute is the Sargent-Welch Cat. No. 2816.

EXPERIMENT 26

A variety of equipment for this experiment is available from both Cenco and Sargent-Welch. The Sargent-Welch slide-wire bridge kit (Cat. No. S-68780) just mentioned for the previous experiment includes a galvanometer and a 1000-ohm decade box that may be used here, but a wide choice of decade boxes and galvanometers is available from all major suppliers of physics laboratory equipment. The comments on meters in the Notes for Experiment 24 apply equally well to the present experiment.

EXPERIMENT 27

In this experiment 200 to 300 grams of water are to be heated through a temperature rise of 10°C to 20°C in some easily measurable time interval. For this purpose about 4000 calories or approximately 16,000 joules will be required. Assuming a five-minute heating time, we see that we are talking about an electrical power input of around 50 watts or 5 amperes at 10 volts. These are very approximate numbers, but they give the idea of the sort of current and voltage the immersion heater must be run at. Such heaters are available from a number of supply houses, units used to heat coffee water in a car being quite satisfactory. These heaters are, of course, designed for 12-volt operation and usually have a line cord provided with a plug that fits the standard cigarette lighter receptacle. This must be cut off and the wire ends prepared appropriately for connection in the circuit of Fig. 52. Immersion heaters of this kind have a fairly large heat capacity themselves, and a value for the average specific heat must be determined by the instructor (usually by running the experiment and seeing what the heat capacity of the heater must be to make the results come out right) so that it can be given to the student for entry in the data table. A better solution is to use a small 10-watt wirewound resistor as the heating element. Attach no. 20 plastic insulated wire leads to the terminals and carefully insulate the entire assembly with an epoxy coating so that the water will not contact any metal parts when immersed. Note that the resistor should have a value of about 2 ohms and that the 10-watt rating is adequate because the resistor will be immersed in water. The 10-watt rating applies for continuous duty with free-air cooling. Although the instructor should run the experiment with such a heater as a check before giving it to the class, the chances are that negligible error will be introduced by omitting the resistor's heat capacity.

The 5-ohm rheostat may be the same one used in Experiment 23, but attention should be paid to the NOTE in Procedure 3 about using a power supply with a voltage control and omitting the rheostat. The Cenco Cat. No. 79550-000 supply has proven very satisfactory for this experiment as well as for several of the others in this book requiring D.C. voltages of the order of 12 volts at currents not exceeding 5 amperes. It has even been used to power the current balance in Experiment 23, although it is extremely marginal in this service, 8 amperes being the absolute maximum at which it can be run. Best results in the Boston University laboratories have been obtained using a 6-volt, high-current supply controlled by a "Powerstat" in Experiment 23, the Cenco Cat. No. 79550 for low voltage D.C. applications requiring $\frac{1}{2}$ to 5 amperes, and one of the pieces of equipment discussed in these Notes for Experiment 24 when well-filtered D.C. at less than 500 milliamperes is needed. It is also possible to use the Cenco supply as a 12-volt storage-battery substitute for all applications requiring 5 amperes or less. However, do not trust the meters on this unit. Use separate meters chosen in the manner described in the Notes for Experiment 24.

EXPERIMENT 28

The electrolytic cell illustrated in Fig. 53 is the Hoffman type, obtainable either from Cenco under Cat. No. 81200-000 or from Sargent-Welch under Cat. No. S-29125. Note that any of the low-voltage D.C. power supplies discussed earlier will provide the 500 milliamperes required for this experiment. Again, if an adjustable-voltage supply is used the rheostat may be omitted, but in general meters provided in commercial power supplies are not trustworthy; separate metering should be provided. Thus in Experiment 28 the 0–1 D.C. ammeter called for in the apparatus list should be used even if a power supply with a built-in output current meter is substituted for the storage battery. This meter should be accurate and easy to read as its readings are an important part of the data.

EXPERIMENT 29

The potentiometer shown in Fig. 55 is the Cenco Cat. No. 83412-000, but it is not used in this experiment. Instead, the slide-wire, galvanometer, protective resistor R_1 and shorting switch K_1 are taken over from Experiment 25. The 100-ohm tubular rheostat is the same as the one introduced in Experiment 24. Both Cenco and Sargent-Welch supply appropriate standard cells, and the meters are selected as discussed earlier.

EXPERIMENT 30

Any inexpensive audio signal generator can be used in this experiment. The Heathkit Model IGW-47B has been used with good results. To save cost, a 6.3-volt filament transformer may be substituted as the A.C. signal source, but because telephone receivers and communications headsets are not very sensitive at 60 hertz, results will be considerably less precise. A typical telephone receiver is the Cenco Cat. No. 80770-000, and a typical headset is either the Cenco Cat. No. 80785-000 or the Sargent-Welch No. 2615M. However, considerable cost savings can be realized by purchasing inexpensive headsets from a local electronic supply house.

Typical standard capacitors are the 1409 series made by the General Radio Company of West Concord, Massachusetts 01781. These are precise but expensive units, and their use is not mandatory. Any good commercial capacitor will serve as the standard, and accurate results may be obtained if the instructor determines a precise value with a good capacitance bridge before putting the capacitor out in the laboratory. Commercial paper or oil-paper capacitors work well for the unknowns.

EXPERIMENT 31

Most physics laboratories are equipped with wall-mounted or free-standing galvanometers suitable for use in this experiment. Typical units are those in the Leeds & Northrup Type 2239 series. The Types 2239-d and -e have damping appropriate for ballistic service. They are available through Cenco or directly from Leeds & Northrup, whose central office is in North Wales, Pennsylvania 19454. The company has a large number of field offices that are listed in its catalog and from which deliveries can be made. Accessories needed with the Type 2239 galvanometer include the Type 2127 wall mounting bracket, the Type 2174 telescope arm, and the Type 3375N scale. Less expensive scales, made in the laboratory and mounted so that a light reflects to them from the galvanometer mirror, can be substituted for the Leeds & Northrup arm and scale at a substantial saving in cost.

EXPERIMENT 32

There are many power supplies whose output voltage is variable up through 300 volts and that supply 100 or more milliamperes for powering vacuum-tube circuits. A typical example is the Heathkit Model IP-17. The current rating is higher than that required for the present work but will prove useful if Experiment 36 is to be done. Because 300-volt power supplies are usually intended for vacuum-tube circuits, they often include a source of 6.3 or 12.6 volts for powering the tube filaments. Again, this feature will prove useful in subsequent experiments even though it is not needed here.

The electronic voltmeter (EVM, sometimes called a vacuum-tube voltmeter or VTVM when vacuum-tube rather than solid-state circuitry is used) may be any of a great number of such instruments available on the market. A very ordinary unit intended for radio service work (as opposed to a sophisticated ultrahigh-input-resistance research instrument) will be perfectly satisfactory. Such units almost universally have an 11-megohm input resistance (10 megohms in the instrument proper and 1 megohm in the probe), and Experiment 32 has been designed on this assumption. The authors have found the Model WV-98C ''Senior VoltOhmyst'' manufactured by RCA Electric Components, Harrison, N.J. 07029, to be accurate, convenient, and reliable. In addition, an inexpensive yet satisfactory EVM is the Heathkit Model IM-18.

Ten-megohm resistors of the deposited carbon type are available from a number of manufacturers such as IRC (International Resistor Company, now a division of TRW) with a $\pm 1\%$ tolerance. A more deluxe glass-encapsulated resistor called the ''Hi-Meg'' RX-1 is supplied by the Victoreen Instrument Division of VLN, 10101 Woodland Avenue, Cleveland, Ohio 44104. These resistors are excellent when really high (up to 10^{14} ohms) resistances are needed, but in the present case we note that 10 megohms (10^7 ohms) is the *minimum* resistance available in the Type RX-1, that the tolerance is again $\pm 1\%$, and that the extra cost of this type is not really justified. Care should be taken in handling any 10-megohm resistor to avoid getting dirt, moisture, or fingerprints on it that could provide a conducting path (even a poor one) between its terminals. Ten megohms is a high resistance, and even a small amount of foreign material can provide a parallel parasitic resistance low enough to put the unit out of tolerance.

The 4-microfarad capacitor must have a high-quality dielectric, *i.e.*, one with excellent insulating properties, so that its leakage resistance will be many orders of magnitude larger than the 10-megohm resistances to be connected in parallel. Electrolytic capacitors cannot be used and, in fact, should never be used for capacities smaller than 10 microfarads. In the present case the 4-microfarad capacity can be made up by paralleling the eight $\frac{1}{2}$-microfarad capacitors on the capacitor board of Experiment 31. These capacitors should be high-quality paper or oil-impregnated paper units and should have a 600-volt rating if they are to be used in the present experiment. A 600-volt rating is recommended in any case as it guarantees a dielectric thickness that permits very little leakage. Of course, a good 4-microfarad oil-paper dielectric 600- or 1000-volt ''bathtub''-type capacitor or a couple of 2-microfarad units in parallel (some of the unknown capacitors from Experiment 30, for example) can be used perfectly well and, if available, will probably be more convenient than the capacitor board called for in the apparatus list.

EXPERIMENT 33

The primary and secondary coils (Apparatus items 1 and 2) used in the development of this experiment form a set supplied by Sargent-Welch under Cat. No. 2399. The Sargent-Welch catalog recommends Cat. No. S-30300 as the associated portable galvanometer, which is the same as that suggested for use with the Wheatstone bridge and the potentiometer (Experiments 25 and 29). The primary–secondary coil set includes a soft iron rod that can be used as a core for these coils and will serve as the ''straight piece of iron'' (Apparatus item 4). A soft iron bar can easily be obtained and bent into the shape of a U of a proper size to touch the ends of the Sargent-Welch rod and thus complete the magnetic circuit as directed in Procedures 4 and 11.

The ballistic galvanometer (Apparatus item 6) may be the same as the one chosen for Experiment 31.

The present experiment calls for a multimeter (Apparatus item 8) for the measurement of A.C. voltages. Although ordinary A.C. meters can be used, a multirange meter is extremely convenient in the laboratory, and a very useful purpose will be served by familiarizing the students with such instruments. Students should in particular be trained to be careful about setting multimeters to the proper ranges so as to avoid damage. The classic standby for teaching and research laboratories alike is the Simpson Model 260, which, although expensive, is accurate, rugged, and fairly easily repairable. There are a number of other good multimeters on the market, those supplied by Triplett and Heathkit being representative examples; in addition, several very inexpensive brands are available. The Micronta line distributed by the Radio Shack Corporation is typical, but the instructor must realize that these are cheap, fragile units. In deciding to furnish a student laboratory with such instruments, he or she must understand at the outset that the life of any one meter will be short. In addition, attempting to get one of these meters repaired is a waste of time, effort, and money. If such a meter gets broken, the best procedure is simply to throw it out and get a new one. A decision must be made as to whether this or the purchase and maintenance of better instruments is more economical. Our experience at Boston University has been that the more rugged multimeters justify their extra cost in the long run.

Apparatus item 9 calls for a transformer delivering 6.3 volts at 3 amperes. This is a standard filament transformer specification, and a wide selection is available at electronic supply houses from such manufacturers as the Standard Transformer Corporation (Stancor) Triad, and the United Transformer Corp. (UTC, now a division of TRW). A good-quality transformer of this rating will work well not only in this experiment but in several subsequent ones, namely 34 through 37, 52, and 53, even though the full 3-ampere capability is not always needed. The UTC Type S-55 was used in the development of these experiments and appears in Fig. 146.

The model generator called for in Apparatus item 10 and shown in Fig. 66 is the Cenco dynamo analysis apparatus, Cat. No. 79860-000. The Cenco No. 79550-000 power supply discussed earlier in these Notes works well as a substitute for the storage battery in this experiment. The 100-ohm rheostat and 1000-ohm decade box may also be the same as the ones used in earlier experiments.

EXPERIMENT 34

Most of the apparatus called for in this experiment is the same as that used in previous experiments and has already been discussed in these Notes. The unknown resistor (Apparatus item 5) is an ordinary 1-watt carbon unit with a 5% or 10% tolerance rating and may conveniently be selected from the list specified under item 12 of the Apparatus list for Experiment 37. However, Experiment 34 is the first one for which a cathode ray oscilloscope is required. The author of this book cannot overemphasize the advantage of equipping even the most elementary undergraduate laboratory with a high-quality oscilloscope. Bargain-store instruments, including the cheaper Heathkit models, almost invariably give badly synchronized, distorted patterns that disappoint the student rather than inspiring him or her with how great electronics, even in a student's "do-it-yourself" hands, can really be. Moreover, the cheap instruments tend to be unreliable and soon eat up the cost saving in "down time" and maintenance charges.

Another important point to consider in the selection of a laboratory oscilloscope is that high response speed (broad vertical amplifier frequency response) is unnecessary for the kind of demonstrations and measurements made in the experiments in this book and in undergraduate physics courses generally. In short, the so-called "fast" oscilloscope is a waste of money. Moreover, many inexpensive instruments are advertised as "fast" because it is easy to build wide-band amplifiers. Although they nominally have the advertised frequency response, it will be of no use in the present or subsequent experiments and will mean that other far more useful features have been neglected. Such advertising should therefore be categorically ignored. Instructors should understand that *the most desirable feature in a student oscilloscope is a good sweep and synchronization system*. The ideal example of the above precepts is the Hewlett-Packard Model 120-B. This research-quality instrument is not very expensive because it embodies only the most basic features and has a frequency range from D.C. to only 100 kilohertz. However, it has a fine, well-calibrated sweep and a precise synchronization system that includes triggered or slave-sweep capability. Unfortunately, the Model 120-B is currently out of production, although used instruments are often available. A refurbished Model 120-B will outperform anything else on the market for student laboratory applications. Failing this, other instruments may be investigated, including some of the less expensive Tektronix oscilloscopes or the better Heathkit models. It is imperative, however, to forget about frequency response, look closely at the performance of the sweep and synchronizing circuits, and try before you buy.

EXPERIMENT 35

Most of the items on this experiment's apparatus list have already been discussed. The Type 6AU4GT is a diode vacuum tube intended for use as the damper rectifier in the horizontal sweep circuits of many television receivers. As with most commercial tubes, its oxide-coated cathode has a very low work function, meaning that satisfactory operation under temperature-limited conditions is hard to achieve. The instructor might want to try it by using a "Powerstat" purchased for Experiments 37 and 53 in conjunction with the filament transformer to reduce the heater power to the point where temperature limitation of the plate current can be observed. The author has found that the 6AU4GT exhibits a pronounced Schottky effect (extraction of electrons from the low-work-function cathode by the electric field of the plate), which tends to mask the temperature limitation of the current. Special tubes have been developed by the major suppliers of undergraduate laboratory equipment, made specifically for the demonstration of vacuum diode characteristics, but we have not tried them.

The filament transformer called for in Apparatus item 4 has been discussed in the Apparatus Notes for Experiment 33. We would like to emphasize the need for a good-quality transformer in the present and immediately preceding experiments, as its output waveform will be viewed on the oscilloscope. Cheap transformers often put out a distorted sine wave, which detracts from the quality of the results in these experiments. Note that although the 3-ampere rating is not needed here, it *is* needed in other experiments, and smaller transformers (standard 0.6-ampere rating) rarely have the desired quality. For the present experiment the transformer secondary must be centertapped, but practically all 6-volt filament transformers have this feature.

EXPERIMENT 36

The Type 6AU6 and OA2 vacuum tubes are both of the modern 7-pin miniature style and continue to be available despite the swing to solid-state electronics. The 300-volt power supply may be the same as that selected for Experiment 32 and discussed in the Apparatus Notes for that experiment. A power supply with a built-in source of 6.3 volts for the tube heater is required if the filament transformer is to be used as the signal source as shown in Fig. 96, but the instructor may elect to use the signal generator from Experiment 30 for this purpose. Doing the latter has the advantage of providing a signal frequency higher than 60 hertz, and the instructor can then ask the students to measure their amplifiers' frequency response. This has not been done at Boston University because the experiment is already quite long. Nevertheless, a frequency response curve is easily taken with the amplifiers of Figs. 96 and 98 and is an excellent learning experience for the student. The low-frequency roll-off is best observed if the coupling capacitors C_g and C_p, which have to be large when a 60-hertz signal is used, are reduced to a value between 0.01 and 0.1 microfarad.

EXPERIMENT 37

This experiment was originally designed for a 60-hertz input signal to be obtained from a filament transformer as in the preceding experiment. However, the transistor amplifier can only handle relatively small voltage swings, hence the transformer output is cut down with a variable line transformer. Only a low-current unit is required, but a higher current one may be used if it has been purchased with an eye to its application in other experiments. See the Apparatus Notes for Experiment 53 for a discussion of variable line transformers ("Variacs" or "Powerstats"). The substitution of a signal generator is a real advantage here because such instruments have output controls that allow the output to be set at any desired level. See the Apparatus Notes for Experiment 36 in this regard.

EXPERIMENT 38

This experiment is designed around the Type 741 operational amplifier, a standard, general-purpose unit available from most electronic suppliers along with the majority of the other components called for in the apparatus list. The units used in the original version of this experiment were made by Texas Instruments. Detailed characteristics are given in data sheets published by that company.

The relay used to discharge C_f in the sawtooth generator circuit (Fig. 112) can be any of a wide variety of SPST (normally open), SPDT, or even DPDT or 4-PDT units having 12-volt D.C. coils. The selected relay should be as small and require as little coil current as possible, as the transistor cannot deliver more than about 70 milliamperes. Cat. Nos. 275-214 and 275-004 from the Archer line of electronic parts distributed by the Radio Shack Corporation are known to work in this application. The Cat. No. 275-214 is a 4-PDT relay that happened to be available. Its coil resistance is 160 ohms, so that it requires about 60 milliamperes to close. The Cat. No. 275-004 is an SPDT sensitive relay with a 500-ohm coil requiring only 12 milliamperes and thus is ideal for the present work. The 500-microfarad value of the delay capacitor C_o is, however, too high for this relay, holding it closed for too long a time and thus distorting the time measurements made in Procedures 12, 13, and 14. The 10-microfarad capacitor from Experiment 35 (Apparatus item 13) will be appropriate in this case. In general, the instructor may find that the value of C_o must be adjusted to suit the particular relay selected. Remember that the relay contacts must stay closed long enough to discharge C_f so that V_o returns to zero but not long enough to introduce a significant delay before recharging starts. Tests of the circuit with an oscilloscope (which shows the return of V_o to zero better than the meter) will serve to determine the best value of C_o.

The feedback capacitor C_f must have a low-loss dielectric to prevent leakage, just like the capacitor used in Experiment 32. See the discussion of this point in the Apparatus Notes for that experiment. Either the capacitor board from Experiments 31, 34, and 35 with four of its capacitors paralleled (four switches closed) or a separate 2-microfarad oil-paper dielectric capacitor (possibly one of the unknowns from Experiments 30 or 31) will be satisfactory.

EXPERIMENT 39

Several of the items needed for this experiment can be taken from Klinger's Blackboard Optics Basic Kit, Cat. No. KO 4100. Individual items that can be ordered separately are the following:

Plano-convex lens	Cat. No.	KO 4105
Semicircular lens		KO 4107
Plane mirror		KO 4110
Curved mirror		KO 4111

Note that the curved mirror has two reflecting sides and thus can be used as both a concave and a convex mirror (Apparatus items 4 and 5).

The Klinger kit also includes a triangular prism that is useful for some internal reflection demonstrations. It is not satisfactory for the work in this experiment, however, because it is a 45° right triangle. This makes the apex angle 90°, which prevents operation of the prism in the setup of Procedure 17. Klinger Cat. No. 336619 (crown glass) or 336649 (flint glass) 60° (equilateral) prisms are appropriate choices for Apparatus item 8. Alternatively, Cenco lucite prisms 85511, 85510, or 85336 work very well.

It has been found helpful to file a small nick in the exact center of the bottom edge of the semicircular lens's flat face. The pin in Procedure 15 is made to engage this nick, thus preventing the lens from sliding sideways while it is being rotated.

Any small, low-power, inexpensive helium–neon laser may be used for Apparatus item 1. Typical examples are the Spectra-Physics Model 155 and the Metrologic Model ML-600.

EXPERIMENT 40

When this experiment was originally developed, a Cenco optical bench and accessories were used. The pertinent catalog numbers are as follows:

One-meter bench rod	Cat. No.	72215-001
Bench rod support (2 required)		72220-000
Accessory carrier clamp (3 required)		72322-000
Illuminated object		86027-000
Lens holder		72288-000
Ground-glass screen		86005-000

An excellent series from Sargent-Welch is listed under the following numbers:

Optical bench assembly	Cat. No.	3618B
Carrier support clamps (3 required)		3617J
Lens holder		3629
Ground-glass screen		3643

Note that accessories such as the lens holder, object, and screen may be interchanged between optical benches, the Cenco units fitting the Sargent-Welch carriers and *vice versa*. Appropriate lenses are available from either Cenco or Sargent-Welch. They can also be obtained at lower cost from the Edmund Scientific Company, Edscorp Building, Barrington, N.J. 08007.

EXPERIMENT 41

The optical bench and accessories from Experiment 40 are again used here. As noted in the apparatus list, the vertically-mounted scale may be the one used with the viewing telescope in Experiments 9 and 23. The unit recommended as a companion for the current balance in Experiment 23 is the Sargent-Welch Cat. No. 2738. The telescope magnification scale is made up as described in the apparatus list (item 9) using paper tape from the kind of roll used in old-style adding machines. Although the instructor can make up these scales ahead of the laboratory period to save time, we have let the students make their own, merely providing the paper roll in the laboratory. The scales are taped to the wall for Procedures 4 and 6.

EXPERIMENT 42

Typical apparatus for this experiment is supplied by Cenco. Representative catalog numbers are:

Sodium light source	Cat. No.	87300
Helium discharge tube		87215
Mercury discharge tube		87260
Hydrogen discharge tube		87235
Discharge-tube power supply (including discharge-tube holder)		87208
Spectroscope		87015-002

The prism normally specified for use with this spectroscope is Cat. No. 87019-003, but if a flint glass prism has been selected for use in Experiment 39, it may be used here also.

EXPERIMENT 43

The sodium light source (Apparatus item 1) may conveniently be the same one listed for Experiment 42.

If a mirror with a hole in it is used in Apparatus item 3, it may consist simply of a polished metal plate in which the necessary hole is easily drilled. An inexpensive metal shaving mirror is a convenient choice. If a glass plate is used, it may be another one like the two listed as Apparatus item 2. In this case you do not really have a mirror, but there will be enough reflection from the glass surface to make the experiment work. Refer to the NOTE in Procedure 4.

EXPERIMENT 44

The laser (Apparatus item 1) may be the same as the one chosen for Experiment 39. It may be placed on a stand and aimed along the optical bench or fitted with one or two rods for mounting on the bench with standard optical bench carrier supports. A simple modification that has been very successful with the Spectra-Physics laser is to attach an aluminum block $5\frac{1}{2}'' \times 1\frac{1}{2}'' \times \frac{1}{2}''$ thick to the bottom surface. Two $\frac{3}{8}''$ diameter rods are screwed into this block and serve to mount the laser on the optical bench via the standard carriers. Appropriate rods with $\frac{1}{4}$-20 threaded ends and flats to permit tightening with a wrench are available from laboratory suppliers. Typical of these is the Cenco Cat. No. 72165.

The tungsten lamp bulb (Apparatus item 9) should not be frosted and should have either a single long filament wire or a very small filament that can simulate a point source. In Procedure 7, a very simplified form of spectrometer is set up in which there is no slit and only one lens. This lens forms an image of the source on the screen, and if this source is a vertical-wire filament, the image will be a vertical line. A small 6-volt dial light (operated from the filament transformer used in earlier experiments) works well as a point source, but in this case the image is a small spot. Note that because only one lens is used, the rays from the lamp are not made parallel by a collimator lens before they strike the grating and then brought to a focus by a telescope lens afterward as in the normal spectrometer. Instead, light from the lamp is focused on the screen by the one lens, which means that the rays are not quite parallel when they strike the grating. This does not, however, introduce a detectable error in the present experiment. If the lens is adjusted to bring the central spot into focus as described in Procedure 7, the displayed spectra will be slightly out of focus due to the longer path length. Critical observers may want to sharpen the ends of the visible spectrum by a small readjustment of the lens position, but negligible error will result from omitting this refinement.

EXPERIMENT 45

An appropriate light source (Apparatus item 3) is the 6-volt dial light discussed in the Apparatus Notes for Experiment 44. Note that the 6.3-volt filament transformer specified for earlier experiments will be needed to power it. The converging lens used in Experiment 44 may also be used here. Placing the light bulb so that its filament (which approximates a point source) is at the lens' principal focus will result in the lens producing the necessary parallel beam.

Various photocells are usable for Apparatus item 5, some of which are built into a complete photometer, which includes the meter. If a separate cell is adopted, any ordinary 0–100 microampere D.C. microammeter may be used with it to indicate the intensity.

EXPERIMENT 46

Typical apparatus for this experiment is supplied by Cenco. The Geiger counter tube is Cat. No. 71227-000; the probe, Cat. No. 71203-000; and the count-rate meter, Cat. No. 71201-009. All these items and an appropriate calibrated mounting board are included in the Radioactivity Demonstrator Set, Cat. No. 71201-003. A companion source kit is Cat. No. 71201-004, but individual sources are also available. A satisfactorily sealed cobalt-60 source of β and γ radiation is Cat. No. 71206-503.

EXPERIMENT 47

The same apparatus used in the previous experiment is used again here. Note that when these experiments were first developed, the calibrated mounting board and the set of aluminum plates (Apparatus items 4 and 5) were included in the Cenco Experiment Set, Cat. No. 71297. This set does not appear in the most recent Cenco catalog, but some units are still available. As mentioned in the Notes for Experiment 46, a suitable mounting board is included in Set Cat. No. 71201-003, and the aluminum plates are easily cut from standard .004-inch thick aluminum sheet stock. Note that .004 inches is so close to 1 millimeter that this standard aluminum sheet may be taken as 1 millimeter thick for all practical purposes, as will be seen when the measurement with the vernier caliper called for in Procedure 1 is carried out.

EXPERIMENT 48

The Geiger counter, probe, count-rate meter, calibrated mounting board, and beta–gamma source are the same as those used in Experiment 46. The short-lived radioisotope may be iodine-131 (I^{131}), a solution of which may be obtained from a number of suppliers. A typical example is the New England Nuclear Corporation, 549 Albany Street, Boston, Mass. 02118. Cat. No. NES-051S is a 5-milliliter vial of I^{131} solution with an initial activity of 10 microcuries. Note that I^{131} has a half-life of 8.04 days, which means that the five measurements made at two or three-day intervals called for in the Procedure section will give an adequate spread for plotting the decay curve.

Radiations from radioactive isotopes may be obtained from two types of sources: a sealed source and an unsealed source. A sealed source contains a definite quantity of a radioactive material, usually embedded in a plastic material. This type of source can be easily handled and contains only a few microcuries of the radioisotope. An unsealed source such as the one suggested above usually comes in a vial, in either powder or liquid form. But the sample to be used in making measurements has to be put in the form of a very thin film of the material. To handle such a source, the cap of the vial should be unscrewed carefully. If the radioisotope is dry, a small quantity of distilled water is added to the vial to dissolve the material. The solution is then poured into a volumetric container and diluted to the desired volume by adding distilled water. If the sample is already in liquid form, a definite amount may be obtained by pipetting. Use a pipet equipped with a rubber bulb, draw some of the liquid up from the vial, and expel it into the measuring vessel.

Samples of radioisotopes to be used in activity measurements are usually prepared on planchets. Planchets are small, metal, cup-like containers used in counting radioisotope samples. When preparing samples on planchets by evaporation from a solution, it is essential that the deposits be of uniform thickness. To ensure even spreading, the planchet must be very clean. Wash it carefully in alcohol using a thin cotton swab; a very thin film of alcohol is left on the surface so that the solution will spread out uniformly. By using a rubber-bulb pipet, a 1-milliliter sample of the solution is transferred to the planchet. The water in the solution is evaporated by placing the planchet under an infrared lamp, while the solution is kept spread out as evenly as possible. If the solution collects in spots, a few drops of alcohol may be added to make it spread out uniformly over the planchet's surface. The liquid should then evaporate in a few minutes, leaving a uniform, almost invisible deposit of the radioactive material. The planchet is now ready for use. Suitable planchets are included with the Cenco sample holder, Cat. No. 71207-000. The sample holder itself provides a convenient way to hold both the planchet and the Geiger tube and may replace the mounting board specified for Experiment 46. However, it accommodates the end-window Geiger tube, Cat. No. 71218-000 in the Cat. No. 71204-000 probe, rather than the side-window tube (Cat. No. 71227-000 and probe Cat. No. 71203-000) previously specified.

EXPERIMENT 49

The apparatus shown in Fig. 132 is supplied by Cenco under the following catalog numbers:

Hydrogen discharge tube	Cat. No.	87206-000
Tube holder for above		87201-000
Power supply for above		87207-000
Spectrometer		87015-002
Grating (15,000 lines per inch)		87019-002
Grating holder		87018-000

Note that hydrogen spectrum tube, Cat. No. 87235-000, and combined mount and power supply, Cat. No. 87208-000, can be used in place of Cat. Nos. 87206-000, 87201-000, and 87207-000 to reduce cost. This tube and power supply have already been listed for Experiment 42, along with spectrometer Cat. No. 87015-002. Less expensive spectrometers such as Cat. Nos. 86970-000 or 72720-035 can also be used.

EXPERIMENT 50

Several manufacturers sell an e/m tube and an appropriate arrangement of Helmholtz coils. The unit dealt with in this experiment and illustrated in Fig. 133 is made by Sargent-Welch and listed as Cat. No. 0623B. Also listed is most of the required auxiliary equipment such as meters, power supplies, and rheostats, but adhering to their list is not necessary. Thus a so-called radio "B" battery supplying 22.5 volts works well as the accelerating voltage supply. Several varieties of 45-volt "B" battery are available with a half-voltage tap at 22.5 volts. With such a battery it is possible to try the higher accelerating voltage, in which case (as Equation 177 shows) the magnetic field and thus also the current in the Helmholtz coils will have to be $\sqrt{2}$, or 1.4 times, as great for a given beam path diameter. As no improvement in beam focusing is obtained by going to the higher voltage, this practice is not recommended. In addition, a large number of small, solid-state power supplies delivering 0–30 volts D.C. are now on the market. An inexpensive example is the Heathkit EUW-17; see the Apparatus Notes for Experiment 24. More sophisticated units employ precise voltage regulation, which is useful in maintaining V at a preset value throughout the experiment, but this refinement is unnecessary if V is read each time the coil current is read as indicated in Procedures 8 and 9.

Filament power may be obtained from the same power supply, rheostat, and ammeter specified for Experiment 23. Because the e/m tube filament operates at a low voltage, a 6-volt storage battery is specified for E_2. The 12-volt source (storage battery or power supply) is then available for the Helmholtz coil supply E_3. However, a 12-volt source can perfectly well be used for E_2 provided that it can deliver 10 amperes. The Helmholtz coil current will never exceed 5 amperes, and a number of power supplies are available with this rating. A typical example is the Cenco Cat. No. 79550 mentioned in these Notes for Experiment 27. With such a supply, whose output is continuously variable, the 10-ohm rheostat R_3 may be omitted. However, note that the 200-ohm rheostat, whose purpose it is to provide a fine, low-current adjustment for determining I_1, the current required to buck out the earth's magnetic field, must be retained in any case. As this rheostat's resistance wire cannot stand the higher currents used to bend the electron beam in the appropriate circles, close attention must be paid to the instructions of Procedure 8. Some instructors like to add a switch in parallel with R_2 to short it out of the circuit after I_1 has been determined.

Many power supplies, including the Cenco Cat. No. 79550, are provided with output current meters, but only the more expensive units have meters of sufficient accuracy to be used in measuring the coil current. This measurement is the most critical one in the experiment, its square entering the calculation of e/m as Equations 177 and 178 show. Therefore, a really good 5-ampere meter should be used at A in Fig. 135. Some instructors like to substitute a 1- or $\frac{1}{2}$-ampere meter in order to obtain a more precise measurement of the low current I_1, but care must be taken to prevent students from running the Helmholtz coil current up to high values with this meter in place. One setup that has been used successfully is to leave the 5-ampere meter at A and add the low-current meter in series, and next in line with, the 200-ohm rheostat R_2. The shorting switch mentioned above is then connected across both this meter and R_2 to cut both out of the circuit when the higher current measurements are being made. Again, care must be taken to ensure that the shorting switch is closed before R_2 is reduced to the point where the meter goes off-scale.

EXPERIMENT 51

This experiment is one of the most interesting that can be performed in the undergraduate laboratory and also one that requires the greatest attention to detail. It is an excellent test of the student's ability as an experimenter, results being highly dependent on the refinement of technique that can be brought to bear on the work.

Although in recent years a number of special apparatus sets have become available for the study of the photoelectric effect, Experiment 51 was designed around a standard high-vacuum phototube, the RCA Type 929. Good results are obtainable with this tube, and the fact that the experimental setup is not a "black box" contributes to the student's feeling of actually arranging simple components so that the phenomenon of interest is produced and appropriate measurements can be made.

The use of a high-vacuum phototube provides a cathode surface of sufficiently low work function so that the threshold frequency ν_0 is in the near infrared; a collector wire, or anode, and a sealed high-vacuum envelope remove the necessity of having a high-vacuum pumping system in the laboratory. The Type 929's sensitivity is maximum for light of 4000 Å wavelength, and decreases to zero for wavelengths longer than 7000 Å. In appearance, the Type 929 looks like a radio tube of the "GT" sort. It has an octal base whose standard connection diagram appears in Fig. 141. For this experiment an octal socket is mounted on a 6 × 6 × 1.5 cm wooden block equipped with a metal rod 14 cm long and 10 mm (⅜ inch) in diameter to fit the optical bench carrier supports. The phototube is plugged into this socket and covered with a light shield S_1, which prevents stray light from reaching the cathode. S_1 can be any cardboard cylinder such as a mailing tube or paper towel tube of a diameter large enough to fit over the phototube. A length of 12 cm and a diameter of 4 cm are satisfactory dimensions. The shield slips over the phototube and rests on the socket or supporting block. An aperture 1.5 cm wide and 2.5 cm high is cut in its side at a point directly in front of the tube's cathode. This serves as the wide slit through which light enters the phototube. The shield may be sprayed with a flat black paint to make it light-absorbing.

Satisfactory examples of Cenco items needed for this experiment are listed below.

High-pressure mercury arc light source	Cat. No.	87268-000
Iris diaphragm, mounted		86105-000
Lens holders		72288-000
Screen holders		86020-000

Available from Sargent-Welch are:

Auxiliary optical bench, 50 cm	Cat. No.	3618A
Carrier supports for optical bench		3617J
Screen (S_3) with adjustable aperture		3640

Various manufacturers make picoammeters suitable for the measurement of the photoelectric current. Typical is the Model 414A made by Keithley Instruments, Inc., 28775 Aurora Road, Cleveland, Ohio 44139. Most such units are quite expensive, however, partly because all commercial instrumentation is expensive and partly because the instruments have many more ranges than are needed for undergraduate laboratory work. The authors have found that voltmeters with input impedances higher than the 10 or 11 megohms found in standard EVM's are rarely if ever needed, and the building of voltmeter ranges into a picoammeter requires expensive circuit design. For the experiments in this book requiring low-current measurements, a simple homemade picoammeter has proved to be both inexpensive and highly successful. Full details of its design and construction are given in the *American Journal of Physics*. Vol. 36. No. 11 (November 1968), page 969.

The following light filters were originally chosen for this experiment as being best suited for its purpose.

5770 Å line: No. 22 Wratten Gelatin Filter, 3 inches square
5461 Å line: No. 74 Wratten Gelatin Filter, 3 inches square
4358 Å line: Primary Ultraviolet Filter, Coleman UV-2
3650 Å line: Primary Ultraviolet Filter, Coleman UV-1

The Wratten filters are supplied by Special Sensitized Products Sales, Eastman Kodak Company, Rochester, N.Y. At the time this experiment was developed, the Coleman filters were listed by Cenco as Cat. Nos. 29842-1 and 29842-2. Cenco's current catalog no longer shows these items, but some stock may remain. The Wratten filters are also available as part of the Sargent-Welch Filter Set, Cat. No. 3663, in which No. 21 can be substituted for No. 22 if desired.

In addition, filters for the 5461 Å and 4358 Å lines are sold under Cat. Nos. KO 3129 and KO 3130 by Klinger. These filters are on the list of equipment for the Klinger photoelectric experiment, which utilizes the mercury line at 4047 Å instead of the two ultraviolet lines selected in this book and thus results in three rather than four points on the graph of Fig. 138. The Klinger Cat. No. KO 3131 for the 4047 Å line can be used if the Coleman filters are unobtainable, with the disadvantage that the graph of Fig. 138 must be determined by only three points.

Before use, the Wratten gelatin filters must be properly mounted. Cardboard mounts for 35-mm projector slides are suitable and can be obtained from any photographic supply house. These mounts have 35 × 23 mm apertures, and the gelatin filters, which come in 3-inch-square pieces, must be cut to the proper size. A filter should be handled by the edges or at the outer corners only. When being cut, it should be placed between two sheets of clean, fairly stiff paper, and sharp scissors should be used. The cutting line can be marked on the paper. The filter is then carefully placed in the cardboard mount, which is thereupon folded over and sealed as described in the instructions supplied with it. The mounted filter is held in place on the optical bench by one of the Cenco Cat. No. 86020-000 screen holders.

The ultraviolet filters UV-1 and UV-2 are supplied already mounted on glass plates 2.7 × 4.9 cm. This type of filter, mounted in this manner, can be conveniently held in place on the optical bench by one of the Cat. No. 72288 lens holders. The optical bench already provides horizontal alignment, and vertical alignment is achieved by moving the lens holder up or down as required before clamping it in its carrier support.

An adjustable-aperture screen (Sargent-Welch Cat. No. 3640) has been listed for S_3, but a satisfactory and much less expensive substitute may be made from a piece of black cardboard 7.5 × 7.5 cm with an aperture 2.0 cm wide and 3.0 cm high cut in its center. This screen may be held as is in a Cat. No. 86020 screen holder or may be mounted on a clear 7.5 × 7.5 cm glass plate for rigid support. In this case a Cat. No. 72288 lens holder is used to mount the screen on the optical bench.

The system may be tested prior to use by plugging the mercury-arc lamp cord into the 115-volt A.C. line with no filter in place on the optical bench and no voltage applied to the phototube (switch K_1 open and/or R_1 turned all the way down). For this purpose the aperture of the iris diaphragm should be made very small. The lamp reaches high intensity in about 3 minutes. Adjust the lens to produce a parallel beam and aim the lamp down the axis of the optical system. The beam of bluish-white light will be limited by screen S_3 and in particular by the small aperture in the iris diaphragm. Under these conditions the current from the phototube should be less than 2 microamperes. Measure it with the picoammeter, first turning this instrument on, allowing it to warm up, and carefully zeroing it by turning the range switch to the zero position and setting the zero control so that the meter in fact reaches zero. Then turn the range switch to the appropriate scale, note the photoelectric current, and carefully adjust the iris diaphragm until exactly 2.0 microamperes is obtained. After testing the setup in this manner, return the picoammeter range switch to zero. It should be left in this position except when actual readings are being taken.

EXPERIMENT 52

The Hoag-Millikan oil drop apparatus used in this experiment is supplied by Sargent-Welch. The current Cat. No. is 0620B. In this model the spray chamber is about 6 cm high and slips over the viewing chamber. The oil droplets can be produced, seen, and controlled with great ease. One can produce and see oil droplets of widely different sizes, and then control any selected one in the field of view for a long time. The charge on a droplet can be changed to other values by means of a built-in ionizing source. Either this or the older model (Cat. No. 0620) can be used, and the same instructions apply to both. Both have the advantages of ease of setting up, simple operation, and a relatively low potential requirement (about 200 volts).

A calibration scale, an atomizer, and a bottle of high-quality watch oil are supplied with the oil drop apparatus. The calibration scale is 10 mm long with 0.1-mm divisions on a glass disk cemented into a metal frame. During the calibration of the microscope the scale is supported in a vertical position on the capacitor at the center of the upper plate. The atomizer is a specially designed non-flooding oil aspirator. Usually Nye's watch oil, which has a density of 0.890 gm/cm^3, is supplied.

The Sargent-Welch catalog also lists other items, such as the power supply, voltmeter, and stop watch, needed to perform this experiment. However, these particular units are not mandatory. Any 0–300 volt D.C. power supply with good filtering, such as the one specified for Experiments 32 and 36, will be satisfactory. The 0–300 volt D.C. voltmeter may be the electronic voltmeter (EVM) used in Experiments 32, 35, 36, and 37 provided that it has been accurately calibrated. The stop watch or stop clock used in many of the earlier experiments is appropriate for the present work also.

The oil drop apparatus should be assembled as shown in Fig. 142. This may be done by the student if time permits, but for large classes and crowded laboratory schedules, setup and focusing is best done in advance by the instructor. The toggle switch at the base of the apparatus is initially placed in the vertical position to disconnect the power source and short-circuit the capacitor plates. The plates are then connected to the two binding posts on the switching assembly marked ''condenser.'' Next, the high voltage supply is connected to the two binding posts on the switching assembly marked ''high voltage input.'' The voltmeter is also connected to the latter binding posts so that it will not be reversed when the polarity of the plates is reversed.

If the power supply unit does not include a 6.3-volt A.C. outlet, a separate voltage supply must be used for the light source. This may be either the 6-volt storage battery or D.C. supply used in Experiment 50 or the 6.3-volt filament transformer listed for Experiments 33 through 37. A single-pole, single-throw switch is also included in the circuit.

The light source and the microscope are arranged on the stand as shown in Fig. 142. The capacitor is oriented so that the small glass window is facing the light source and the larger one is facing the microscope. The entire apparatus should be leveled so that the capacitor plates are horizontal.

For a determination of the distance separating the capacitor plates, one of the glass plates on the capacitor should be removed and the distance between the inner faces of the plates measured using a suitable metric caliper. If the glass plates appear foggy or misty in any way they should be removed, cleaned, and then replaced.

The proper illumination of the oil drop is very important. The initial adjustments of the light source are as follows: First, the lamp housing is raised slightly so that the beam of light passes just over the top of the capacitor plates. Then the nut on top of the light source is loosened and the lamp moved to focus the beam of light at a point exactly above the center of the capacitor. This is done by holding a small piece of white cardboard, perpendicular to the light, at the center on top of the upper plate and focusing the light on the cardboard. The spray chamber must first be removed from the top of the capacitor. The lamp housing is now lowered to a position where the beam of light passes through the small window and illuminates the edges of the upper and lower plates equally. This face of the capacitor should be oriented perpendicular to the light beam. Now the bright beam of light is directed at the center of the viewing chamber.

The initial focusing of the microscope may be done as follows: The arm supporting the microscope should be oriented at an angle of about 30° from the direction of the beam of light from the light source (see Fig. 142). The height of the microscope is adjusted and the microscope oriented so that it is pointing directly at the center of the viewing chamber. First of all, make sure to short-circuit the capacitor plates by setting the toggle switch vertical; the power supply should also be disconnected. All of this is done to ensure that high voltage is not accidentally applied to the plates. Now with the bulb lighted, a small pin is inserted vertically through the hole in the center of the upper plate and the microscope focused on the point of the pin and, if possible, on the bright streak of light reflected from the pin. The focusing is done by turning the knob of the focusing device. When the point of the pin and the bright streak of light from the side of the pin are both in sharp focus, the microscope is focused for viewing the region where the oil drops will appear. The eyepiece of the microscope contains at its focus a graduated scale by which measurements are made. The eyepiece should be rotated about its optical axis until the scale is vertical, with the highest number on top. The microscope being already pointed toward the dark corner of the chamber, roughly at the orientation suggested above, its angular position is now varied slightly until the background of light just permits the scale to be clearly distinguished. The focus of the microscope and of the eyepiece should then be carefully adjusted so that there is no parallax between the image of the pin and the scale. The focusing of the microscope should not be disturbed for the remainder of the experiment.

The next step is to calibrate the microscope scale so that correct measurements may be made. In calibration, the scale, divided to 0.1 mm, is supported in a vertical position on the capacitor at the center of the upper plate. Without moving the arm that supports the microscope, or disturbing the focus, raise the microscope enough to point just over the top of the capacitor and in the direction of the center of the plates, where the small pin is still inserted. An additional light source is used to illuminate the scale in the microscope. A lamp with a 40-watt incandescent bulb, placed directly in front of the microscope and about one foot away, is a satisfactory source. The microscope is next pointed at the small pin and adjusted so that the image of the pin appears exactly on top of the microscope scale. The small pin should now be removed. The calibration scale is now placed on top of the capacitor, with the side of the glass on which the scale is etched facing the microscope, and with the part of the frame marked ''down'' touching the capacitor top. The scale is thus in a vertical position, with the face perpendicular to the optical axis of the microscope, and as near to the center of the top plate as possible. The exact location of the calibration scale is checked by making sure that the scale itself is sharply in focus in the microscope and superimposed on the scale in the microscope. Only the moving of the calibration scale slowly and carefully, a very minute amount, along the direction of the microscope for focusing and perpendicularly for lateral adjustment, will do this. When both scales are sharply in focus and superimposed, the calibration can be carried out.

Select two points on one of the scales, such as 6.00 and 10.00, and read off the corresponding values on the other scale to three significant figures. From these measurements, the scale in the microscope can be calibrated in centimeters. Thus the measurement of the distance that an oil drop falls will be in centimeters. Now the calibration scale should be removed. The small pin is again inserted in the hole in the center of the upper plate of the capacitor, and the microscope is lowered to its original position without moving the arm that supports it and without disturbing the focusing in any way. The height of the microscope is adjusted, and the microscope is oriented so that it is pointing directly at the pin, with the image of the point of the pin exactly on the scale in the microscope. The pin can then be removed and put away. With this orientation, the scale will be at the point where the image of the oil drops will appear, the oil drops will be sharply in focus, and the background of light will just permit the scale to be clearly distinguished.

EXPERIMENT 53

Although a number of apparatus suppliers advertise Franck-Hertz tubes and associated ovens, the present experiment was developed using the equipment provided by Klinger. The tube is Cat. No. KA 6040, the oven is Cat. No. KA 6041, and the thermometer is Cat. No. KH 2235, although any laboratory thermometer whose scale includes 180°C and that fits through the hole in the oven will do.

The variable line transformer is a small ''Variac'' or ''Powerstat'' and may be the same one used in Experiment 37. The one shown in Fig. 146 is a Superior Electric Company ''Powerstat'' Model 116B, but its 10-ampere rating is much larger than needed for the present application. A more appropriate model is the 10B, which requires mounting in a cabinet and addition of a cord, plug, and output socket, or the model 2PF10, which is a 10B already equipped with the above-named accessories.

The filament transformer may be the same as that used in Experiments 33 through 37. Any standard 6.3-volt, 3-ampere unit will be satisfactory. Note that although the 3-ampere rating is more than enough for the earlier experiments, it is a minimum for the present work, as the Franck-Hertz tube draws a rather large filament current. The transformer illustrated in Fig. 146 is a United Transformer Corp. (UTC) Type S55. It is shown mounted on a small aluminum chassis with terminals for the 6.3-volt output and has been provided with a cord and plug for connection to the ''Powerstat.''

When Experiment 53 was originally developed, a large 45-volt ''B'' battery was used for the accelerating voltage V_a. Adjustment of the voltage was provided by a 5000-ohm rheostat as shown in Fig. 145. This is a perfectly good arrangement. If it is adopted, the 5000-ohm volume control is used for V_a, and the 200-ohm rheostat from Experiments 24 and 50 is connected across the 1.5-volt dry cell to set V_r. Fig. 146 shows an alternative setup, in which a Heathkit Type EUW-17 power supply is the source of V_a. As this supply has its own output voltage control, a separate rheostat to control V_a is not needed, and the volume control has been used to set V_r. In this case the 200-ohm rheostat is not needed. Note that the value of the resistance across the dry cell is unimportant as long as it is not so low as to draw excessive current from the cell (200 ohms will draw only 7.5 milliamperes) or so high as to present a significant resistance to the collector current. Because this latter is less than 10^{-8} ampere, 5000 ohms will appear completely negligible.

The 10,000-ohm protective resistor is shown in the apparatus list as having a 2-watt rating to indicate that one of the resistors called for in Experiment 35 may be used here. Actually, any resistor between 4700 and 10,000 ohms with a power rating of 1 watt or more will be satisfactory. Its sole purpose is to limit the current drawn by the Franck-Hertz tube's grid in case a discharge develops in the mercury vapor. Such an occurrence will quickly destroy the emitting surface of the filamentary cathode and should be avoided. No discharge will develop if the outlined procedure is carefully followed, but should a student complain that he or she is seeing overly large collector currents, the filament power and accelerating voltage should be shut off immediately on that apparatus.

Because of the protective resistor's presence, the voltmeter used to measure V_a should have a sensitivity of at least 20,000 ohms/volt. For this reason the one used in Experiment 50 may not be suitable for the present work. However, if the multimeter used in Experiments 33 and 34 to provide A.C. voltage measurements is of high quality, its D.C. ranges probably have the necessary sensitivity. The Simpson Model 260 shown in Fig. 146 is a popular example that meets all the requirements of this and the earlier experiments. The EVM specified for Experiments 32, 35, 36, 37, and 52 may also be used here provided that its ''low'' side may be isolated from the common ground point.

An appropriate choice of picoammeters has been discussed under Experiment 51, which has a similar requirement for the measurement of small currents. The unit described in the referenced *American Journal of Physics* article is pictured in Fig. 146.